John Badcock

Hind's Farriery and Stud Book

Farriery, taught on a new and easy plan: being a treatise on the diseases and

accidents of the horse: with instructions to the shoeing-smith, farrier, and groom

John Badcock

Hind's Farriery and Stud Book
Farriery, taught on a new and easy plan: being a treatise on the diseases and accidents of the horse: with instructions to the shoeing-smith, farrier, and groom

ISBN/EAN: 9783337237295

Printed in Europe, USA, Canada, Australia, Japan

Cover: Foto ©berggeist007 / pixelio.de

More available books at **www.hansebooks.com**

A
B
C
D
E
F
G
H
I
J
K
L
M
N
O
P
Q
R
S
T
V
U
W
X
Y
Z

A
B
C
D
E
F
G
H
I
J
K
L
M
N
O
P
Q
R
S
T
V
U
W
X
Y
Z

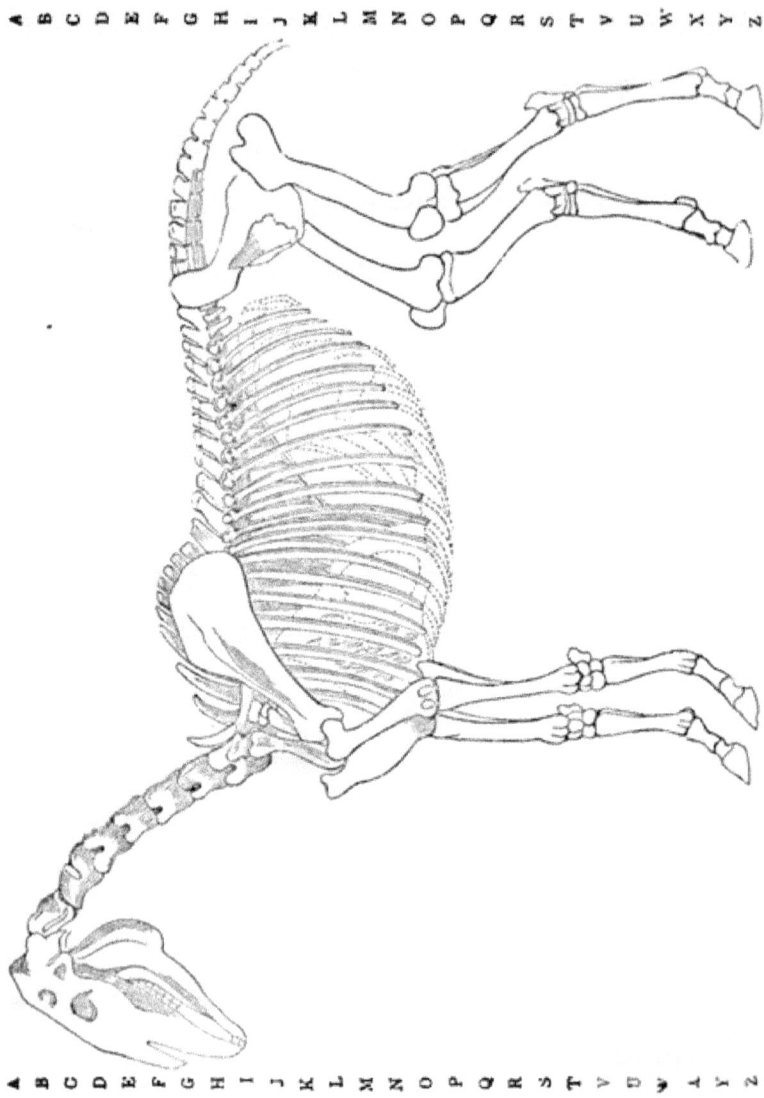

Near Side View of a Horse's Bones; showing, also, the Situation of some principal Internal Parts.

N B. The liberties here taken are explained in the proper places.

FARRIERY,

TAUGHT ON A NEW AND EASY PLAN;

BEING A TREATISE ON THE

DISEASES AND ACCIDENTS OF THE HORSE;

WITH

INSTRUCTIONS TO THE SHOEING-SMITH, FARRIER, AND GROOM.

PRECEDED BY

A POPULAR DESCRIPTION OF THE ANIMAL FUNCTIONS IN HEALTH, AND HOW THESE ARE TO BE RESTORED WHEN DISORDERED.

BY JOHN HINDS,

VETERINARY SURGEON.

WITH CONSIDERABLE ADDITIONS AND IMPROVEMENTS, PARTICULARLY ADAPTED TO THIS COUNTRY.

BY THOMAS M. SMITH,

Veterinary Surgeon, and Member of the London Veterinary Medical Society.

WITH A SUPPLEMENT:

COMPRISING

AN ESSAY ON DOMESTIC ANIMALS, ESPECIALLY THE HORSE,

WITH REMARKS ON TREATMENT AND BREEDING;

TOGETHER WITH

TROTTING AND RACING TABLES,

SHOWING

THE BEST TIME ON RECORD, AT ONE, TWO, THREE, AND FOUR MILE HEATS;

PEDIGREES OF WINNING HORSES, SINCE 1839; AND OF THE MOST CELEBRATED STALLIONS AND MARES;

WITH

USEFUL CALVING AND LAMBING TABLES, &c. &c.

BY J. S. SKINNER,

Editor now of the Farmers' Library, New York; Founder of the American Farmer, in 1819 and of the Turf Register and Sporting Magazine, in 1829: being the first Agricultural and the first Sporting Periodicals established in the United States.

PHILADELPHIA:

J. B. LIPPINCOTT & CO.

1867.

PREFACE.

WHATEVER person would consult these pages with profit should previously read the first book with care ; for in it he will find laid down the principles upon which all the subsequent details are founded, how the process of nature is carried on in health, and the cure is to be effected in every species of derangement. Indeed, he should study it hard, if he would become proficient in "the Art of Farriery," and not rely implicitly upon other people's prescriptions for the cure of any alleged disorder, which have been composed for the most part without any such preparation.

From this neglect, also, symptoms of one disorder are confounded with those of another, when the proposed remedies can not possibly effect the cure. If he be imbued with the proper thirst after knowledge, be his station in life about the horse what it may, he had best to comply with the advice strenuously urged at the very outset, to examine the internal parts of dead horses, as often as opportunity presents itself, which, in the neighbourhood of large towns and hunts is frequent enough. For this is the manner in which I was myself mainly instructed; as well as by noting down whatever then appears worthy of observation, connected with the previous disease of the deceased subject.

Such was my manner of proceeding for several years. And next about the present volume, how I came to write it, and what were my views in the manner of executing the task that was rather imposed upon me by the booksellers than sought after by me; and which was, in effect, occasioned by the nature and quantity of veterinary facts and observations I had a long time been in the habit of heaping together. But I had already been an author nearly a quarter of a century, having partly translated the manual of La

1 *

Fosse, at the request of another bookseller, Mr. Badcock,
of Paternoster-row. I claim no credit for that performance,
and have already stated my present opinion of its degree
of usefulness, at pages 133 and 135. Proceeding with my
"literary history," I may here add, that a few communications in the (old) Sporting Magazines,* to the Monthly
Magazine,† to the Weekly Dispatch newspaper, and other
such publications, on topics connected with animal medicine, preceded the essays on the structure of the horse,
which comprise the first book of this volume, and found
place in a newer and much more brilliant publication. An
accumulation of materials for these pages lay by me, with a
latent hope of publication, when the mammon of a "ten
pound prize," for their insertion in the Annals of Sporting, and some cheering commendations that attended the
appearance of those essays, from time to time, induced me
to finish the design of a complete pocket manual for owners,
grooms and aspirants after the knowledge of horse-medicine, of every degree.

Like all other practitioners of the old school, or rather
no school, my late father had long amassed together and
preserved, in an immense and shapeless volume, entitled
his "Receipts," all the alleged remedies recommended as
eligible and found good in every variety of case: I believe
he may have tried the efficacy of each, though I am now
tolerably well convinced that some must have failed of complete success. Yet was the manuscript preserved like a
family treasure; and destined to fill my pockets at some distant day, its contents were secluded from vulgar eyes,
though it contained nothing but prescriptions. As usual
with all similar accumulations, the proper remedies were
therein stated, without a word as to symptoms or those
anomalous cases that frequently baffle the utmost skill, for
the practice of medicine in any of its departments is but an
imperfect science, even when we can ascertain the precise
ailment under which the patient labours. This necessary
preliminary is not always possible in veterinary practice
we are more frequently baffled than assisted in our inquiries. Notwithstanding all this, my revered parent sus-

* For November and December, 1820, on "Fever in the foot," in refutation of Mr. Cherry, in which was described the successful treatment of a nag
belonging to Mr. Bowley of Covent Garden.
† January, 1821, &c.

tained a high character for successful practice; his close observation of the symptoms and attention to the operation of his physic, supplying the want of a " regular education," which no one farrier could at that time boast of: indeed, few of them could even copy their own receipts, which they preferred to carry in their memory. At a very early period I endeavoured to repair this apparent defect by study; with what success the reader may judge, and I will endeavour in the next pages to make him comprehend how my task has been executed.

The reputation of our name induced the bookseller just named to ask my father's opinion and mine (among others), of a certain manuscript he held in his hand, which upon inspection turned out to be a treatise on the rationale of horse-medicine, with very plain directions for ascertaining the true symptoms of dieases before attempting to apply any remedy, however estimable. As the expositions of the writer agreed mainly with our own ideas, it was impossible to withhold approbation. Finally, Mr. Badcock also consulted with W. S. Rickword, of Moor-lane, and other veterinary surgeons of the college, and resolved upon the spirited publication of his new purchase, notwithstanding he had received the uncheering disapproval of Bracy Clark, of Smithfield, who gave for answer that " no one could learn the treatment of horses' diseases from printed books." Yet has Bracy Clark since then printed many books. The great success of the publication alluded to, which was James White's " Compendium of the Veterinary Art," justified our opinions of its merits, and gratified my vanity at the early share I took in its promulgation, and the revision of many passages with a view to simplifying the terms (in particular); in which commendable quality, by the way, Mr. White is not deficient, though, in other respects, a lapse or two which have since fallen out, come under notice in the course of the following pages (viz. pp. 39, 83, 111, and 154). No man can be perfect: how few among us know every thing that pertains to themselves.

Even at this moment preceding the birth of my volume, I am not certain but I may be found similarly tripping—to have expressed myself obscurely, when I fancied my language most completely understandable by the meanest capacity; and I doubt that my familiar style may frequently appear vulgar to more polished eyes and ears than mine. But I take credit for having sedulously avoided the use of technical phrases, terms of science and learned dissertation, as well as the crime of over-refinement with which I have rebuked two cotemporaries, whose laughable sublimations are idealized at page 166.

Candour and ability for the task are not always found combined with willingness, even among our best friends, to amend certain slips of the pen, or to curtail such exuberances as the more animated writers are liable to fall into; and I am free to aver, that the friendly assistance I have obtained in this respect, the nature of which may be inferred from the note at bottom of page 50, has not always seconded my plain meaning, nor adequately fulfilled my wishes, though I am grateful for these and every act of kindness. After all my care, repetitions have crept in, and owing to the length of time occupied in the composition, or rather the manner in which the various particles of information were collected together, and digested into form, great variety of style may be discovered, though unity of purpose, and the desire to *instruct*, pervades every page. The arrangement is at least obvious; the principles being taught in the first book, the details of practice follow in natural order in the second and third books, and seem to arise out of the preceding "observations on the animal system of the horse, as regards the origin of constitutional disorders." The references from the latter chapters to the former, operate as exercises with those students who may have neglected to acquire and retain sufficient intimacy with the principles laid down in the pages so referred to.

The diseases of brute animals are few and simple, and easily cured when the symptoms can be distinctly traced up to their causes; for the remedy then consists in little more than putting the animal upon a direct contrary course to that which brought on the disorder (though not too rudely), and health follows. For example, heat, inflammation, fever, is the most general cause of constitutional derangement in the horse: in a state of nature, he seeks out and employs the remedy himself; when domesticated and pampered, or at least denied the use of green food, we judiciously set about reducing the heat by cooling medicines and factitious regimen, and the fever subsides. Again, hard work occasions lameness, rest restores the feet to their wonted state in incipient attacks, topical applications effect the remainder in bad cases.

For the same reasons few medicines are necessary in veterinary practice, but certain modifications of these add to their efficacy in particular cases; though the school in which I was first initiated, as well as the modern writers, White, and the Lawrences, quite overwhelm their readers with the quantity and apparent contrariety of their prescriptions, that frequently possess no essential variation from others that may be applicable to a whole series of disorders.

Under such circumstances, I have been extremely chary of puzzling the reader by merely altering the vehicle when the active material of the prescription had been already compounded for a similar disorder; therefore I have avoided repetition of such (mostly purgatives) by referring the reader to the page where these may be found. Notwithstanding the apparent difficulty of this mode, yet has it certain advantages that outweigh the trouble, and compensate for the moments thus expended. During my noviciate, and long intercourse with persons employed about the horse, in almost every capacity, I noticed that all those who consulted the books respecting any actual disorder, did little more than turn to the prescription which was recommended in their particular case, and it was made up and given to the animal without once more reading over and comparing "the symptoms," and notwithstanding they already had the same medicine upon the shelf. By this blind manner of proceeding, they did but adhere more closely to the old system of their "book of receipts," to the entire neglect of the anomalous symptoms, and risked the mistaking of one disease for another, in many cases. To compel the inquirer to study his case before he applies the remedy, I at one time thought of adopting the method of La Fosse, and others, who have thrown their prescriptions all together, and referred to each numerically; but, after due consideration, I adopted the middle course, and simply avoided repetition in this respect, as that which best suited with my views of instruction. In some cases, the remedy is mentioned in general terms only; for example, at page 170, I said, "blistering ointment may be applied," &c. The reader will of course, in this and all similar cases, consult the *Index;* and under "Blistering," he will find himself referred to page 76.

Throughout the volume, though I naturally evaded all
controversy, yet in a few instances it seemed necessary to
advert to certain existing errors and authorized mistakes;
to disabuse the public mind, to negative the mischiefs these
were calculated to spread of themselves, and to assure the
reader that I was not wholly unmindful of the dissonance
of opinion betwixt the authors mentioned and myself.   To
the "Annals of Sporting," a monthly publication much
devoted to the natural history of animals, I have frequently referred, and often quoted; because in the course of its
earlier volumes many desirable facts, some good and useful
hints, and valuable suggestions, appeared from time to time;
some new opinions and statements were started, and met
with repulse, or were more securely placed upon their proper bases.* In these respects a favourite project, first communicated to me by Mr. Badcock in 1802, and partially
acted upon,† was therein realized, viz. of collecting together
the scattered opinions, remarkable cases, and fugitive suggestions that should occur to various isolated practitioners
throughout the kingdom, in the same manner as had long
effected so much progressive good for human medicine. He
had engaged me and Mr. Rickword to assist him in this
undertaking, and wrote to Mr. White and others for their
contributions; but it failed at that period, like many other
projects of a similar nature; and I observe that the last-
named gentleman, in every successive edition of his "Compendium," constantly inserts his correspondents' letters on
various topics at length, though it was clear to me that dis-

* In that useful publication ordinary passing events are recorded monthly,
under the head of "Horse Intelligence," with brief comments, accompanied
at intervals with exhortations to veterinarians to contribute their experiences
to the same stock. In one instance, a vivid appeal, in the number for September, 1824, page 191, produced several valuable communications concerning hydrophobia, that are embodied in the present work, and acknowledged
at page 162–3. The intelligent papers of Mr. Perry, of Swaffham, and
others, also owe their origin to the same stimulus to publicity and the desire
to establish a name for ingenuity in their profession to the writers.

† I took occasion to advert to that project in my preface to La Fosse's Pocket Manual, and to lament that "the want of a more liberal practice is felt as
an insuperable bar to improvement in the art of farriery, which would be best
served by communications of the discoveries made, and the mode of treatment
most successfully followed by various practitioners. This it is which of late
years has done, and is still doing, so much for other branches of medicine, and
which, for the sake of humanity, it is devoutly to be wished could be extended
to this branch also." Page vi.

cussions like these rather belong to periodical publications,
such as the "Annals" professes to be (where they admit of
refutation), than to a "Compendium." For my part, I
was early induced to enter into the spirit of those periodical investigations, and the inquiries set on foot in that work,
and occasionally to furnish the materials for an article, or
the argument in point for a controverted doctrine, or dis-
puted "improvement." An offer of two premiums of ten
and five pounds for the best and second best of an "Essay
on the Structure of the Horse," had first induced me to labour in the pages of the Annals. The award of the highest
premium to my paper* encouraged me to hearken to pro-
posals for its enlargement, and the present volume is the
result.†

The volume has been a long time at press, and in October
last was fully announced by advertisement. The author
could not, therefore, satisfactorily account why his title
was adopted by another in the month of April of the present year

* Divided into magazine-like portions, and inserted as convenience offered
in many successive numbers of the Annals of Sporting, for the years 1822,
3, 4, 5.

† The second premium was followed by the like result: the writer of it,
Mr. Percivall, (I presume) having since then published his volume on the
Principles of the Veterinary Art." The utility of such periodical works
that devote their pages to the promotion of useful arts, is thus manifest in the
fact that to those premiums the public owe two volumes at least on animal
medicine.

London, July, 1827.

INTRODUCTION.

As the value of the Horse is daily becoming more mani fes*, it is presumed that any attempt to reduce into a system, the art of preserving it in health and of removing diseases will not be unacceptable.

It is certain that at no period in the history of this country, has the horse stood so high in general estimation, or by the display of his various powers, rendered himself an object more worthy of our consideration.

As greater attention is now paid to the breeding of horses for the different purposes of the turf, the road, &c.; so should our anxiety for their preservation increase.

The object of this publication, is to render as plain and familiar as possible, a subject that has for a length of time remained in obscurity: the want of a work possessing practical facts and illustrations, has long been severely felt and acknowledged.

Under this conviction I am induced to lend my aid, in bringing forth the present volume, with such alterations and additions as an extensive practice in this city may warrant.

To remove long standing prejudices, I am aware is a difficult task; still I venture to hope, that a careful perusal of these pages will excite in some degree, the feelings of humanity, in respect to the many sufferings to which the generous animal is frequently liable from unmerited cruelty and injudicious treatment, and that mankind may be induced to view his sufferings with an eye of sympathy and tenderness, and have recourse to a rational mode of practice, when accident or disease may require it.

I am not aware of any publication having issued from the

2

press in this country, in which the Veterinary Science, or Art of Farriery, has been laid down in such a manner as to be clearly understood; the present work is so familiar in its composition, as to render it at once interesting and intelligible to every one who may think proper to peruse it.

To such persons who are removed at a distance from those places where the assistance of a farrier can be had, in cases of emergency this work must prove highly useful, as such rules for the discovery of disease, and such a plan of treatment is recommended, as, if judiciously followed, will rescue from the danger of blind experiment, the noblest and most valuable quadruped in creation.

THOMAS MOORE SMITH.
VETERINARY SURGEON

Philadelphia February 1 1830.

CONTENTS.

-ᐁᐁᐁ-

BOOK I.

PLATES.

FARRIERY

ON AN IMPROVED PLAN

BOOK I.

THE ORIGIN AND SEATS OF VARIOUS DISEASES IN THE HORSE EXPLAINED, WITH A VIEW TO THEIR CURE OR MITIGATION.

INTRODUCTION.— *The necessity and* **advantages** *of veterinary* **knowledge,** *and the means of acquiring it, as* **regards** *prevention and cure.*

ALTHOUGH it can not be denied, that "'tis better, in a humane point of view, to prevent diseases than to cure them;" yet, looking at the fact as a veterinarian, without forgetting my feelings as a man, I do not hesitate to say, "this is a consummation we can not reasonably hope to arrive at, whilst the horse is compelled to exert himself to the utmost of his power for our daily profit," whereby he acquires a constant disposition to create disorders. Nor would I be thought to maintain, that "preventives ought never to be employed:" the succeeding pages fully disprove such a conclusion. I merely mean to inculcate, that, under existing circumstances, they can not be resorted to generally: and this I say, notwithstanding it will be found I have here noted very many occasions, when rest, alteratives and regimen, might be often substituted for active medicines, more economically, (in my opinion,) both of time and expense. The hour is not arrived, however, for me to insist too strenuously upon an entirely new mode of treatment of the horse in health and in disease, since that course would appear rather too theoretical for a Treatise designed to be wholly practical.

Those are the reasons which have induced me to keep in view the readiest way of enabling the sick animal to return to his work again, according to the long beaten track of my practice; whilst my main purpose is to show, by an examination of his powers and his parts (external and internal,) that a moderate mode of treatment, in sickness and in health, would be not only more humane but more profitable, as preventive of many of those evils to which thousands of horses prematurely fall victims every year. More conducive, also, to a profitable result to their labours would it be for the owners of horses, instead of studying how to "physic" their property, were they to put themselves in a condition, as near as may be, for rejecting, with some degree of certainty, not only such horses as are offered to them actually diseased, but such also as, by their awkward built or structure, and consequent ill-formation of the internal parts, can not fail to possess some inherent bad quality, and thereby a proneness to its corresponding affliction to the end of their days. This ought to

2 *

constitute every horseman's first step to horse knowledge, whether he undertake it as an owner or as a farrier, the latter most especially; of him I may
justly add, that he can not be said to exercise his calling honestly as he ought,
who sullenly neglects to learn those rudiments of art and practice that teach a
knowledge of the animal economy and the functions of the horse in particular.
I do not hesitate to insist upon the examination of the animal's internal parts,
as constituting one main item of those rudiments; and I would not avoid giving
this operation the proper name of dissection, but that I fear to alarm the general reader with an apparent difficulty where none exists in reality. How
without that previous knowledge, durst he venture to pronounce what parti
cular ailment, out of the numerous catalogue that pertain to the horse, his pa
.ient labours under? How can he ascertain the degree, or quantity and quality
of the attack, so as to know when it may be increasing in malignity, or its
virulence is expended? Least of all can he succeed in the cure, when so much
uncertainty hangs about his means of discriminating between one disorder and
another,—to say nothing of the usually attendant ignorance of the mode in
which medicines operate upon those internal parts that lie concealed from his
view, but upon one or the other of which they are, nevertheless, destined powerfully to act. If it be allowed, that no two horses are ever affected exactly alike
.n those disorders that depend upon the secretions, as I shall show at the end
of this chapter, how is it possible that such neglectful men could ever reduce
the symptoms of any disorder, without reducing, at the same moment, the
power or functions of the part upon which their strange and ever-violent mix
tures expend their force, and thus entail upon the animal a disposition to acquire some other disorder.

Every man who would make himself proficient in the knowledge of diseases
should open his own dead horses, and as many more as he can obtain access
to, and attentively examine the state of the stomach, the liver, the lungs, the
heart, kidneys, and bladder. If the animal be recently dead, this profitable
inquiry will be far from disagreeable, unless the cause of death has been of
the putrid kind, spoken of in Book II. Chap. I. as Typhous, but which rarely
happens. In the pursuit of this necessary first step to veterinary knowledge,
he will proceed in this manner. The horse being on its back, two legs on the
same side are to be elevated by a cord passing round the fetlock of each, and
fastened to a nail in the ceiling or elsewhere aloft. Then with a sharp knife,
of the common shoemakers' kind, he will draw a straight cut all the way from
the first rib or breast bone, at the intersection of O with 21 in the picture, to
the sheath, or thereabouts. If the cut be not too deep, the skin will recede a
little, and expose the membrane; cutting through this the intestines will protrude, and drive forth a thin expansive membranous sac, apparently unattach-
ed, being designed for holding the guts, and preventing friction. This soon
bursts, and the blind gut (or *cæcum*), described at section 48, appears. He
will slit open this pouch, and examine its contents before he quits the subject,
probably; but his first business is with the stomach, which is depicted in the
annexed plate, as situated at the conjunction of IKL with the figures 26—29.
Herein will be found the last drench that sent him out of life, or the last food,
that gave hopes of a prolonged existence; and on its surface, vulgarly termed
the coats of the stomach (when turned inside out), may be discovered the havoc
committed by the farrier's unskilfulness: according to the strength of the poisons so administered, will the coats show the dilapidation, or at times a hole
will have been perforated, that is the cause of instant death.

The young operator will keep in mind what is said of those parts at sections 45, 46, &c., if he do not turn to and read them over once more before he
takes up the knife. With the same precaution as to re-reading section 52,
&c. he will proceed to examine the state and appearance of the liver and kid

neys. The description of these will be found at sections 52 and 53 respectively; and they are delineated as situated in the picture, the liver between the parallels of J—N, 22—28, and the kidneys at H, 29, 30. Returning forwards, the operator will find his way to the heart and lungs obstructed by the midriff, (see plate at 22 to 28, ascending slantwise from L to H) that divides and keeps asunder these from the first-named parts, lest the guts and liver should obstruct the action (functions) of the heart and lungs, and *vice versa*. Its appearance has been described (sect. 35.) as resembling a drum-head; and like it, if pricked with the knife, the cavity of the chest is instantly laid open—an immense vacuity, that proves to what a vast extent the lungs must fill at every inspiration of fresh air, to occupy so great a space, and further spread out the ribs to the utmost extent of the intercostal muscle that holds them together. In the plate the lungs are depicted in a quiescent state, at J to N, and 15 to 22; but when filled they occupy all the vacant space above, in addition to their lateral width. Hence, the importance of this *viscus* (as they call each of the vital parts above named), to which I have attached such high consideration in the sequel, will at once be seen and appreciated. See sections 31—36.

Concerning the Heart, its structure and functions,—so much has been said in another place, and so minute is the description of each, that I shall add no more here, than refer to the sections, where the reader may find ample instructions for examining this main-spring of animal life. See sections 37 to 40. In the annexed picture, it is delineated as lying near the lungs [LMN, 19—21], to the upper part whereof it is attached, as described hereafter.

By pursuing this course of inquiry, the operator will discover what is, or ought to be, the healthful state and appearances of the main functions of the animal system,—he will perceive the auxiliaries and their uses,—he will have informed himself (it is hoped) of the treatment any horse has received previously to its death; and he may thus store up in his mind, or better still, upon paper, what dread effects may be produced by the drenches, cordials and diuretics that stimulate but to destroy the vitals of the animal. [*] He will see and compare the animals that die in health (accidentally), or after a short illness, with those which die after protracted illness; upon the healthy ones that are doomed, a few hours previously, he may try the experiment of some favoured farrier's celebrated mixture, and subsequently send him the stomach to prove its efficacy in "killing all disorders."

Happily, the cause of humanity may be served, and the interests of his owner promoted at the same time, by our (first) ascertaining the nature and amount of the horse's powers by his make, shape, or built; and, thereupon, demanding of him no more, in the way of service, than is clearly proveable to lie within his power, or putting him to those labours only to which his capabilities are best adapted. In the neglect of this plain rule lies the root of all error as regards preserving the health of horses. Some materials for making a tolerably good estimate as to this head of information, are arranged in the first chapter: the second being well pondered, and the facts and observations it contains rightly stored up in the reader's mind, he will learn what functions belong to each part of the animal in health; or, these being deranged or obstructed, he will know in how much the horse is affected: and the third chapter being read with reference to both, I entertain the well founded hope, that this course will enable the general reader to form tolerably accurate notions of the nature, origin, and tendency of the animal's internal and constitutional diseases, upon which all the others depend, but which have hitherto received but little attention any where here, and, consequently, are but imperfectly known among us. Not only so, but the reader may, by these means, by study and close observation, enable himself to demonstrate nearly to a certainty, when a cure is hopeless; and further the cause of humanity and the interests of its owner at the same time, by ordering the horse to be destroyed at once, rather than by fruitless delay, and at a heavy expense, prolonging the animal's sufferings to no worthy purpose.

I have not confined my researches to disease only: in the first chapter, the *shoeing-smith* will find explained the principles upon which depend deformities of the hoof, and he may fashion his work accordingly; whilst the choice of a *horse* may be undertaken with some confidence, if *the purchaser* keeps in mind the practical advice and information here collected together from various sources, and added to my own observations, and long, extensive, and successful experience, in all matters of this nature.

Explanation and practical use of the Skeleton annexed.

THE references that are made to the annexed plate, and which will necessarily be found rather numerous in the chapter on conformation, are so made by means of letters and figures, corresponding with similar letters and figures upon the plate. The letters direct the reader's eye across the picture, the figures from top to bottom; when he is referred both by letter, and figure, the place of intersection is the point to which his attention ought to be directed. Thus [G. 37.] which, by placing a flat ruler, or a piece of paper, across at "G." and running the finger downwards from the figure "37," would be found to intersect each other at the insertion or commencement of the horse's tail; whilst [Y. 40.] would bring us to the hindermost pastern. Again, [K. L. M. N. 14, 15, 16.] or [K—N. 14—16.] directs the reader's attention to the shoulder-bone; at [M. 20.] is his heart, and at [H. 29.] his kidneys are placed.

The reader will please to observe, that the Frontispiece is meant to be, less what is termed "a pretty picture" than a practically useful one, calculated to facilitate his comprehension of what is said in this treatise about the living horse, his structure, and internal formation; of his capabilities, and all of the diseases arising from their misapplication. To this end, a mere elevation of the skeleton was requisite; and, that this should be rendered more practically useful, it is divided into squares, for more ready reference. The figure itself, is that of a rather long bodied horse; the blade-bone having been lowered to show the continuity of the vertebræ, or backbone, between the shoulders, and the elbow being bent forward for that purpose, so that the shoulder-bone is brought to form its sharpest angle. This position of the limb, of course, rendered the subject of the plate lower before than he would be were those bones more straight up and down than they are. See Section 8.

He will observe, too, that the situation only of some internal parts was required for the purpose of elucidation; thus, the heart seems unsuspended by its vessels, as its pericardium and part of the lungs are removed; and it follows, that whoever expected to find a delineation of every viscus, perfect, has deceived himself,—if any such there be. Respecting the poll, or bones of the head and neck, the reader will find some remarks in Section 16.

Further, the references my readers will meet with in the midst of the text are necessarily as brief as they are useful, and are made to the sections, or parts, into which the first two chapters are divided after the manner of verses.

This mode of reference will be found highly serviceable in his inquiries by the attentive reader, who is unused to study things of this nature, but who must soon perceive the great practical advantages to be derived from so intimate an acquaintance with the subject as this method of learning it will furnish him the means of acquiring. If, in the prosecution of his studies, he happen to forget what has been before said, tending to the same point of in formation, or he be at a loss whereabout he should look to refresh his memory, these references supply him with the ready means of overcoming the difficulty. By adopting this method, I have likewise avoided the repetitions inseparable from a work of this nature, and have thus saved room.

CHAPTER I.

External formation or structure of the Horse, and the disorders originating therein.

Section 1.—Scarcely any man who is in the habit of seeing many horses perform their labour, and observing their capabilities of several kinds, but acquires, thereby, some insight of the properties conferred on the animal by such or such points of conformation. He can tell, at first sight, nearly from this habitude, "what a horse can do;" but few men reduce their observations to writing, least of all to principles, upon which we may afterwards reason, or draw conclusions with any degree of certainty, as to what duties a horse can not perform properly, when wanting those points of excellence, and which duties ought, therefore, never to be required of him; or, being so imposed upon him improperly, are productive of certain disorders that invariably attend such misapplication of his powers. No doubt it has happened, that a horse with a radical defect,—in the shape of his hind quarters, for example,—yet having a corresponding defect before, the one makes up for the other, and such horses may occasionally perform well for a short time, but then they are no lasters; all the while they may thus be at the full stretch of their physical powers, straining to the utmost the immediate coverings of the bones, some thing or other is going to wreck—of muscle or tendon, of ligature or sinew. Sooner or later so much excessive fatigue of the deformity runs along the solids, and reaching the vitals, occasions constitutional disease, or leave behind it an incurable malady of the limbs, mostly descending to the feet. Equally true is it, that we find out new properties, or hidden powers in a horse, which had never hitherto been known to his owners; but, then, as I shall particularize by and by, no such latent powers were ever discovered in any horse, without his possessing certain just proportions of the bones taken altogether.* What these proportions are, as well as what they are not, I come presently to lay down: the integuments (or coverings) ever adapting themselves thereto, in one case produce what is called symmetry; but if the limb be disproportioned, the coverings adapt themselves to that particular defect, and enlarged muscle at these particular places becomes visible to the common observer.

The acquiring a ready mode of discovering when a horse of the one or the other formation is presented to our notice, forms the perfection of art in purchasing a horse.

2. But the horses's achievements, or "what he can do," under certain circumstances of shape and make, would ill employ my pen at the present moment—valuable as the investigation must always be in itself—were it not for the practical application I mean to make of it shortly, by way of illustrating the direct contrary, or defective shape and make, as being the harbinger of

---

* Eclipse, a horse whose very name is used as synonymous for speed, had none of the proportions generally deemed indispensable to great speed, and he was cast, by the Duke of Cumberland, for his apparent deformities when a colt, but his defects in one particular were amply supplied by excesses in another, and, taken altogether, composed the very best bit of bone, blood, and muscle ever produced. His lineage, lateral consanguinity, and the kind of cross by which he was got, demand the breeders' serious attention.

several radical disorders of his frame. Nor is this all; some are so evidently ill-formed in the chest and carcase, from the moment they are foaled, that no art of ours is equal to preventing the return of certain disorders which are sure to attend a horse of that particular formation all his life time. As the one is known and inevitable, so the effects of the other may be foreseen, and, in some degree, alleviated, if so much trouble and expense be not greater than the value of the horse. This is all that can be done for such an animal; and since the resources of art are not equal to the obstacles of animated nature, so no man ought unreasonably to expect, least of all, to force his beast, to perform any species of labour or exercise for which nature or the accident of birth hath rendered him anywise unfit; although it must be allowed, as a general axiom, that it is only by pushing the animal to the extent of his powers, that we can find out the most he is capable of performing at any given work. In this way it was the fast-trotting powers of the Phænomena mare (which was before then a butcher's hack) were discovered; for people of this trade generally try the utmost their nags can perform in the trot.

To be able to judge of a horse's defects as to what he can not do, undoubtedly it seems necessary to ascertain what constitutes a fine figure, or a perfect one, that can do every thing; but when it is considered that the exposure of those defects is intended to apply wholly to the origin of disorders for which he will require medical treatment, if he does not deserve rejection *in toto*, I shall find less occasion for adverting to any known horse, entirely without error in his form or built. In most cases, however, good symmetry being accompanied not only by the power of achieving great feats, but a good portion of health also, or, at any rate, the absence of the diseases incident to a bad form, I may be allowed, while exposing his faults, to deviate a little, and to contemplate some few of his perfections also.

3. The most obvious physical truths are those which can be explained upon the principles of mechanics; upon such a basis, even the most abstract can be securely grafted : that intelligence which is derived from experience, from observation, experiment, and acute reasoning, is rendered more easily understood when conveyed with mechanical precision; and however strange it may appear to some, the gift of speed, if not of all progression, depends more upon mechanical principles than is commonly understood to be the case. See farther onward at Section 9, where the details are given. In all compound bodies, whether animate or inanimate, intended for our active use, it is above all other things requisite that they should stand well upon their bases or legs. A horse, or a joint stool, evidently defective in this particular quality, would be shunned as insecure; and the one is sometimes endued with movements as little suited to one's ideas of getting on safely as the other, both being indebted to their original bad built (or *charpente*, as Lafosse calls it) for the defect. Cover them both, the one with muscle and skin, the other with drapery, how you will, the faulty legs are faulty still. A good stable aphorism has it thus —"a horse that does not stand well can do nothing well; and by natural inference, the horse that walks well can perform other paces well."

A much better example, however, may be found in a four legged table, of which every horseman knows there are many of different sizes and of various workmanship, some for heavy or rough usage, others more for show and to sustain light weights. But, if the fore and hind legs bend towards each other upon the ground, any carpenter may see that this first element of an ill-formation must sooner or later, produce a fall; he will know that more strength for supporting great weights would be found by making all four legs perpendicu 'ar. But a horse not being like a table, immoveably fixed upon its legs, but being required not only to bear up but to proceed with his load,—which is sometimes effected with difficulty on account of its weight; then must his powers ot pressing onwards be estimated by the positions in which he can place the bones of his hinder part, the legs particularly, since it is to these the propul sion of his body forwards is chiefly indebted. In his efforts to accomplish this duty, the position of his hind legs will resemble those of the second table in the margin, stretched out constantly as these are, and each leg alternately twice as much beyond his body; while his fore legs will bend under him alternately also, like those in the first table. In both movements his legs are stretched to their utmost when the drag is up hill, because the resistance to be overcome is then greatest, and we can thus form an opinion how much "he has the free use of his legs." When this is the case, all horses step short; but, upon even ground, the hind leg, to be perfect, should come finely forward in the walk, and occupy the identical spot which the fore leg had just quitted. See further at Section 8. As the horse gets old, is tired, disordered, or over-much laden, he ceases to do this as usual, in the exact ratio that he is affected the one way or the other.

4. Mares, occasionally, and skittish horses, frequently bring their legs to-gether, much resembling the first figure, and are insecure roadsters as well as poor draught horses. The second sketch is the walking motion of an unladen cart-horse or a coach-horse standing still; these, as well as hunters, take the same position, which indicates that they have the free use of their limbs. In the drag, the former bring their fore legs under their bodies, the principle be-ing applicable to any quadruped performing the like task; and such a horse would consequently fall down forward but for the resistance of the load he draws. But this accident seems provided for, by the power the horse has of contracting the muscles (see Section 10), and drawing up quickly the lower part of his limb, in time to get it out of the way of his hind leg, both motions forming each a separate effort toward progression. I still have in view a walk ing pace, all other paces being no other than modifications of the walk; and, in fact, "a horse that walks well can do any thing else well," an aphorism that is a-twin with one equally well founded in the preceding section.

With some horses, the hind foot, instead of coming forward, as described at the bottom of the last section, upon the spot of ground marked by the fore one, falls short of the mark.—These never turn out fast ones, although their fault does not always consist in the shape or disproportion of the bones, but in the contraction of the muscle or tendon (see this tendon described under the head of "Foot"); at times it is owing to the relaxation of the immediate coverings of the bones, described at Section 16. Such horses may be well enough to look at, but can not perform properly. The extreme of this misfortune is termed stringhalt; but every approach towards it, however trivial, is good cause tor rejecting the animal. In case of the hind foot coming too far forward (in the walk still) and striking the fore one, the fault lies in want of sufficient strength (or quickness) in the fore leg; besides which see further at Section 10. If the hind foot comes down sometimes inside, at others outside, the just quitted situation of the fore foot, the animal has a disagreeable rolling in his gait from side to side, the fault being as often in the fore leg as in the hinder one, some-times in both. Such horses commence a journey with much apparent confi-dence, but tiring soon, they fall into their old error, and the security they have inspired is found to be deceptious:—many accidents are the consequence This fault I hesitate whether to ascribe to the fore leg or the hind one but it certainly originates in a disagreement between the fixing of the two upon the body, either as to the situation, or want of muscular strength at the place of joining.   Such a horse is a stumbler, and when he trots away from us, we can see nearly as much of his fore legs as of his hind ones; in the straight-built, well-set limbed horse, the fore legs are then concealed from our sight by the hind ones.   I own this is with me a grand criterion for judging as to a horse's capability of going over the ground.   In racing, or indeed any running, the fore legs are then brought closer together, the hind legs rather wider (so in leaping), as we see in greyhounds, hares, deer, and all other fleet creatures.

Such as I have described is the act of progression with all horses, but in various degrees, according to their sizes (as with the coach-horse, saddle-horse, poney); four such efforts having called into action all the bones of the body, including more or less that of the head, tail, and neck, according to the pace or other circumstances.—See Section 11.   Hence it must be clear, that to perform this duty of progression, or getting forward, properly, as regards either the length of time he sustains it, or the quickness of performance, weight, or velocity, the limbs must be adapted to the kind of work the horse has to perform and to each other, whether that be in harness, on the turf, the chase, or the road.

5. We do not find this adaptation of the limbs so much in the amount of covering the bones may have on them, as in the size and proportion of these, and the suitable manner in which they are fastened together; as may be seen in those horses (blood) where tendon supplies the place of muscle, and most strength resides in the smallest compass; and, as may be proved by the obstruction to his paces, which is always observable in the horse burthened with very muscular shoulders.   Equally true is it, that, after we have approved of the proportions of a pair of horses in respect to bone and built, certain powers of going or lastingness are frequently discovered to be possessed by one so much beyond his match, that we are compelled to admit those powers do reside in something else than in his built.   Superior health, sound wind, courage, give this strength, with speed, and lastingness ; the bones being then well cased together and strongly supported by their immediate covering, have full and fair play.*   But wherever they be fundamentally ill-adapted to each other, in whatever degree this escapes our observation, the muscles and tendinous parts adapt themselves in some measure to that lamentable kind of form, but which no filling up, or after-accommodation of the parts to each other, can completely eradicate, though it may be concealed from our view.   The muscle that is so perverted rises up in the middle preternaturally, as if some sprain or other had caused that appearance : the contiguous parts, consequently, undergo greater fatigue than, in the event of finer symmetry, would have fallen to their share : and the extraordinary friction or working thereof, occasions, at a day more or less remote, the exhaustion of its powers (see Section 21), and the lodgement of acrimonious matter in the cellular membrane, which appears in tumour, abscess, &c.   This protuberant appearance of the muscle is most visible at the stifle [N. 30], and on the shoulder [M. 16], just above the elbow.

A more minute inquiry, however, on those points would lead me away— too far from my main purpose, at present ; I therefore return to notice, in the first place, the structure of the legs of such horses as, by their untoward posi-

---

* Firing is supposed to restore derangement of the integuments, by causing inflammation and retraction thereof upon the bone, so as to embrace it more tightly. This is effected by much of the muscle being taken up into the system, or sloughing off in the cure ; as well as the contraction of the flexor tendon (back sinew) and its sheath.

tion, entail on them the chances of producing some one or other of those evils that are known to afflict certain horses, incurably, to the end of their days. Thus, some are known to tread on the inner quarter of the hoof, others on the outside, without the real cause being ever ascertained, and remedies are frequently applied that have not the remotest chance of achieving any good, on that very account. Some horses "cut" in consequence of treading on the outer quarter; on the contrary, by punishing the inner quarter in treading, others contract a disposition to "quittor and ringbone;" both instances of mal-formation, or bad built (as I call it), produce splents, diseases of the frog, of the sensible sole, and of the coronet, as the case may be: how the various modes of wrong treading are brought on remain to be examined into hereafter. Meantime, it may not be amiss to observe that the right mode and make may be discovered by noticing the proportions of those horses, that, by the acknowledged just synimetry of their bones, the agreement in size of one limb with another, and the faultless manner in which these are attached to the body, go tolerably free from any such diseases, until old age, accident, or the misapplication of their powers, brings on disease.

6. There are, then, three kinds of mal-formation, or bad shape, attendant on the limbs of horses, which I consider original faults, those others to which they give rise being but secondary ones. 1st. That wherein the leg is ill-formed in itself. 2d. When it is badly joined to the body. 3d. When the fore legs disagree with the hind ones in length or quantity. Each, being attended by its respective defect in going, as to safety, speed, or strength, and liable to incur one or other of the ills enumerated, as appearing on the legs and feet—is worthy of the reader's separate consideration; although it frequently happens that an individual horse is afflicted with all three faults at the same time, the two first being found together, subsequently producing the other also. But I have generally noticed that one of those faults sometimes accommodates itself to the other, amending it considerably; as, when a limb that is too long is set higher up on the body than is esteemed right construction, in the same manner as a horse lame of a leg may be passed off for sound should the corresponding leg of his body also fall lame.* Much the same is it with the third kind of disagreement, in the opinion of many people; because it has existed in some celebrated horses, and they would have us believe that this very disagreement was itself the cause of the celebrity those individuals arrived at. This, however, was not the fact.

7. The *Phænomena* mare, unquestionably the first trotter of her inches in our days, never did her work in style: nobody could account for her achievements upon the view, and I had always my doubts whether hers was a fair trot, though I won upon her. In the trot she had an unaccountable shuffle. She was low before, but had the gift of taking her fore feet out of the way of the hinder, which fell (in the walk) about half a shoe beyond that of the fore ones, the feet reaching the ground in succession.

Laertes, a grey horse, hunted in Leicestershire, 1818, 1819,+ of no particular powers any where, and confessedly clumsy in the forehand, without much fire, was yet in the habit of taking the ordinary six-feet leaps with ease, and clearing a ditch of twenty-five feet with pleasure, often exceeding those admeasurements by nearly a fourth. *Eclipse* is known to all of us (as matter of history) for having had a low shoulder, which gave his fore quarters an awkward appearance: but this was compensated for by the fine form of his hind quarter, which, being particularly strong and muscular, threw his body

* Certain dealers are known to have inflicted lameness on the foot with this view. Horrid and disgusting as is the relation, 'tis no less true.
+ At that time the property of Mr. Maberly.

3

forward at every leap, in despite of his low fore quarter,—for running is no other than the leap reiterated. One leading characteristic, however, denoted all three horses to be of the right stamp in the main : they stood even on their leg-bones and the soles of their feet; that is to say, straight up and down, nearly, from the *elbow* [N. 16] to the ground before, and from the *stifle*-joint [N. 30] to the ground behind, respectively; both these parts, viewed sideways in the plate, being placed nearly horizontal, as regards each other, on the line [N]; at least, this was the relative position of the stifle and elbow, in the two first-mentioned animals, and of the third I do not presume that he was so, for "the history" of his form in this respect leaves us a little in doubt.

But "the shoulder of Eclipse was a low one," say the published accounts of him; yet, as this defect, real or supposed, consisted in the inclination of the shoulder-bone [K to N] above the elbow, by reason of the great freedom of the muscles which held it and the shoulder-blade in position, he would, when stepping out with the fore leg, rise higher than when he stood still; a particularity that is reversed in horses whose shoulder-blades are set on more nearly upright than those of Eclipse were. This accounts for the vaulting manner he had, as we read in the printed accounts of his exploits; and his running greyhound fashion, with his chest close to the ground, for he would thereby keep off the ground longer betwixt each leap, until the impetus received from his hind legs was nearer spent than it would have been but for thus holding up his fore feet. On referring to those parts in the annexed plate, they will be found thus drawn.

8. Viewed in front, the fore legs, upon which the safety and ease of the animal's going chiefly depends, should, to be perfect, be widest next the chest,

Fig. 3.

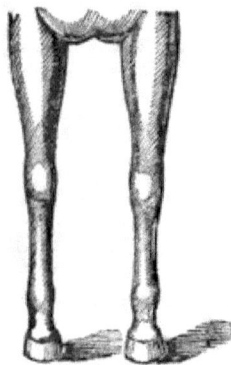

approaching each other gradually, until the eye, having compared that part with the pastern, scarcely perceives the difference. Here, the leg, taken by itself, is smaller, though the interval between the knees and the feet does not differ, on account of the width and flatness which ought to exist in the well-formed knee, yet, taken on the outside, considerably more breadth will be found above than below. Such a knee, when flat and finely marked at the joint, is always well covered in a healthy horse, (see section 15), he then throws it out with great freedom, and takes a firm step fairly on the entire bottom of his hoof; but, should the leg be ever so good a one in itself, yet placed too high upon the chest, where it is held, not by a socket or insertion of the bone, but by strong elastic muscle only, this throws the feet too near together upon the ground; the horse then treads on the outer quarter of his hoof, and wears away the wall; and, when tired, is most commonly given to cut. Endeavours are used in shoeing to amend this fault, by paring away the inner crust; but it is one of those defects in the built which no art can completely eradicate, and has been termed "pigeon-toed."

Nor is the matter rendered any better when, by reason of the knees turning in, the toes turn out, and the horse then treads on the inner quarter; and, however those of the one or the other description may have the reputation of great speed, it can be for a short distance only, because the action of such horses must be laboured and imperfect, particularly one of the latter kind of make. He must, consequently, fatigue himself more at every step, and tire sooner than one of the same size, and formed in every other respect similar but having legs that come nearer in shape to those in the annexed sketch [fig.

3]. That such knock-knee form is occasioned by weakness, is evident from the position of the knees, when the animal stands at rest. This he does by supporting himself at times like a dancing-master, with one foot before the other; and, no doubt, the twist with which his pace is always attended when going, occasions certain disorders of the feet, which he seeks to ease by shifting the weight alternately from one, to the other foot. He will, moreover, sooner "knock up," and ultimately "get done for" earlier in life, by reason of the origin of this species of malformation being seated high up on the limb, thereby incommoding the action of the shoulder-muscles: the elbow, at N. 11, by being pressed close to the ribs, having thrown in the knees, receives, at very step the leg takes, a kind of double motion, which, of course, doubly affects the action of those parts; and much fatigue, pain, and anguish succeed each other, until it communicates to the cavity of the chest, or other internal parts. Such animals have frequently the shoulders unusually muscular, hiding, in a good measure, the original defect from the eye and touch of a common observer; but it may, nevertheless, be ascertained to exist, by the symptoms just now mentioned, as well as by the appearance of the protruding muscle before noticed at Sect. 5. To knocked knees and inside tread, let me add the circumstance, that such horses have a broken pace, kicking loose stones before them, with a certain rolling from side to side, to the great annoyance of the rider. All this arises from awkwardness, by reason of the shoulder's bad position, whereby the leg being thrown sideways removes the foot in an increasing ratio from the centre of gravity, and, instead of its being thrown straight forward, describes part of a circle, more or less curved, according to the amount of the original defect.

Fig. 4.

The straight dotted line shows the space a well-formed foot, such as belong to the leg in our preceding sketch (3), would take, being on paper just one inch; the curved line shows the course, or nearly so, the foot is thrown which belongs to an ill-formed shoulder, contracted at the elbow: as this line is an inch and an eighth (1 in. ⅛) in length, the horse so formed does an eighth more work than one with straight legs would do on going over the same ground.

In addition to his other evils, a horse with such a shoulder (being muscular) is most liable to contract "fistula in the withers;" but, if not so muscular, "strain of the shoulder" is likely to attend his twisted manner of treading, when hard worked. With such a built horse, "splents" are usually more tedious than with a straight-limbed one; and strains of the sinews, i. e. of the tendon, as well as those of the coffin-joint, happen oftener, and appear with worse symptoms, in proportion as the limbs are more or less cross-built*.

9. Long and sloping pasterns [Y, 13—16, and Y, 34—39] partly denote the Arabian, are handsome to view, and make easy goers; but such horses soon tire, and, I may say, are generally weak, having the flexor tendon, or back sinew, considerably relaxed. The small pastern, or bone inserted at the hoof, always rises in a direct line from the hoof, both being about 45 degrees for saddle-horses, as at b, (fig. 5.) and the large pastern is then several degrees nearer to upright. These hoofs stand of an oval shape, and have small frogs. But some, as draught horses, have large frogs, the hoof round, and more upright by nearly ten degrees in early life, as at (c) in the annexed scale, in which case they are liable, if no change takes place, and they get older and weaker in the joints, to "knuckle over." But, getting aged, and the supply

* I reserve until a latter part of the volume what I shall have to say, respecting strain of the back sinew and of the coffin-joint, which I have thus named in conformity with the general vulgarism, in order to make myself intelligible to the meanest capacity.—See F ot. a section of

of nutriment for repairing wear and waste falling short, the horse becomes
pummice-footed. The wall or crust is then lower; and as the bottom of

Fig. 5.

the foot grows convex, causing the animal to slip about, so the front of the
hoof (a—e) grows concave, the toe (c)
almost turning upwards. Horses kept
for heavy draught have short pastern
bones, the small one entering the hoof at
the coronet (c) in early life, but afterwards changes, as I say, to (a—d); and
this new inclination, it will be seen,
must depress the bone, as the animal
acquires the sort of hoof called pom-
mice-footed, and causes a constant straining upon the coronet; hence, the crippling, insecure gait, horses of this description acquire, even before they get
old; and hence those numerous disorders to which the feet are liable from
this one origin.

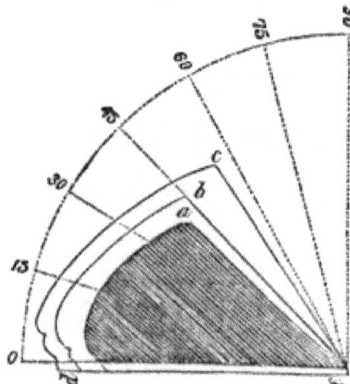

Contracted heels of this or any other species of horse, being destructive of
h.s capability of going, should be guarded against as much as any other individual misfortune to which he is liable: a disposition thereto constitutes sufficient ground for rejection. When this is the case, the interval or cleft between the heels, at (d) in the annexed scale, is found to be more or less tender,
according to the progress of the disease; the cleft will, in health, receive two
fingers lain in, the part having in it nothing unusual in the feel. Soon, however, the heat increases, the part hardens, and the cleft scarcely admits of a
small finger; the horse flinches as if you touched a sore, and nothing but time
and proper treatment can restore him, if any thing can. Most commonly,
however, the disease proceeds until the clefts of the heels meet and become
rotten. Pressure upon the frog, is the certain preventive of contraction. See
" Foot."

When the pastern-bones (great and small) rise one above the other too up-
rightly, the small one receives the whole concussion, and communicates the
jar to every minute construction of the internal foot. See Foot, section of.
The jarring of his pace is then very great, both to the horse and his rider.
Such horses are very liable to go lame occasionally, but they recover by rest.
The ass and wild horse (poney) are thus formed; but being hardy, and having
less blood and less weight of body to carry about, suffer less by it than the
horse.

The just form or elevation of the hoof in front, upon which mainly depends
its form behind, has been discussed by various writers, but remains yet awhile
uncertain and unsettled. Mr. B. Clarke judges 33 degrees of elevation from
the ground to be the best form of the hoof, and Mr. White quotes him with a
portion of approbation, but most unaccountably refers to his "plate iv," on
which an inscription tells us the fact is not so, but 45 degrees is the best possible elevation of the hoof: whilst those which are higher (lower he writes it,
or "33"), "approach too near the perpendicular;" the figure on the plate
itself differing with the diagram on the page of his book (305).

My ideas, however, on this subject are not so general; for I have found the
best form of the hoof differ, according to the shape of the two pasterns, as
they regard the hoof and each other; deeming that the best, in its particular
case, where the small one follows the same declination as the hoof, and the
large pastern ascends twenty degrees nearer to the upright, as before stated.

The preceding figure (No. 5) shows the outline of three feet of different degrees of elevation: *b d* describes the line of the coronet, or orifice, into which the thickest end of the small pastern-bone sinks, and rests upon the springy substance attached to the inside of the hoof, and which bone, we naturally expect, should ascend out of, and take the same direction as, the hoof, whence it springs. Any departure from this rule of nature is clearly an approach towards disease. In the paragraph above, I showed what mischief might be derived from an upright small pastern, such as would suit the outline hoof (*c*); of course, this elevation, or a greater, would be a mis-shapen hoof as well as pastern. In like manner, we knew that the pommice-foot is out of point and diseased, and it follows that the best possible elevation of the hoof must necessarily lie in the medium of those extremes, which we know to be diseases in themselves: this it is to determine a contest mechanically, without once adverting to the well-known circumstance of the health and free use of its heels, which attends the horse whose hoof is, at any time of life, near 45 degrees of elevation or depression. Did we require more arguments to prove this to be the proper elevation, a conclusive one could be found in the well-known circumstance of those hoofs of horses which are very upright in early life becoming the lowest when the animals get old; whilst those hoofs which come near the standard of excellence in youth (45 degrees), retain the same form, as nearly as the injuries of shoeing admit of, to an extreme old age.

---

10. So far as the foregoing observations on the fore-legs apply, they do belong, in every particular, and with equal reason, to the hind legs also; with the exception, however, of what is said concerning the elbow of the fore-leg, and its adhesion to the chest, for which we must now substitute the stifle of the hind-leg [N. 30]; and add, instead of the kind of defect described as being occasioned by the contraction of the part, it is here owing to the expansion or spreading of the stifle from the sides. This throws the houghs together, and forms "cat-hammed horses," as they are termed; the mode of going such animals are constrained to adopt, the circle their hind feet describe, at every step, the additional fatigue they undergo, the awkwardness of their tread, and the consequent diseases communicated to the sole, lately described (in sec. 6.) as pertaining to the fore-leg,—most undoubtedly afflict the hind-leg also, with the additional fact, that this one is more liable to "grease." At rest, if an animal so built does not place one foot before the other, his houghs not unfrequently touch each other,—poney's and low horses more particularly so; and it seems worthy of remark, that this species of mal-conformation seldom appears on the fore and hind-legs of the same animal. Indeed, I can not recollect having seen one instance, and I am thence led to conclude that this twist of the legs is a contrivance of Nature to accommodate itself to the disproportionate length of legs before or behind. But, when it so happens that the strength of the parts resists this bending of the hough or of the knee, such horses walk higher behind than before, and *vice versa*, *i. e.* when one pair of legs seem to have outgrown the other pair; a defect which, though

Fig. 6.

3 *

often overlooked, is no less deserving of notice. The wound termed "over-reach" is inflicted by the hind-leg of this formation upon the fore one. "Forging" is, likewise, occasioned by the hinder toes striking the shoe or shoes of the fore-feet; and is sometimes brought on by injudicious shoeing on feet of the very best construction, and a loose rein; it is, therefore, to be corrected only by the contrary practice, keeping the hind-toes short, and the heel of the fore-foot low, and driving with the reins borne up. By these means, the fore-hoof will spread at the heel, and the animal be enabled to take it out of the way in time for the hind-foot to occupy the identical spot on the ground it had just quitted; for very few horses have the greyhound tread of Eclipse before-noticed, wherein the hind-feet tread much wider than the fore-feet. Neither is such a gift desirable to any but racers, perhaps: nor is it, indeed, compatible with the duties the generality of horses have to perform.

The great additional labour horses with houghs so formed undergo, added to the pain and anguish of continuing it, occasion irritation of the whole hind quarter, that communicates itself to the region of the kidneys and intestines, and superinduce inflammatory complaints, which frequently terminate unfavourably. Constitutional diseases appear on the leg and foot behind oftener than before; and those of the Coronet, with Curb, Thorough pin, spavin, strains, windgall, scarcely fill up the catalogue of evils caused by, or receiving aggravation from, too much expansion of the stifle, with its attendant, the cat-hammed hough, and, consequently, a twisted tread of the hoof. No doubt exists in my mind that Eclipse would have been a cat-hammed horse had he been raced at two or three years old, as our practice now is: both he and Flying Childers were five years old before they started on the turf. Heavy long-legged children of our species, in like manner, become knock kneed men, by being put on their legs too soon; this form of their knees deprives them of calves to thin ill-formed legs, and the thigh, too, seems wasted, when the deformity is great.

11. When the fore-legs are shortest, the horse, whilst going, nods his head up and down a good deal, as he does when these are either weak, tired, or tender of foot: when they are very feeble, without any other ailment, he carries the head high constantly; but he works his head from side to side when the same subjects of complaint assail the hind legs and feet. Poneys being ever out of point in one or other of these respects, afford unerring proofs of those remarks. The value and advantage of the straight position of a horse's houghs are never more apparent than when he rises upon his haunches to take a leap, a service which never was performed satisfactorily by a cat-hammed horse, because he seems to hesitate about what shall be the distance between his feet at the precise moment they are to leave the ground: a blunder which is most visible in the standing leap, when the feet are seen first to straddle to their utmost; in an instant they are brought so close together as to lose all purchase, and he goes over from an intermediate spot, the whole transaction occupying as much time as does the counting of one, two, three, and away!

The motions of the head are always good indications of pleasure as well as pain. A horse will frequently throw up his head, almost in his rider's face (as if to rebuke his barbarity,) when he has been hit on the head or ears. (See Section 16.) He looks at his flanks dolorously when affected by a dull pain in the intestines; if it be sharp pain, he turns about quicker: he thrusts his nose towards his chest, when pain assails his lungs generally; but when one lobe only is affected, he turns his head only to that side. If a horse be girthed too tight he will sometimes (justly) bite his tormentor, for this operation retards the action of the muscles between the ribs and of the ribs themselves, so that the lungs do not get room to play. (See Section 31, and Introduction, page 4.) Old horses contrive to avoid this punishment by "holding their

wind" (keeping the lungs filled) during the girthing; a fine proof this of Nature's dealings, for which they usually either get kicked under the belly, or
hit about the head; but both kinds of punishment are the harbingers of further
disease, viz. the first of the blind gut, as described at Section 48 and 49; and
the other leads to poll evil, as described in Book 2.

A horse is frequently found to have contracted lameness in the fore-leg
without showing any visible sign of its exact situation, and applications to the
shoulder is the usual remedy in the hands of the generality of common farriers. Some of them imagine the strain is situated lower in the leg; but they
are no nearer the fact, though they are to the spot. A defect in the conformation of the limbs occasions the foot which leads to come upon the ground with
more force than its fellow: the concussion of the hoof is greater, and is un-
equally placed when the leg is a-twist than in the upright form; the leading
tires sooner, and the sensible sole becomes inflamed when the horse is constantly urged to step out with it, the affliction barely showing itself between
the frog and the toe, if any where. If a horse receives the impulse to proceed
from the right hand or heel, he will step out with the fore-leg of that side, accompanied by the hind-leg of the near side; but his rider, or driver, should
early teach him to change the leading-leg, by sometimes touching him upon
the contrary side. It is worthy of note, too, that the horse which executes
this change with the least trouble, and oftenest, has most power and command
of his limbs. [See Index—Fever in the feet.] When both legs before are attacked, the horse exhibits a crippling uncertain gait, not unlike that of a
drunken man, whence the term "groggy" has been applied, and, if he is not
timely indulged in rest and a run at grass, he is a ruined horse, and becomes
soon what is termed "foundered," of which disorder there are several kinds.
The mistaking one kind of founder for another generally costs the animal his
life, sooner or later, and the studious inquirer had better turn to the next
Chapter (at sect. 21, paragraph 3), where he will find a few words on chest
founder, many of the symptoms whereof are not unlike this of the feet.

Horses full of feed, and requiring purgative physic, stand with the legs
stretched, more than our second cut, at page 7,—inordinately at times. Old
Gibson attributed it to vice, and a disposition to kick, when a horse holds his
toe scarcely resting on the ground; this is not always the case, for his fore-
leg is as frequently so held a-trip as his hind one; and I consider it the token
alike of either sore feet, or of incipient founder.

12. Besides the disproportion the fore and hind legs bear to each other,
another series of defects in construction exists between the length of the fore
limbs and that of the trunk, being sometimes most apparent at the belly and
flank, at others on the back, its tendency always depending on the turn taken
by the latter. Although this is the old English way of judging of long car-
cased horses, Lafosse (an old French farrier) took the measure of proportions
more properly from the breast-bone to the buttock, in the annexed plate being
from the parallel line 11 to 38; then comparing this with his height, he tells
us "a good horse, as we can learn from experience, should be a *tenth longer*
from the breast to the buttock than he is *high* from the top of the shoulder to
the ground." The latter admeasurement will be found upon the annexed
plate to extend from the line [D to Z] and, with the former, will compose a
square rather wider than high,—the integuments being removed from the
bones on all sides. My notions of just proportion, however, differ from the
French standard, though they do not run into the contrary extreme; for I can
not help thinking inordinate length of body, as compared to a horse's height,
a very great defect as regards his health, that form being invariably attended
with meagre, washy flanks, and a painful manner of going. But the Flan-

ders and Norman breeds have all this tendency; and they are invariably of a sluggish nature, when the belly, also, hangs low.

The major part of our horses of this built have their sides falling in, more or less, towards the hind quarter, some few of them to such a degree that the flank appears as if it were fastened to the loins.   These are remarkably poor feeders, have a good deal of short-lived vigour, without the gift of keeping it up at any kind of thing.   Nutritious food, but less in quantity, does for horses which are out in the first-mentioned point all that can be done, and that is very little : those of the second species of bad form can not bear long journeys, nor long privation, or they contract flatulencies and spasmodic cholic.

Another species of disproportionate length, as compared to height, consists in what is called " high mounted," the limbs having then much more length than the body ; a defect that is rendered still more apparent when (as generally happens) the horse is also roach-backed, like the first sketch of back

Fig. 7.

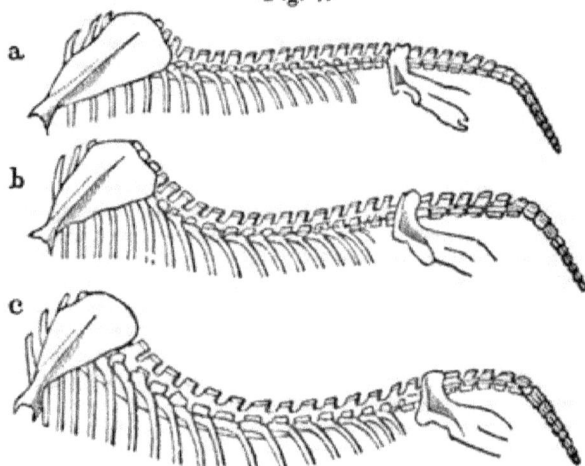

bone in figure 7; and it is still more striking when a little man is mounted upon it with a saddle that is ever sliding forward upon the withers.   Such a form always denotes weakness of limb, and want of freedom in the fore-hand; nor can a horse of this built take a long step, or trot well, or thrive in the field, by reason of the difficulty he has in reaching the grass, which induces him to bend one knee forward, whilst the other leg is drawn back under him.   A ludicrous story is even told among horse-dealers of a horse so formed having starved itself in the fields, while the food lay within an inch of its nose; and though such stories are no argument, they, nevertheless, convey the general feeling of the narrators, which is seldom completely wrong.

13. But a horse may be short in the carcase, which is not exactly "high-mounted," in my view of the term; since much will depend upon the shape (or bend) of his back-bone, to bring him under the one or the other description.   We have seen what sort of character a roach back bestows on a horse, the direct contrary form, or hollow back-bone, [see the lowermost sketch in the last cut (c)] is no better, though built upon long limbs, horses with this shaped back being in all cases weak in the loins ; and, therefore, are they more liable to contract inflammation of the kidneys, and to resist the cure longer than those of any other shaped back whatever.   Yet are they preferred by

timid horsemen, principally on account of the easy seat a hollow back affords
Great caution in administering strong repeated diuretics should be impressed
upon us at the sight of a very hollow-backed patient. See Sections 53—56.

When the bend in the back-bone, or "hollow back," is restricted to the fore
part of the animal, the loins being well filled up, his built in other respects is
less material, to be "short in the carcase" being then an advantage; and it is
much greater when the bend is confined to a gentle curve, scarcely distinguishable, just behind the withers. [See middle sketch in the last cut (b)].
This is considered a straight back, belonging to a light made, compact horse;
he is invariably ribbed home ; and, as there then exists but a small space between the last rib and the hip-bone, as seen in the plate at 30 to 32, so is it
always accompanied by the deep chest, good hind quarters, and wide loins
(i. e. not pinched together), and his ribs finely curved. Horses so formed are
always healthy, and esteemed at first sight super-excellent, being supposed,
with justice, capable of doing more work than those of any other built whatever. But the gift of leaping or of great speed do not always belong to horses
of this form, however perfect in other respects, though health, vigour, strength,
and lastingness do. Such horses always feed well and retain their condition.
The inquiring reader would do right to turn back to what is said under Section 10, and draw his conclusions from what is there stated.

14. Low-buttocks generally accompany roach backs [see sketch a.] and are
always attended with another fault—"hind legs too straight," and incapable
of stepping out. A horse so formed can execute no pace tolerably, and trotting worst of all. In the drag, such a horse steps short, and is always upon
the bustle. as if his legs were tied. A large head, with short thick neck, denotes a sluggish horse, heavy in the hand, and usually "carrying low:" these
are faults generally attending his entire breed. When the neck is longer, the
case is not bettered, for then the animal is of the long-bellied kind, with thin
flanks and washy. As a small light head, but wide at the forehead, with expanding nostrils, and bold prominent eyes, denotes (blood) strength of body,
and vigour of constitution, so the contrary may be looked for in horses which
have narrow foreheads, small or sunken eyes, and small arid nostrils. I
never saw a fine well-turned head that did not belong to a good set of legs,
well fixed upon the body; the correspondence goes still further, inasmuch as
the quantity of white in the face is commonly attended with a proportion of
white upon the heels, thus : a star, one white foot; a blaze, two white feet;
white face, four white legs, &c. Horses with large jaws are given to keep
open their mouths while at work ; and, when aged, grind their teeth more
than is necessary in feeding. The manner of breaking his food being, with
the horse, different from that of other animals, viz. by rubbing his under teeth
from right to left against the upper ones,—a motion to which the term "grinding his corn" has been applied,—an old horse will sometimes continue it when
he has nothing to eat, thereby wearing away his teeth ; a circumstance that
occasions imperfect mastication and its consequences,* besides subsequently
leading us into error in examining his age. Hard-mouthed horses, and those
which champ the bit much, fall into this idle habit.

Flat, or narrow-chested horses are subject to those attacks which lead to
consumption (see Section 36), and, consequently, are liable to show bad con
dition ; or, it may be, that disorders of the chest do contract its capacity. In
some horses on the contrary, the cavity of the chest seems too great for its
contents; they are short-winded horses of one description (there being several)

---

* Indigestion, flatulency, cholic, &c. are all produced by animals swallowing their food un
broken.

that are afflicted with these kinds of mal-conformation, or disagreement in size between the parts containing and those contained.

15. My purpose in making this exposition of the ill-effects produced by mis-shapen limbs, &c. on the horse's health and usefulness, would be incomplete, were the original causes thereof left unnoticed. The most remote, or more general one, resides in the breed, or the manner of breeding the animal, whence we are sometimes led to say, "what is bred in the bone will never go out of the flesh." As regards the kind of stock from which to raise a supply of young ones, breeders may undoubtedly suit their own fancies; but it must be seen that a brood mare which receives too much of the horse for her capacity, will produce a foal *all father*, as it is called, being at the same time larger than she can conveniently carry; it then bids fair from the beginning to be a mis-shapen animal. This happens oftener than is commonly imagined; but it is easily prevented by adopting a horse for her whose strength comes tolerably near that of the mare. Disregard of this precaution is found to produce the first foal much smaller, though more lively, than the next and subsequent ones, especially if care be taken in the latter case to give her a horse more and more vigorous as she becomes more roomy. For it must be clear to any body [upon mechanical principles again] that if the fœtus, growing too large for the cavity in which it is generated, originates too much bone, it must determine towards some particular part of the young animal; and the colt will be brought forth with that deformity, and carry it through life, after plaguing two or three of its owners with fruitless endeavours to physic off its ills.

I say nothing whatever of the cross to be adopted; that, being contrary to my plan, would carry me too far away from my main subject. But I may observe, in passing on, that no breeder in his senses would think of employing a horse to raise stock that has served half a dozen or more mares in the course of the day; and yet nothing is more common, nor more inevitable when the payment for covering is low (say a guinea or two), than that the smallness of the sum must be made up by the number of mares served, the price, keep, and attendance, upon stallions being expensive. This error must be so palpable to any man who calls himself a father, as to render any further argument upon that topic utterly unnecessary. Some twenty-five years back into the last century, I recollect reading a well-attested account of a celebrated horse's dying in consequence of twelve or fourteen successive efforts in procreation; and if such be the deplorable case with the parent, what strength, bottom, or lastingness, can be hoped for in the progeny so begotten? Nevertheless, I am of opinion that a vigorous horse which may have been freely engaged (if early in the season) may be in a better condition for raising large and lively stock, than under any other circumstances whatever, except recent exhaustion. Aged stallions produce hard-mouthed foals, and further proclaim the ill-adapted ages of sire and dam by extraordinary hollowness over the eyes.

Much depends on the country, the climate, or kind of land, in which the gestation or breeding may be carried on; and it may be presumed, that no one in his senses would choose such a situation as is known to be disadvantageous to the particular kind of breeding he may have in contemplation to pursue; whilst those who may already be so placed, have no right to complain when they engage in a branch of business thus ill adapted to their plan of farming, and they get disappointed. As both objections lie at the option of the parties concerned, they require no further remark; but another point of consideration well worthy our careful attention is, the treatment the mare ought to experience at our hands while she is breeding; this being a matter of some moment, and within every one's control, should not be neglected. Though a brood mare in foal requires no pampering at any period, yet it is clear that, from the third or fourth month, she should not be worked so hard as usual, and from this period to the day of her foaling, the duty to be required of her should be less and less every week. Nor, on the other hand, is complete idleness befitting her situation: in cases where she has not been used to hard labour, a run at grass, in a paddock, with access to an outhouse or stable, as it leaves to her option the quantity of exercise her strength is capable of sustaining, would be found most conducive to the best purposes of nature. Her food should be of the first quality, and regular, and, though full enough, should not be too much. Occasionally, she may be off her feed, during the "time," but she does not therefore require "physicking," nor coaxing to eat. Great care should be taken that her body is emptied regularly, that no derangement take place either way: and that if opening physic is required at all, aloes is not in her case the best that can be prescribed for that purpose, since they act mostly upon the intestines lying immediately in the vicinity of the foal. An opening draught or drench should be substituted for the pill, as its operation begins sooner.

A very general cause of mis-shapen limbs is the placing upon younkers too great weights at first, whereby the houghs or the knees are thrown together particularly when the animal is constructed with the fore and hind legs disproportioned to each other, as noticed at sections 9 and 10. Splents and sprains are the inevitable consequences of mounting colts, &c. too early in life; and hollow back is oftener induced by this premature error than existing originally. As if all this were not enough, many breeders nearly starve their young ones until they are brought into use; whereby they become deficient in solidity of bone and quantity of muscle, if they do not imbibe some internal or constitutional malady, and the event of their limbs growing mis-shapen is no longer left to chance.

16. Notwithstanding all that has been said and done, little would avail the finest proportions of the bones towards the formation of fine-shaped limbs, least of all to symmetry of the whole horse, but for the seemingly adventitious circumstance of the covering with which they are immediately invested; and which, embracing tightly several bones, and connecting them together, constitutes a limb. Some of these coverings are confined to the joints only, holding them in position as near as the Creator designed them, unless accident (of parentage, of birth, or misusage), as before described, should induce them to a perpetual strain, and they enlarge at these joints in spite of the next or universal covering of the bones: this is membrane (of which more shortly,) the uses whereof on the bone may be illustrated by taking a stocking of good length, and having filled it with pebbles of its own size, and tying the end tightly, a stick or club is produced of some degree of flexibility resembling a limb and its joints. If the tying be not performed well, by bracing the stocking to its utmost, the flexibility of certain parts (or joints) of the limb will be greater: it will possess less strength at the joints when bent, and be liable to give way or break unless supported by some other covering. It is easy to perceive that the horse which has those coverings in the highest perfection would move his limbs more correctly after the fashion they were designed for, than he which constantly strained them out of their places. He who was endowed with the first-mentioned quality in perfection would be considered a sinewy tight-built horse; the second kind I have already depicted in section 10, where the houghs are described as keeping those integuments in a perpetual state of derangement, straining or twisting them in such a manner that constitutional enlargement at the joints is the consequence.

At the ends of all bones, a yielding substance, in appearance like bone itself, prevents friction, and by its elasticity gives a spring to the animal's steps. The ease of a horse's going mainly depends upon this substance, which re ceives the name of cartilage, and is liable in some measure to be absorbed or taken up into the system, or, in cases of diseased joint, to become stiff and bony. Consult sect. 23, &c. on those points of information. We may notice this absorption in very young animals, whose bones are all substituted by cartilage, until the blood furnisheth the means of forming a more substantial frame, such as I have been describing; and teaches the validity of some remarks I made in a preceding section (15) on the kind of attention we ought to pay to our brood mares while the *fœtus*, or unborn animal, is being formed in the womb.

Not only between bones, and embracing ever joint, but at the termination of the four legs in their horny feet, is this springy substance to be found, the whole being liable to wear out, to contract or to harden with age or disease. Besides this casing of the joints in cartilage, the ligaments connect or tie the bones together. These ligaments are seldom troubled with any ailment but that of great lassitude when the animal is tired, and occasionally to sprain. This accident takes place when the horse steps aside upon uneven ground, and the ends of the bones press laterally upon the ligaments. It follows, of course, that mis-shapen horses whose feet are always constrained to take an uneven tread must be subject to a constant strain, and must be more liable than others to incur permanent accident,—every step forming a trivial one.

But the ligament demanding the student's most serious attention is that which suspends the neck bones, on the same principle as our old fashioned lamp-irons are suspended by a small one from above, only that the ligament lies closer, and covers the intervals of the upper side, as at *a—b* of the annexed sketch.* So placed, and passing from the skull to the backbones, to both

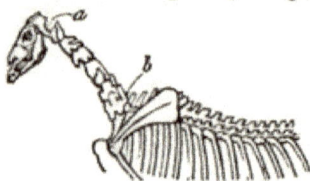

of which it is fastened, it has the power, at the will of the animal, of bending down or drawing up the head, which would, in fact, but for this support, fall to the ground. Horses in their last moments, when that will may be supposed to have left them, always cast back their heads considerably, by reason of the contraction of this strong ligament during the paroxysms of departing life. At *a*, however, where is the seat of poll-evil, it is usually thin, the cavity there found between the bones being mostly filled with muscle (s. 27); but this does not happen invariably, as some horses have little or no cavity to be filled with ligamentary substance, or with muscle. Our frontispiece is the portrait of a subject of this latter kind; but the reader is referred to some subsequent observations and cases on " poll-evil " for more detail on this hitherto-neglected point of conformation.

17. At the joints formed by the bones and covered by cartilage, the whole are surrounded by a strong membrane, which wraps the bones tightly, and secretes an oil at the joints for its further defence from the effects of friction. Of this secretion, and of the membranes generally, some further notice is given in the second chapter at section 22.

This strong membrane is not, however, confined to any particular part, but continues its close attachment, or embracement of the bone, over the entire frame of the horse. Throughout its extended course it serves as an excellent holdfast for the sinewy ends of the muscles (see sect. 27), which are attached to it above and below joints, whereby they act as levers to raise the lower bones of the limbs, as described hereafter.

---

* Called by the learned " cervical ligament " and " the cervicular." In operations for the poll-evil this ligament is frequently divided by the unskilful farrier cutting it across rather than lengthwise, which is the only right practice.

According to the parts this membrane may cover, it has received from the learned in hard words and many, a separate name for each, as it that course would further the cause of science; and whenever they speak of it as being found upon the joints, and skull, or the bones generally, they term it perichondrium, pericranium, and periosteum, as the case may be: why, no one explains. It has been considered insensible, because in health it has not t ie sense of feeling so fine as other parts of the system, which are furnished with more nerves (s. 30); but, the very few of these fine organs with which the membrane of the bone is furnished, renders the pain occasioned by disease, whenever it may be attacked, the more acute; when flying from one nerve to another, those well-known shooting pains are felt (by us) that are universally mistaken for pains in the bones themselves. We do not go too far in inferring that the horse is similarly affected. This takes place in splents and spavin, when the bone enlarging forces its way through this tightly-braced membrane, and causes inflammation, temporary lameness, and, at length, those well-known appearances I have just named. In the living horse this membrane is red, by reason of the fine blood-vessels with which it abounds; but in the dead subject, the supply of blood being withdrawn, it then turns white.

CHAPTER II.

Concerning the Horse's Inside, of its Conformation, the Functions of the Organs of Life, and the Diseases to which each is liable: together with Outlines of the Principles upon which the Cure is to be effected.

18. Such, as I have endeavoured to teach, being my view of the external frame or structure of the horse, which I have termed its built, I come, in the next place, to speak in a more particular manner of his inside; noticing, as I pass on from one part of him to another, the seats and causes of his diseases, with a view to their cure, but referring you to the second book for the separate treatment each requires. In the third chapter will be found my reasons for following up the principles herein laid down, by a line of practice, at variance, in some material points, with the present mode of treating the animal in health as well as in disease.

Organs.—But, before I proceed to describe those several parts of the horse's inside, there appears to me an absolute necessity for previously making the unlearned reader better acquainted with a few general topics, that we may proceed with the details smoothly and more intelligibly together; viz. the names, uses or offices and powers, of that infinity of small organs which lie spread over most parts of the body, and belong in common to several of these parts in nearly equal degrees. The large organs, having the power of carrying on the animal system, first, as regards digestion, secondly, those employed in the circulation of the blood, and third, those of respiration, are too well known to the sight and touch to require explanation here; yet are they (the heart, kidneys, lungs, liver, &c.) composed or made up entirely of those minor organs I mean first to describe. But the precise way in which these act in and upon the large ones, the great share they hold in furthering the system of animal life, and the eminent rank their services maintain in restoring health when the system is any way disordered, has not received, in the practice of horse-medicine, that share of serious consideration the importance of the subject imperiously demands. To these points, then, I shall shortly call the reader's undivided attention; meantime, as some cramp words and phrases are

4

applied by most people (writers and others) to those offices of the animal's organs, they stand in need of previous explanation.

19. Each kind of organ, whether small or large, was designed by the great maker of all things to perform some office towards the preservation of the animal in health.   When such office is performed properly, as ordained, the organ is said to "perform its functions well."   For example, the heart is given for the purpose of sending the blood through the arteries, all over the body; but when the pulse beats low or irregularly, that organ is said to "perform its function badly;" when it ceases to beat, this function is lost or gone. So, certain of the organs are said to secrete something or other that is liquid; the doing this is their function; the power of doing so, that of secretion; and the article secreted or collected together, is called the secretion of this or that organ.   Thus, the kidneys secrete urine, and it runs off (sect. 53): the glands, under the jaws, secrete spittle (saliva), which passes off with the food by the intestines; therefore are they properly considered as excretory also, seeing both the secretions are drawn together for the express purpose of being so sent away, this last by the grand canal (or gut), as the first mentioned is by the bladder, and the perspiration is through the pores of the skin.   But some secretions are found that have no outlet visible to us weak mortals, though they find their way through the skin, sensibly enough at times; and this then becomes the sensible perspiration or sweat, but when we do not see it, this third species of evacuation is termed the insensible perspiration; and in health, one of the two is always in action,—in disease not so.

When, however, it happens such functions are obstructed, or, on the other hand, too much of either secretion is furnished to the system, then disease begins; as does, also, our duty of finding out what part of the vast machine has ceased to perform its office properly.   For, without this previous information, no man can possibly know how to apply the remedy in restoring the disordered organ to the proper exercise of its function; nor can any one hope to arrive at this desirable point of veterinary knowledge, unless he has acquired the means of ascertaining where, when, and in what degree the mischief has taken place, by patiently examining the action of those organs while in health, and comparing their appearance, after death, with the particular symptoms which preceded that event.

20. SECRETION.—Although, as I say, the secretions just spoken of are important in themselves, and of several sorts, as bile or gall by the liver, urine by the kidney, &c. yet the chief object of our present notice is the secretion of a fluid, more or less watery, which pervades the whole system.   It differs in quality a little, and very little any where, being adapted to the nature of the parts requiring its aid: 1st, In softening and enabling them to move freely over each other (as, between the ends of bones); 2d, Acting as a defence against injuries from extraneous bodies (as on the inner coat* of the intestines); and 3d, To prevent the parts from growing together (as the liver to the midriff), &c.   Misfortunes these which invariably happen when the supply of this fluid falls short of the quantity required for a long while together; and this is the case whenever the animal is worked until the fluid, at some part or other, is exhausted: a circumstance that strongly bespeaks the propriety of allowing the worn-up poor creature more frequent supplies of water although this be done in smaller quantities.   Inflammation, or fever, which is occasioned by suddenly checking the secretion, eventually exhausts this moisture by its great heat.   Both those disorders are therefore referred in the

The surgeons of human practice will observe, that I here transgress the doctrine of the surfaces; but they will please to recollect that my object is to make myself understood to a certain class of readers, of which they compose a very small art.

sequel to the same origin; the first being local, or pertaining to some particular organ or part, whilst fever pervades the whole system, and the solids in particular. The total absence of perspirable matter marks both diseases

On the other hand, when too much of this fluid is secreted, and remains unabsorbed, disease ensues : upon the heart it forms "dropsy of the covering of the heart ;" on the covering of the lungs it becomes "dropsy of the chest ;" in the membrane of the belly it forms "ascites," or dropsy of that part, and usually falls into the scrotum. The powers of medicine have hitherto proved of no avail in the first description of ailments; and are but partially applicable to the last mentioned ; the operation of tapping too frequently disappoints our hopes, to induce us to rely upon it as any other than a temporary relief, and it is, therefore, seldom or ever applied to the horse. Thus, in whichever way we view this important secretion, its eminence must strike us as quite equal to any other. Whenever obstruction in this part of the system takes place in the horse, the consequent adhesion of the parts being invisible, he gets worse used by his inexorable master for his inability to perform his usual work, and he soon falls a victim to the lash, the spur, and the bit. At the joints, this fluid is considered to be an oil (cynovia); at the heart it is confessedly nothing but water : whilst it partakes of a mucous, or slimy nature at some other parts of the body. This is the case with the membranes of the throat and gullet ; on those of the nostrils, the heat of the horse's breath converts it into a "*viscid mucus ;*" when the secreted watery particles come off by sweating, it assumes a white or milky appearance, after a little time appearing thicker and more slimy as the sweating continues, and the watery particles becoming less and less, its fluidity is also lessened. See membranes, sect. 26.

21. In all animals, the secretion of this watery fluid is carried on by the membranes, which are thin films placed between the various organs, over the bones and among the fleshy parts. These not only secrete, but sustain the fluid in its place, for the purposes above mentioned, and being of various texture or fineness, the fluid that is so secreted and held to its purpose by each, partakes more or less of water, is more or less slimy, or consists more or less of an oily nature, according to the use it may be designed for. Each kind of membrane, and its proper secretion, has received a learned name,—the first being called serous, the second mucous, the third fibrous; but, having resolved to abandon learned words, whenever the thing can be understood as well without them, I find less occasion for introducing them here than is generally practised. For, the peculiar nature of the horse having assimulated together, by its action, the three kinds of secretion more so than is the case with other animals; and its habits contributing as much more to the hasty calling off of one kind of fluid from certain parts to the assistance of another part, which may have been exhausted of its kind ; and as the treatment of the horse in all cases of a disordered secretion of these fluids is the same throughout, the action of medicine upon one always affording the assistance to another (as I shall prove shortly), there is no such necessity for carrying the distinction farther in horse-medicine, although it may be so in the human practice.

Perspiration is always at a great height in the horse; it is one of the chief means of cure in most of his disorders, and consists in drawing the watery secretions from all parts of the body. These pass to the surface readily, coming through the membranes from the joints, the solids, the bowels, and their coverings; as may be noticed in the case of hide-bound, upon opening the animals that die in this state of exhausted nature; the mesentery canal (hereafter described) is invariably discovered with yellowness, being, at times, almost orange colour; but I have as constantly found the lacteals of a fine coated horse shine through as white as milk. Again, on over-working the horse, so much of the joint-oil is sometimes drawn off by perspiration, that he becomes stiff in the knees, for want of that softening quality which kept the parts supple; we feel the same ourselves upon such occasions; and in taking off the knee or the hough of a permanently "stiff-jointed" horse, I have invariably found the joint-oil affected; in very bad cases it no longer existed. During life, the escape of this oil, by reason of wounds (as bad broken knees), leave the joint stiff. Further comment on its uses is unnecessary; but those facts should teach his owners a practical lesson of moderation.

On the subject of absorption of these secretions, I noticed many years ago, a very ingenious reason assigned for "lameness of the fore legs, of English horses particularly," in the great work of La Fosse, the elder, on what he calls "Hippo-pathology," or the diseases of horses. He says, "The fluids which did lubricate the parts (the shoulders) and keep them supple, being reduced in quantity, the food flying off by sweat, the remainder gets thicker in consequence, and the solids of his limbs become stiff and dry." It happens, mostly in the fore limbs, and he calls it a cold or chill, and says, page 267, it resembles a "stroke of the shoulder,"—" *Cheval froid et pris dans les epaules.*" A species of founder, that is clearly not to be cured by external applications, (as the oils, firing, &c). but by restoring to the part the function of secreting a sufficient supply of the fluid which had been so exhausted. In these few words are included the whole secret of my method of cure in such attacks; and, in this case, gentle sweating is that remedy which is best calculated for restoring the function.

22. When the skin does not permit evaporation, and sends forth the secretion by perspiration, disease has begun, the hair looks staring near the part affected, and not a stable-boy exists, who, when he sees a horse with a rough coat, can not tell that "something or other is the matter with him." This arises from want of moisture within; the skin itself not having the power of secreting or drawing towards it, by effusion, the moisture which is necessary to keep it supple, it shrivels up, and this important evacuation, which is second only to the urinary, is then stopped, so that even the insensible perspiration ceases. Some idea respecting the amount of this insensible evacuation may be formed, by placing a horse, that has been exercised, between ourselves and a well white-washed wall upon which the sun shines: when the shadow of the insensible perspiration may be seen upon the wall ascending in tolerably thick volumes, something very like steam from a boiling pot. Indeed, the insensible perspiration is, when compared to sweating, the same as warm compared to boiling water.

Yet, although we do not know the exact workings by which this internal effusion (as it is called) of the watery particles from one part of the animal to another takes place, we do know, accurately enough for our purpose, that abundantly perspirable matter lies in and upon the intestines; as any affection of the heart, arising from the organs of sense (sect. 30), causes a sudden suffusion of blood in the skin, and induces heat and irritation there;* so do the intestines send forth their watery particles upon the slightest occasion, to the same place of exit, in order to moisten and render it more supple. Whether the very transparent membrane, called *peritoneum*, which sustains the bowels, or that other large part of it which covers these and all parts of

---

* rear, for instance, of the dealer's whip often occasions the skin to contract and expand, so as to cause the tail to shake with every alternate vibration of the heart; and I once rode with a Jew, a right-out journey of forty-four miles, who whipped and spurred his horse to such a egress, that the hairs actually fell off from his tail, except a few at the end, an occurrence that is usually ascribed to scrophula on the horse's hide; a disorder it might have also laboured under, — aught I know to the contrary.

he inside, is most concerned in this secretion and effusion, is not worth the trouble of inquiry here. But, in addition to what is said in the last section respecting the colour of the lacteal duct, as it passes along the mesentery in cases of hide-bound, I may be allowed to observe, that we may daily witness the sensible perspiration from young and healthy horses to contain more of water than is found in feverish, old, or generally unhealthy animals; and that with these the sweat is more frothy, or becomes so much sooner, his mouth gets clammy, and his tongue dry and hot underneath, with less work than they; and that horses so affected are always found insatiably craving after water. Moreover, as regards the connexion that subsists between one part of the animal and another, I have many times found purging physic, given in the usual doses fail of the effect intended, and come off in the shape of profuse perspiration. Not only in those large and decided doses that are intended to produce much effect, but even milder ones, as alterative-laxatives often turn out of their course, and, as well as diuretics, not unfrequently disappoint us in the same way, the latter also coming off by the skin instead of urine.* It follows, of course, that the less sweating a horse has got, the more he must stale, and accounts for the profusion of the latter kind of evacuation in winter, when he scarcely ever sweats, and perspires, but little, comparatively speaking. As a farther proof of this connexion between the secretions and evacuations, let any one notice a horse when he first stales in consequence of taking a diuretic, and he will find a transparent water hanging in little globules at the end of each particular hair of his coat all over his carcase.

23. We come now to speak of glands, nerves, membranes, absorbents, (being 1st, lymphatic, 2d, lacteal,) and muscles, which are the names writers and practitioners of eminence have agreed upon to speak of those numerous minor organs that are employed throughout in carrying on the functions of animal life, and the uses whereof I shall come shortly to explain. The reader is already aware of the sinews, of three kinds, that more immediately cover the bones and keep them in their places (sect. 16, 17), to which if we add the bare mention of the muscular, or fleshy parts, and refer to the "circulation of the blood" (sect. 37—44), for a description of the veins and arteries, he will have before him the names of all the integuments of a horse's body beneath the skin. Detailed particulars respecting all these follow next in their order; the larger organs of the inside being reserved to the subsequent sections of this chapter. By this course of proceeding he will be better enabled to comprehend, as we study those things together, why and wherefore these were given to the animal, and what functions each has to perform in health; or these ceasing, or being obstructed, we shall be led to consider in the next place, what species of remedy is proper to be applied for removing such obstruction, and thereby of restoring health; for he may rest assured, that not the least atomy of matter has been conferred upon the animal form without intending that some good and demonstrable end should be answered by its creation. In addition to all which, there are many causes, incessantly operating towards the simply grand purpose of prolonging life, and of providing for the waste which is constantly going on in the animal system, that are far removed from our sight, and others almost surpassing our comprehension, but which are nevertheless known to exist by their effects; but, of all these several matters, more in their proper places; one instance of the insufficiency of human knowledge having been already adduced in the preceding section, as regards the unknown mode in which the watery secretions penetrate from one part of the body to another.

* Tears or any other evacuation of the water that moistens the animal system, are liable to the same kind of comparative remark. In man, when excessive salivary secretion attends the toothach, the glands of the mouth and jaws carry off so much water as to affect the quantity of urine voided, and we may infer that a diuretic would reduce the inflammation of the jaws. So much for the comparative practice: but not worthy of rejection on that account alone.

4 *

24. All those important points of knowledge in the first principles of our art lie within the compass of every man's capacity, who can read; they are certainly open to his inquiries; and he who is constantly among horses can not fail to learn (after studying the subject in the manner I now propose) to make himself as well acquainted with the symptoms or signs of approaching disease as the generality of veterinarians. He certainly may render himself much superior to the old, ignorant set of farriers, who were bred up in the days of stupidity that are just gone by—never to return. Let such an inquirer after knowledge bring to the task industry, patience, and good common sense, and he may soon acquire knowledge enough of the outlines of the art to be able to pronounce when a pretender is at work, or when it is that a man of judgment and real sound learning in his art has undertaken the treatment of this valuable animal in the distressful hour of sickness. On this head I am not ashamed, after the lapse of nearly half a century, to own that I once wept over the sufferings of a sick animal which died of the medicines administered by a stubborn self-willed farrier, who could read, and write, and talk, give a drench, and drink himself—and nothing more: he could not *think*, of course could not compare one disease with another, nor mark the difference that exists between two or more that are frequently and fatally mistaken for each other. And here, once for all, I can not refrain from thus early insisting most strenuously on one point, which therefore I shall not have to repeat when I come to notice certain barbarous practices perpetrated by some such men, and the not unguilty practice of other physic-giving horse-doctors; and this is, in short, whoever of them dares to undertake the administering of medicines to this incomparable animal without paying especial attention to the subject matter that is handled in this chapter, commits an unpardonable act of inhumanity on his suffering patient, and of gross dishonesty towards its owner. The remedy for a disease is not always to be found in medicine; preventives never. Purgatives are not only the most obvious means of cure, but the best, the least dangerous, and those which promise in the readiest manner to dispose the most vital function to resume its wonted action. Alteratives are the safest and most effectual remedy for valuable horses, and those which can not be spared from labour; they are indispensable in all cases of vitiated blood, and where found ineffectual nothing else can be of service. Bleeding is the very best, or the very worst auxiliary we can employ; its efficacy and precise periods of utility may be learned in the sequel (see sect. 37 to 44), where "the circulation" comes under consideration, also in the first pages of Book II, where the pulse is justly made a subject of primary consideration.

25. For the sake of making myself more clearly understood, I shall, when explaining the formation and functions of the horse's inside (i. e. as much of it as will answer my purpose), consider it under two distinct heads; namely:

1st. The fore part, or throat part, as it is called from its neighbourhood to the throat, or gullet; and,

2d. The hinder part of him, being his belly, properly speaking, &c.

Both of these parts have obtained learned names; but that is no business of ours.

The fore part of a horse is that which lies between the rider's two knees, within the chest and true ribs. To the farthest of these is attached, as well as to the middle of the back bone, a natural division of the two parts, stretched tightly across his inside, like the head of a drum; and it is also fastened to his breast bone, but admits of the gullet to pass through, as it does of the great vein and great artery which carry on the circulation of the blood of the hinder part. With these exceptions it is air-tight, and it bears resemblance to the head of a drum in another particular—it is membraneous, except round the edge next to the ribs, &c. where it is found somewhat fleshy. From its situation in the middle, this natural division is termed the midriff, or skirt, and appears to have been designed for keeping back the stomach and bowels of the hinder part, which as it is, when full, press it out of shape, not unlike that of a watch glass, and would, but for this barrier, interrupt the action of the heart and lungs. But by the present contrivance, as we shall see presently, this pressure from behind soon recedes, the midriff returns to its level, and the ribs, no longer contracted towards each other by the aforesaid pressure, expand, thereby enabling the lungs to perform their function, of drawing in a fresh supply of air. Upon this principal agent in the function of respiration, see more in detail in the 35th section of this chapter.

26. After this necessary preamble, let us proceed, as before proposed, to consider the construction of

> The Membranes,
> Muscles,
> Glands,
> Absorbents, (i. e, 1st lymphatics, and 2d lacteals,)
> Nerves.*

Of these the most universally dispersed over the frame, those which occupy, defend, or embrace every part, are the membranes. As well behind as before the midriff, not only inside, but on the outside and every part of the animal are these skinny films placed, for the purpose, 1st, of keeping those parts which they encompass in a compact state; 2d, to secrete a fluid for protection (see sect. 21); and 3d, to prevent those parts from rubbing against and injuring each other, or adhering together. The better to accomplish these purposes they are admirably calculated for the secretion of a fluid, as I observed before (sect 20); but whenever the property of secreting such fluid is suspended, then disease begins, and according as the secretion may prevail, being either too little or too much, will be the kind and quantity of disease. Hereupon may be calculated the importance they hold in the animal system: but of those matters I have already spoken higher up.

MEMBRANES.—To appearance they are nearly transparent, web-like, and of a strong texture; some are simply film, having more or less of feeling according to their uses, and are those which, being interposed between one organ, or part, and another, prevent the interruption which would otherwise ensue; as the midriff, for instance, which I have just above adverted to (see sect. 31 and 35), which is the thickest of all, or the loose membrane that covers the lungs and divides them into two parts, so as each may act separately (see sect 32). The second species of membranes are finer, more transparent, and paler than the first mentioned, and possess the quality of containing in their cavities, resembling sponge, the matter deposited within them by the arteries for the purpose of repairing waste and adding new flesh; these we term cellular membranes, from their sponge-like texture, and they are, moreover, so infinitely thin as to pervade all over the solids, or fleshy parts, without being in every case visible to the eye. They are nevertheless proved so to exist, from the circumstance of those being greatly distended, when the subject dies of being "blown," as I have shown lower down (sect. 35). Then, not only the forehand, but the hinder quarter, even down to the hocks, become inflated with the wind, which, by reason of the animals being strangled, the lungs had no power to discharge, and the cellular membrane admits it into its cells or cavities. Any one may perceive this membrane and its numerous cells to advantage in a buttock of beef, or leg of mutton, after being dressed; upon taking a slice between the fingers and straining it nearly asunder, the mem-

* For ligaments, cartilages, tendons, &c. see sect. 16 17.

brane appears, but more evidently at the corners where two or more muscles meet; and in summer time, particularly with over-driven beasts, the membrane between the muscles will be found charged with a dull brown sort of matter, that may be, and frequently is, scraped away with the knife. Another familiar illustration of the uses of the cellular membrane, first mentioned by old Dr. Bartlett, of Windsor, in 1764, is that of "the inside of a shoulder of veal, which butchers blow up with a tobacco-pipe, or quill, to delude their customers." When the animal becomes adult (or full grown), the membrane that is so capable of being blown up is filled with meat, and shows the impropriety of pushing young animals in their work before those solids have reached maturity.

When once divided, membrane of either species never again unites, but in case of a healed wound the granulations of new flesh hold the divided parts of membrane to their respective places; the obstruction thus occasioned in the deposite of blood causes pain upon change of weather, when the new flesh either expands or contracts, as it may be effected by heat, cold, or humidity. In the human physiology, another kind of distinction is made between the kinds of membrane (as I said before), tending to show whether their respective secretion is more or less watery, slimy, or oily; but this view of the affair is not applicable to the physiology of the horse. My reason for abandoning that course was given at section 21. I may, however, here aptly observe, regarding that species (the slimy or mucous) which lines the nostrils, throat, and intestines, that its chief disorder is a cold, which shows itself in the cessation of the secretion; soon after this, the parts being inflamed, throw forth a thin acrid discharge, which is greatest when the inflammation arrives at its height, sometimes producing a little blood, either upwards or downwards; when the inflammation wears off, these appearances are also lowered by the mucus becoming more and more thick, until it reaches its usual consistency. The cure is to be effected by lowering the inflammation; but this is most frequently effected by the natural discharge of the mucous matter just spoken of.

27. MUSCLES are fleshy bodies of various sizes and shapes, according to their uses; reddish, of a fibrous texture, easily separated, but more stringy at some places than at others: the last-mentioned are termed "coarse parts," or pieces, in the animals sent for our sustenance, and are those where the greatest strength lies. These fibres formed into bundles, and surrounded by the cellular membrane, are visible to the eye, if there be not attached to each fibre a continuation of the same membrane that is not visible. Several of those bundles, being further enclosed by a stronger membrane, form a muscle; each whereof is attached by its two farthest extremeties to some other, or, to two different bones, upon one or the other of which it acts as a lever. A muscle accomplishes this motion of the bone by expanding its belly or middle part, and contracting it towards the centre; whereupon the bones to which the muscles' ends are so attached are drawn towards each other, and that which is farthest from the trunk is drawn forwards or backwards, at will. Thus, if we wish to bend our elbow, the muscle which is situated just above that joint, inside, contracts in length, and expands in breadth, till the fore-arm is brought up to touch the muscle itself. Fighting men (boxers) exhibit this muscle, as indicative of their strength; and horses of good action show the same sign at every movement, whilst with those that are over-fed, the muscles are concealed in fat, that obstructs their movements; whilst, with those which are impoverished, the muscles dwindle away, hang slack, and ill support the wonted action of the bones. When much compulsory exertion, in hot weather, has exhausted the secretions that keep these parts supple, aridity and stiffness follow, and the action becomes impeded, difficult and uncertain.

All muscles of the limbs are long and narrow, when quiescent; those of the body are more wide than long; in a good measure, squarish, oval, or tri-angular, according to their uses. They have been compared, with good reason to the shape of flat fish, some being long and narrow, like the sole, others wide, like the plaice. At their ends, muscles often terminate in a much stron-ger substance, closer in texture, inelastic, bending with facility, and insensible, answering the same purposes, but occupying much less room than muscle. These are tendinous, and the horse which is well kept, having the tendons strong and vigorous, is bold, strong, and "sinewy," moves his limbs with agility, and gets over his work to admiration, by picking his feet off the ground well and replacing them (as you see while he is going) within a hair's breadth of the spot you may mark out for them to pitch upon. On the legs, tendon supplies the place of muscle, wholly so in blood-horses, less in the cart-horse breed. Muscle is constituted of blood deposited in the membrane, innume-rable small arteries, some of which are scarcely visible, terminating within each muscle, by a kind of doubling up, or curl, as shown in the margin; within each of these a correspondent vein is twined, and the whole being covered with the finest membrane, con-stitutes a gland. Herein it is that the veins commence their share in the work of circulat-ing the blood afresh, as we shall see in the se-quel, and the lymphatics obtain the watery particles into which the morbid matter of those solids are converted: those figures receive the name of "glands."

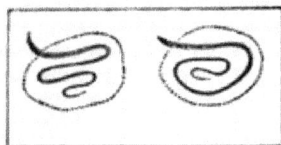

In blood-horses (natives of hot climates), as we have seen, tendon supplies the place of muscle, or flesh, upon the limbs particularly, which are always finer than those of other breeds; this accounts why our fleshy horses in sultry weather, or hot stables, feel the greatest lassitude, even to weakness, whilst those of full blood seem invigorated by the same circumstance. When, how-ever, the atmosphere of the stable be moist as well as hot, both breeds suffer equally in one way or another; laxity of fibre and profuse perspiration, with weakness, follow, and this producing an obnoxious effect upon the excrema-tory organs, occasions in stables those stinking ammoniacal vapours that de-stroy the lungs, by disposing them to contract inflammation.

28. Besides the GLANDS just alluded to, they are situated in and about the solids and more secluded parts, and so small and concealed as to be scarcely exposed to the sight or touch, unless when inflamed and enlarged by disease, other larger and more evident ones occupy the hinder part of the animal, of which I shall speak in their place. They are, 1st the liver; 2d, the kidneys; and 3d, the testicles; the functions of each being tolerably well known. See sections 52—55. All glands, of whatever size or shape, are employed in se-cretion, taking up and separating from other matters that quantity of watery particles which is constantly escaping out of one-part of the system into another, by means of the cellular membrane, as described at sections 21 and 22. The smaller glands, just now described, have each a small tube attached to it, which seems intended to hold the acrid, or otherwise noxious, matter which its lymphatic had refused to take up, as being at variance with its func-tion; here it remains concealed, until the proper occasion arrives for carrying it off, which may be found by one of the three natural evacuations; but these failing, it is clear disease of one sort or other must ensue. Perspiration seems to be its most natural mode of passing off, unless the demand for that kind of evacuation happens to be low, and then it is drawn to the kidneys, (sect. 22). But, if the discharge by dung has been so copious as to afford too little of this acrid matter (essential probably in a certain degree) by means of the absorb-ents of the intestines, then, and in that case, it is taken up once more. When the animal's spirits are low, the absorption imperfect, and this offensive matter

lies a long time in the tubes of these small glands, a general languishment of
the beast takes place (called *lentor* by the old farriers); he perspires upon the
least exertion, becomes unnerved, shows a rough hide, and refuses his meals.
This constitutes "low fever," when the whole animal system is affected. This
state of things, which is very common, points out the impropriety of now re-
sorting to diaphoretics (sweating powders); for it has been neglected so long,
that tired nature, being offended thereat, refuses to part with those particles
which occasion the greatest injury. "By the urine," be it said. This way
offers the same difficulty, and the answer resolves itself into my plan of opening the principal evacuations first. (See what I shall offer concerning bleeding and purging in a subsequent page.) So much, however, seemed necessary
to be advanced here, that the reader, who reads straight an end, should be at
no loss as to what lately passed between us concerning secretion and effusion
at sections 21 and 23. Of all the smaller glands, the best recognised are those
termed salivary, situated near the jaws for the secretion of spittle, wherewith
to moisten the food while descending into the stomach, and thus assisting di
gestion in its first stage. The strangles and vives are disorders of these
glands: the swelling at this part is a corresponding symptom of glanders, and
sometimes attends farcy. But the largest of these minor glands is situated in
the solids, and lies within the buttock, concealed near its centre, into which
passes an immense quantity of blood for its size, since it is found in the dead
subject most disposed to putrify, especially when the animal has been driven
hard, as is the case with all the cattle killed in London for food.

29. LYMPHATICS are one of two species of absorbing vessels; the other
species (the lacteals) being reserved for description under the head of "digestion," at sect. 44. They are small tubes, with mouths that suck up or absorb
the thin watery particles of the solids, one or more being placed on each gland
of these parts. Some idea of the important nature of this part of the animal
system may be formed from the circumstance that mercury applied to a glandular part of the body undergoes immediate absorption by the lymphatics, and
is conveyed by this means through the jugular vein to the blood. Persons
who may be unfortunately ordered to rub in mercurial ointment on the thighs
will feel a fulness under the left ear in the course of a few minutes, according
to the previous state of their bodies. How mercury acts upon the second species of absorbents—the lacteals, remains to be seen hereafter. The tendency
of both is towards the heart, or rather the left collar-bone; increasing in size
and diminishing in number, until the lymphatic duct meeting with the milky
juices of its co-absorbent in the thorax, the mixture soon becomes blood by
the action of air in the lungs, as described at sect. 39. As the lacteals, it will
be seen, absorb only nutritious juices, so the lymphatics absorb none but offensive ones, as the matter of diseases, wounds, spavins, broken bones, ulcers,
and the useless part of the deposite made by the arteries as said at section
27; these being mixed, pass through the heart, there receive fresh vital powers, as hereafter is described, and thence to the liver, there to be purged of its
bad qualities, which, passing incessantly into the intestines is soon eliminated
with the dung. At least, such is the natural course in health; a change takes
place when these organs do not perform their functions aright, and we can
perceive this misfortune in the dung, when the absorbents are at fault, particularly in the yellows. The importance of stimulating the lymphatics in
all disorders of the outer surface, as mange, surfeit, farcy, &c. must be evident: as it is, also, in cases of tumours, as poll-evil, fistula, &c.

30. The NERVES, like the glands, run in pairs, mostly, to all parts of the
body; they are the organs of sense, communicate immediately with the brain,
and are thus principally concerned in the function of voluntary motion. That
the horse entertains likes and dislikes is certain; he has a memory too, both

for persons and places, as every one knows; he must, therefore, have perception, and he is kind and docile in his nature, which entitle him to a kinder return from his master than he usually receives. I have often lamented that he was not endowed with one more faculty, even in the smallest degree, that he might distinguish between those who really love him, and those empirics who make a profit of his sufferings; he would then be induced, probably, to kick some among them, as an example to all the rest; and I never hear of one of those fellows, or their employers, being unhorsed, but I think of retributive justice. So, when the horse is girthed up unmercifully, in such a manner as to obstruct his respiration, he frequently attempts to bite the operator—and, "serve him right," I say. As the nerves of a horse are the seat of no distinct disease, I shall content myself with adding, that they consist of small cords, white and roundish, like thread; and are certainly the vehicles of pain, which vibrates from one to the other, pleasurable sensations being conveyed by the same means to the *sensorium*, or brain.* The nerves are closely connected with the circulation, and with the brain, where they originate.

———

31. RESPIRATION is the act of drawing in the air by the expansion of the lungs, the cells whereof thereby become filled to their utmost, the ribs are distended, and the midriff pressed back upon the stomach, liver, &c. This is inspiration; the expulsion of the air, forming the re-action, being termed expiration; both together constitute what we call breathing or respiration, and the matter was before introduced (in section 8,) when I noticed that powerful auxiliary of this function—the midriff. Now, as I have always attached much importance to the act of respiration, seeing its close connexion with the formation of blood, and the almost constant state of disease in which are found the organs that contribute to this great function of animal life, I shall enter into more minute particulars respecting these, than I have thought necessary for any of the preceding organs. By this course, the reader will be enabled to form more distinct notions respecting the forming and "circulation of the blood," and its concomitant, the formation of *chyle*, commonly called "the digestive powers"—both of them functions most essential to health; but unhappily, both together become, by contravention of those powers, the fruitful source of numberless ills, we thence call constitutional or bodily disease, as fever, abscess, farcy, &c. To this point tends all that I have hitherto said concerning the inside of the horse; and the inquirer after veterinary knowledge will find his labour in studying this portion of it amply repaid, by the just principles upon which he will subsequently conduct his practice.

32. The LUNGS, or lights, are two well-known spongy bodies (called lobes), having at their conjunction a small lobe nearer to where the pipe enters that is to inflate them. At the same place is fastened the ends of a thin membrane, or rather two membranes, that enclose each one of the lobes: this membrane is termed the pleura, and seems designed to admit of one lobe performing its functions whenever the other may at any time be diseased. Between the two

* Conscience (consciousness), which agitates the nerves by the faculty of thinking, when applied to the evils that are in the world, does sometimes cause the accession of fever to those delicate organs in human nature; but brute animals being denied those powers (or of memory, except as regards the means of prolonging life) are little likely to contract "nervous fever;" although that state of fretfulness some high-bred horses are prone to, partake of a good many symptoms of the human ailment, and may be cured by the same means. Sedatives, quiet, and a cooling regimen are those means. The loose stable recommended by John Lawrence, and now much adopted, contributes much to sooth the fretful horse. When the same fretfulness or despondency comes over a horse, one of condition, or whose condition has been recently reduced he acquires slow fever. See what is said under this head in Book II.

lobes, the membrane (pleura) is double, and, from its situation in the middle (*in medio*) is called mediastinum : it forms a passage for the great blood-vessels running near the spine, and it is very liable to contract disease, which shows itself in "thick wind," or rather short wind. Sect. 36. In the pleura, then, are wrapped up, as in two silk handkerchiefs, the two lobes of the lungs, the upper part of that membrane being fastened to the spine and ribs ; and on its surface is generated or secreted some of that fluid I before spoke of (sect. 20, 21), which is designed to keep the parts moist, and prevent their adhering together. His powers of secretion, however, and those of this organ in particular, often fail in the horse, in consequence of his very great exertions, combined with the heat of his blood, exhausting more than the secretory power can supply; and we frequently find the pleura growing to the ribs, the lungs, or the midriff, by reason of its wanting a due portion of this fluid : from the same cause (a defect in the secretion), we sometimes find the upper orifice of the stomach partially attached to the midriff, evidently caused by inflammation of the parts. But whichsoever of those misfortunes attend the horse, he is invariably "hurt in his wind," suffers much pain at the commencement of a journey, and subsequently, if pushed hard, dies of a locked jaw, through excessive suffering. Disorders of this nature were hitherto unknown to farriers of any description, being mistaken for the worms by every one who has written a book upon horse diseases ; and, by the most eminent veterinary author of modern times, the last stages of this mal-conformation are vaguely noticed by the erroneous term of "debility," and "general debility," which may mean anything amiss. Whenever the animals that are slain for our sustenance turn out to have been so affected, their flesh is rejected by the Jews, under the denomination of *trifler;* for the whole animal system is entirely affected by the horrid circumstance ; the secretory functions in general refuse to perform their share in the production of good and sufficient animal matter, and *lentor,* or slow fever, is the consequence, as mentioned higher up, in Section 28.

33. The WIND-PIPE, as its name imports, is the pipe or tube for conveying to the lungs the air which every act of inspiration draws through it. Extending from the throat to the lungs or lights, at their conjunction this tube divides into two branches, one penetrating to near the bottom of each lobe, and these again, having a dozen holes a-piece in their sides, inflate an infinite number of little tubes, or pipes, which compose the lungs much in the fashion of sponge. Except eight blood-vessels, which enter the horse's lungs, the intervals are filled with cellular membrane, and these being also connected with the same kind of membrane in all other parts of the body, accounts for a phenomenon, I shall take occasion to notice shortly (sect. 35), in the case of a blown horse. At his upper end, the wind-pipe is composed of strong cartilaginous plates, connected together by ligaments, and put in motion by small muscles for producing the sounds expressed by the animal. Next to the throat these cartilages, which are there strongest, form a curious kind of chamber, termed *epiglottis,* over which is a lid or valve, placed there to defend the passage into the air-tube, from the entrance of victuals, drink, &c. For, upon the descent of any such substances, this valve shuts down like a trap-door, and they pass over it. No sooner, however, are they gone past, than up rises the valve again, lying back towards the mouth upon the palate, and being very large in the horse, accounts for the gulps with which he takes in water, and his peculiar mode of feeding. For the same reason it is, that the horse breathes only through his nostrils, between which and the wind-pipe there is close affinity in some diseases, and accounts for his incapacity for bellowing like the ox, or vomiting like man. At this spot it is, that certain savages in human shape press the finger and thumb with brutal force, in order, as it is

called, "to cough him." No certainty, however, lies in this imagined test of his wind; for, although a thoroughly broken-winded horse will not cough, yet one which is partially affected will do so in most instances; whilst the soundest horses do most obstinately resist the coughing; and in a few, the cir cular cartilages so well defend the muscle, as to defy the inhuman effort, and seem to rebuke the ignorant attempt "to prove the goodness of his wind."

34. Farther towards its lower extremity, the wind-pipe becomes more membranous, but less sensible of injury, and the cartilaginous rings gradually lose their form: they no longer describe a circle, being composed almost wholly of strong elastic membrane, that it may bend out of the way when the gullet is distended with swallowing. Its internal surface is lined with a membrane, which incessantly secretes a quantity of the mucous fluid spoken of in sections 20 and 21, hereby defending its coats from the action of the air in passing to and from the lungs. But this secretion being exhausted, sometimes by the very great exertions of the animal, he then coughs so as to shake his entire frame, as if to incite the membrane to make fresh secretions of fluid for its defence; or, in default thereof, the cold air still rushing in at each inspiration, he contracts a permanent cold, or catarrh, which, if suffered to continue, increases and runs along the membrane to the lower branches of the pipe, and ultimately communicates its baleful influence to one or both lobes of the lungs. If the attack be trivial, small green spots are found on the surface of the lungs, which afterwards form ulcers, increasing in size and number, according to the number of small tubes or cells that may be affected. These tubes lose their functions in consequence of the first attack, the animal's wind becomes worse every time he is hard pushed, and the cells burst into each other, until, perhaps, one lobe or half of his lungs is rendered useless. In process of time, it turns black as one's hat, infects the other lobe, and mortification ensues, which is rather accelerated by the cordials with which the poor creature is usually punished, and it dies.

But when it so happens, that too much of the fluid is secreted in the windpipe, the animal snorts or coughs it off by a sudden natural effort; wherein, the midriff being made to press forcibly upon the lungs, by the sudden con traction of the muscles of the lower ribs, out flies the wind through the nostrils, carrying with it whatever may have adhered by the way. Whenever this is the case, the membrane that lines the nose inside becomes irritated, and fresh accession of its own secretion, thickened a little, is the consequence; in flammation of the part, ulcers, and a running of foul matter ensue—and this, if the blood be not in a good state, soon becomes that obstinate malady—the glanders.

An instructive experiment may be made upon the pluck of a sheep—the relative situation and functions of these parts in all quadrupeds being the same, except that the sheep's lights, compared to those of the horse are not so long in proportion to their thickness. Take a pair of bellows, and having introduced the nozzle tolerably well into the windpipe, tie it round with a cobbler's end; then, blowing hard with one hand, while the other is employed in squeezing the pipe, to prevent the escape of the air back into the bellows, you may form an accurate notion of the effect of inspiration. The lights or lungs at first give out the whole of the air which has been driven in, and may be inflated to an enormous size; but, if much force is used, the cells burst into each other, some appearing on the surface thin and transparent, and refuse to give back their wind; this forms "broken wind" of one description, and is that wherein the expirations are slower than the inspirations—the pleura being then affected in its thickest part, and the midriff also Out of the first part of this experiment may, likewise, be derived a more accurate knowledge of what is termed "second wind," among sportsmen: when the animal (or man)

5

has made great exertions, so as to fill all the cells of his lungs to their utmost, and then relaxes from the labour, he finds himself renovated, the cells being rendered more capable of distention and expulsion, when each inspiration and expiration also occupies more time and less labour.

Sporting men, who are fond of our bear-baitings, Pecora-fights, and monkey scratches, may daily witness a practical natural illustration of the same doctrine, in the conduct of the bear towards his antagonist. Seizing the dog between his paws, he squeezes him up till he gasps for breath, when Bruin, being muzzled, rams his nose tight into the dog's mouth, and, blowing with all his might, you may hear the wind whizzing : the dog swells all over, by reason of the air entering the cellular membrane, and he dies unless timely pulled off. A dog which has "had a hurt" of this sort seldom regains his proper wind; he must be "a good one" to face the bear again, "as long as he crawls." Such is the polished language at those elegant places of town amusements.

35. The MIDRIFF has been already mentioned (ss. 25 and 31). It is termed diaphragm by the learned in hard words ; and we have seen how materially it is engaged in the business of respiration. But for the action of this drum-head-like membrane, neither the lungs on one side of it, nor the stomach, bowels, and liver on the other, would obtain their full degree of motion, which is thus kept in tune, as it were, by those organs acting alternately upon each other ; the action of the heart, too, is in unison with that motion ; but when through agitation (occasioned by great exercise, affright, &c.) it does not keep time, the temporary disorder, termed palpitation, is the consequence. We may infer that, when the lungs have discharged their contents, the lower or thinnest end of that organ, falling upon the muscular border of the midriff, is by it repulsed and excited to action. Any man can feel, when he has expired all his wind, a kind of throbbing internally, lower down than the heart, until he inspires a fresh portion of air. When the lungs are in such a state of supineness, those of the horse are about three or four inches thick at the conjunction with the windpipe, and ten to thirteen inches from thence to each extremity, according to the size of the subject ; but, when fully inflated with air they together fill up the whole cavity of the chest, obstructing in a trivial degree the vibration of the heart : then do they reach to the enormous difference of twelve or thirteen inches in thickness, and somewhat more in length. At least such were the dimensions of this organ in a horse which was opened by me in May, 1820: he was of the cart-horse breed, under sixteen hands, and healthy in other respects than having been blown by eating too much corn ; whereby nature was compelled to leave the lungs quite full at the moment of his death. The same subject is alluded to in the 26th sect. where I intended to illustrate the formation and functions of the cellular membrane.

"No part of the animal has been formed in vain," as I before observed (s. 23): quadrupeds and bipeds both press the earth which gave them birth and which affords them the means of prolonging life ; accordingly this order of beings is furnished with a midriff, but fishes and insects, having no such occasion for this organ, are without it : neither have birds a midriff ; but Mr John Hunter was of opinion, that the want of it is supplied by the hollowness of the bones, which not only increase their buoyancy, but the air contained in them re-acts upon the lungs in the same manner as a midriff would do.

36 In health, as in disease, the midriff is liable to be affected by its neighbours, both before and behind it, the stomach often communicating its state of feeling to the lungs through the midriff ; for it is by this medium that medicines impart their beneficial effects upon the lungs, as may be experienced upon our swallowing cold water at a time when our lungs are heated—the relative situation of those organs being much alike in man and in the

norse. Immediately hereupon, a sensible difference takes place in the number of respirations, and the quality thereof is also changed from a hot to a cooler temperature; well be it, if the suddenness of the check do not occasion inflammation of the lungs: again, whoever swallows spirituous liquors feels an immediate disposition of the lungs to repel the heated air of the cells which lie contiguous to the midriff; the first breath which escapes the mouth being less heated than that which follows and finishes the expiration, and imparts a sensation wholly different from the vulgar belchings of an overcharged stomach. These come up by way of the gullet, and carry forth a nauseous effluvia; whereas air from the lungs is ever sweet, unless this organ be already in an advanced state of decay. This state of the case leads me to make one practical observation, which shall not be set down, as more curious than useful; out of seventeen subjects, which successively fell to my lot to examine as to the immediate cause of death, only two, tolerably sound at the lungs, presented themselves; which I take to be the fair proportion of sound horses, as respects the lungs of all that live or die. All those cases occurred from February to May, a season when such an affection might not be considered most prevalent. Hence, (my reader may smile!) I conclude from all that has been said, that a tolerably good guess at the state of a horse's lungs may be formed, by smelling at his breath after a canter, in like manner as our Smithfield dealers smell the animal's nostrils in order to detect the glanders.

From what has been said, it follows, that a diseased stomach may be produced by diseased lungs and *vice versa*, and that the midriff suffers in either case: then does the midriff become livid, purplish and inflamed, with dark-coloured stripes, as if thickened at such places, the muscular border thereof assuming a putrid appearance, and sending forth a villanous stench. When this is the case, or any other ailment prevents the midriff from performing its proper function of inhaling and expelling the air from the lungs, that species of "broken wind" takes place which is known by the sort of breathing wherein the expirations are quicker than the inspirations; being thus contradistinguished from that other species of broken wind, which is occasioned by rupture of the air cells. A paralysis of the midriff, or the adhesion of the stomach to its lower side, is equally obstructive of its reaction upon the lungs: and I have this day (May, 1820), cut away an adhesion of this sort as wide as the palm of a man's hand.

Unfortunately for the horse which is affected, either in the midriff, the lungs, in the pleura, or covering thereof, his doctors heal the whole series in the same manner, not unfrequently including in their uniformly mistaken practice, the affections of the stomach, liver, pancreas, &c. In all, the inflammatory symptoms are predominant, and a cooling regimen presses itself upon our notice as more proper than the best of medicines, although having the same tendency; whereas, the direct contrary is the practice mostly followed, and heating medicines, under the fascinating name of cordials, made of spices, ale, wine, &c. are administered daily. Or, at most, if a sedative or opiate, by chance, finds its way (properly enough) into the animal's stomach, this organ is thereby only rendered more susceptible of the heating mixtures which are again had recourse to immediately thereafter.

THE ORGANS OF RESPIRATION are liable to seven or eight several kinds of disease, mostly originating in the horse's having caught cold; they are denominated according to the particular place where he may be affected, and in one respect, according to the degree of attack. A cold (simply) or catarrh, produces that affection which denominates the patient "a roarer." Chronic cough brings on "broken wind," of which there are two sorts; and consumption usually follows the long continuance of either. When either has con tinued awhile, and reduced the animal's strength, he is said to be in a consumption; "worn out" is also a common phrase, as is "rotten;" "debilitated," and "done for," stand a little higher in gentility; but all mean, that there is small chance of his recovery.

Pulmonary consumption is the only kind which may be attributed to constitutional defect; i. e. heated blood, with viscidity, causing over-much action of the parts (see section 32), when the pulse becomes powerful and quickened, and the horse seems anxious and fearful. Should it subside by judicious treatment, or the natural strength of the horse, he commonly retains so much of its effects, as to cause great danger whenever he may catch cold, or be worked too hard. This attack is too frequently neglected, or put off with the remark, "only a little touched in the wind," unless by the addition of a cold, the disorder comes on rapidly, when it is termed "inflammation of the lungs," and the animal goes off in four or five days, if he be not promptly relieved. On dissection, the parts are found spotted with a livid colour, and evident gangrene; every variation indeed, is equally appalling to humanity, and I have often wondered how the horse could have lived an hour under such horrid circumstances.

A cold, simply, or catarrh, commences by inflammation of the lining of the windpipe, which may be confined to some given part of it, or extends itself generally from the nostrils to the lungs both inclusive, much resembling "a cold" in man; in either case, much matter is secreted, and thrown off by snorting: when the attack is confined to the upper part of the pipe, and lasts some time, the horse becomes a confirmed roarer, his groan bearing great resemblance to the roar of some wild animals, and he is equally incurable.

Chronic cough is that obstinate cough which remains and plagues the horse long after vain attempts at curing his cold ought to have taught his tormentors the inutility of their endeavours and the dishonesty of their prescriptions. Broken wind is of two sorts; the first is caused by the rupture of the air cells of the lungs by over exertion, in which the expiration being slower than the inspiration, he is, aptly enough, said to be "broken winded;" the second kind of broken wind is known by a breathing the direct contrary, and is occasioned by the cold having settled upon the perforations of the branches, and enlarged them, so that the air escapes too readily. Thick wind, on the other hand, arises from the secretion of the pipe getting into the cells of the lungs, and affecting the orifices or perforations by thickening them, so that the air passes through with difficulty. Similar symptoms attend inflammation of the pleura (section 15), particularly when it reaches to the *mediastinum*, or double part; as they do when the muscular border of the midriff is affected. But these, though perhaps considered two different diseases at the commencement, in the horse, very shortly become one common affection, more or less, of all the organs of respiration; inflammation predominates over the whole series; and if the performance of his duties does not render them fatal, an injudicious treatment will fix upon him an incurable disorder, until the knife terminates his usefulness.

Lafosse, junior, observes, that "flat-chested horses, are almost always subject to consumption, whilst (again) consumption narrows the capacity of the chest, and re-produces itself—there is no remedy for misconstruction." He further says, "Short wind is either produced by disorders of the chest, or it is a fault of construction, and both are irremediable. Such a horse is generally of less use than one that is thick-winded."

37. THE CIRCULATION OF THE BLOOD is carried on through all parts of the body, to which it affords the means of life and health; or, being ill performed, is the fruitful source of lingering, obstinate and incurable diseases, some whereof almost baffle our skill and care; and, while they induce us to admit that the practice of veterinary physic never will reach perfection, inspire the

hope that, by patient investigation, we may at least find out the means of alleviating their evil tendency. These considerations should excite particular attention to the subject in hand, being that to which all other functions are but subservient or conducive; respiration and digestion being more closely connected with it than any other, and requiring a corresponding degree of attention. The heart, jointly with the lungs, occupies the cavity of the chest, rather inclining towards the left side, against which its point may be felt beating, whenever the animal is agitated by exercise, or internally affected by inflammatory complaints. Its figure is too well known to render description necessary: in size it approaches that of an ox, and, like it, is enclosed in a membraneous sac, but very thin, not unlike a tight purse. Another sac, called the heart-bag, less tight, surrounds the former, but is never so large, nor encumbered with fat like that of the bullock. On opening this sac there issues forth a fluid which turns to water, if it were not so already; in the ox it is confessedly water, and rather more in quantity than in the horse;—the reason for which difference I will show presently. So great is the stimulus of this organ, that its contractile power often exists long after the animal is dead, and the arteries cease to flow: a phenomenon that occurs when the horse is killed in full health, by an accident, or otherwise, while in full possession of its functions. In animals of much mettle or courage the heart attains to a great size, that of the famous horse Eclipse weighing 18lbs. In some horses, the vessels that nourish this muscular organ lie exposed on the surface; with others, they lie wholly concealed; a fact from which I have yet found no opportunity of making any sure deduction, but apprehend it may arise from the breed, and conclude it must have considerable effects on his disposition.

As the heart is the principal organ employed in converting into blood what is drawn from the finer particles of food sent into the system for that purpose, as well as in renovating that which has been exhausted of its vital principle in the circulation, it may well be supposed full of small blood-vessels. Four large ones meet at its thick end, and suspend it, by being attached to the bones of the back and ribs: two of them, bringing the dark vitiated blood from the extremities of the fore-part and of the hind-part, are veins; the other two large vessels are arteries, which, receiving the blood from the heart, just now refined by the process of nature, convey it with rapid contractions to every part of the system, there to give fresh vigour, and to impart health to the whole. These contractions constitute the pulsation, or pulse; a criterion of health as of disease, which we investigate with primal anxiety, as from it may be deduced the best prognostic of the state of disease, especially of inflammatory ones, to which this noble animal is more than any other subject, from causes to be explained hereafter.

38. Two large chambers (as I call them) and two small ones, each of the latter appended to one of the former, mainly contribute to the process of circulation; these, by their co-action, aid the contractile power of the heart, compelling the contents of the larger chambers to issue with much force into the arteries. But, before the blood can be thus again fit for circulation, nature has provided the means of rendering it so, by the action and re-action of these four chambers (or cavities) in the heart, contributing to refine it for that purpose; one large chamber, with its small one, being placed on the right side of the heart, and the like pair on the left side thereof. The blood from the veins flowing into the small chamber on the right side, irritates its inner coats, and they each contract upon its contents with nearly as much strength and quickness as we can open and shut the hand;—but certainly not so much open, although there is a flap on the entrance of each small chamber, which they liken to "a dog's ear," and call by the Greek word *auricle*, that being

5 *

an ear. At each of those closings or contractions, the blood is forced out of
the small chamber into its large chamber on the same (right) side, through a
small door-way, or valve, which opens only inwards, the door being too large
for the door-way. By the way, these valves are very numerous in the animal
system, principally as regards the circulation of blood in the veins; and the
reader will do well to recollect, when I speak of valves in future, the illustration just given by comparing the same to a door-way, which bears a near resemblance to the valve. In the performance of this office, the heart may be
said to have three several motions belonging to it: 1st, a quick one, by the
contractions of the smaller chambers upon their contents: 2d, the contractions
of each large chamber upon its contents, being just half the number of throbs
made by the first mentioned: and, 3d, the vibration or tremor of the whole
heart in consequence of all those motions.

Well, on the right side still, the large chamber being filled with venous
blood through the valve, or door-way, from the small one, becomes in like manner irritated so as to contract upon its contents, and to drive it out somewhere
or other. Back to the small chamber it can not possibly go; the valve does not
open the way, as I have just now said; and therefore it issues with much more
force through another valve into a short artery, which soon opens into two,
like the letter Y, the heart being supposed at the bottom of the letter. Up
rushes the blood to the top of the two branches—where it meets with—What?
What do you think it meets with, gentle reader?

Here, however, let him stop a little, and consider awhile; for, upon his rightly understanding what now becomes of the blood (thick, dark, and unserviceable as it is), and how, in a trice, it becomes healthy, of a bright scarlet colour,
and invigorating, mainly depends his being able to comprehend, by-and-by,
what I shall have to offer concerning the diseases to which a vitiated or corrupt state of the blood gives rise. He will not, otherwise, make out sufficiently clear in what manner the blood of an animal can contract and retain that
morbid state which shall predispose it towards acquiring a constitutional disease that, however differently named according to the parts whereon it may
fix, has but this one common origin for the entire series. To this page, then,
I shall frequently refer him when speaking to these points more in detail, hereafter; and he had best, also, keep the book open at this place, whenever he
may be endeavouring to comprehend what the learned veterinarians of the
present day are striving to say respecting "the circulation," as they quaintly
term it. Another of them, speaking upon the topic I have just brought to a
conclusion, says, "The heart is divided into two cavities, termed ventricles,
each having an auricle, resembling a dog's ear. The blood-vessels proceed
from these [those] cavities, the arteries from the ventricles, the veins from the
auricles, &c." All which is very true, but not very intelligible to the generality of readers; and yet is the author, who thus speaks, (Mr. White, in vol.
i. p. 63) said to be the plainest spoken among the moderns; indeed, were he
any thing else than a good one, I should not have deemed him worthy of this
rebuke.

39. The Blood, as I have said, rushes out of the large chamber on the right
side of the heart into an artery that soon divides into two branches; whereof
one enters each lobe of the lungs, and there disperses, through certain cells,
the blood with which it is constantly supplied. Here lies the the secret! At
this point it is, that health or disease (at least a predisposition to one or the
other) is imbibed and engendered in the blood. The lungs having received
the thick discoloured blood from the right side of the heart, and being the re-
ceptacles of the air we all breathe, do, by means of that air, bestow upon the
blood afresh the principles of life, and health, and vigour. The cells, or tubes,
through which the blood passes in the lungs, termed pulmonary, are eight in

number, being double the quantity given to man, and show, from that circumstance, the immense circulation of which they are the agents. A cruel and almost incurable malady, that attends most horses at this part of the organs of respiration, with many and variable symptoms, was alluded to higher up, at sect. 36; and is what we term from those vessels, pulmonary consumption. But then, it is clear, that the air which is so brought to effect those beneficial changes upon the blood in the lungs, must be fit for the purpose:—that is to say, it should be vital or atmospheric air, uncontaminated by any noxious stench (as the ammoniacal smell of the stable, or the stench and smoke of cities); no poisonous vapours (as burning brimstone, the gaseous fluid, smelting of minerals*), nor infectious effluvia (as of cesspools or stagnant waters, producing fevers, glanders, &c.) can give to the animal's blood that healthy vigour which was designed for his well-being; but, on the contrary, every departure from purity, in the air he breathes, must be an approach towards disease†. Yet, how constantly is this simplest law of nature transgressed! And what, in such a state of things, can be expected, but that the blood will assimilate in character with the kind of air the creature is compelled to inhale?‡ Accordingly, we find in the domesticated horse, that previous disposition tc certain diseases which we endeavour to meet by correctives and repellants, but which by better treatment might have been prevented. Of these evils, abscess, or ulcer, is the most prolific, showing itself now on the head, at others on the heels, under the varied denominations of poll-evil, quittor, fistula, &c.; but more frequently attacking, and making the greatest ravages on the internal organs, where it is generally mistaken for the worms, and erroneously treated with hot and burning remedies, when evidently a direct contrary practice would best assist nature in casting off the evil.§ Ulcer upon the lungs, as it produces a staring coat, is too frequently mistaken for worms, and if the animal be not physicked and cordially too much, a partial cure sometimes takes place, though in what way it is carried off is most inscrutable; but, upon opening the dead subject, I have often noticed spots which had been corroded and gangrenous, where the cure had been effected in this spontaneous manner. As for ulcers upon the liver, also arising from the viscidity of the blood, if

---

* Scarcely any truth is easier proved than this: Horses that are constantly kept in close stables, in large numbers together, very soon become unserviceable, by the constitution throwing off some evil or other upon the surface; as one proof whereof, those which are occasionally placed in the under-ground stables, at the Swan with Two Necks, Lad-lane, show evident signs of distress which subside upon being brought into the air. I hear from good authority, that at New Orleans, in North America, where the atmosphere "exhibits a blue misty appearance," nothing is more common than a disease which affects the knee, hock or pastern joint, with abscess, or near those parts with cancer, and the limb actually rots off. The like kind of attacks were found formerly most common in Cornwall and in Wales, and are attributed by all to the arsenical vapour of the copper mines being inhaled: they are, however, much less frequent at the present day, owing to the higher state of cultivation to which the land has been brought, and to the change of situation horses now enjoy. The free use of sweet oil is a good preventive of this poison in human as well as in cattle medicine.

† This is not the place for a finished dissertation on the communication of the glanders; but I must observe, in illustration of the text, that horses which have eaten glanderous matters without receiving the infection, no sooner smell it than they become diseased.

‡ In cases of much sorrow or grief, our respiration is much increased; and the presentiment of death awaiting them, occasions all animals to take harmful substances into the lungs. Pigs, oxen, and even sheep, show signs of uneasiness, horror, or madness, at the effluvia of blood of their own kind: and I have found in each kind of animal, upon slitting the trachea as far as the bronchia, dust, dirt, or other rubbish, which they had snorted up in the last paroxysms of despair.

§ Four ounces of spirits of turpentine have been given with partial success: but such a dose must go near to destroy not only worms, but the horse into the bargain, whilst it is not very clear to me that the small worms we sometimes meet with are hurtful to the animal, but are designed to act as a stimulus to the intestines: and, if the doctor mistakes for worms some more serious disorder, the horse dies of this monstrous medicine, as sure as fate.

they take place near its thin extremity, the common natural process is, that the part of the liver attaches itself to the gut, and the offensive matter will then slough off into the intestine, and come away by stool: a partial cure is the consequence: not produced by medicine acting upon the part, as is very clear, out by a common natural effort, aided by a more vigorous and healthy state of the blood than when the disease was engendered. To assist the circulation in regaining this state is clearly the duty of the medical attendant, and is the only manner in which he can be of any service to the animal in restoring it to health.

40. In the lungs, then, does the blood receive from the air its invigorating principle, and no sooner does this take place, than it drops into another shor blood-vessel (a vein), and, by it, is conveyed again to the heart: not the same small chamber on the right side, of course, but to a similar one on the left side. Here the contractions go on as before mentioned, only that the blood differs in quality; this being now properly fitted to promote the purposes of life, and for imparting the vital principle, occasions the heart to assume that twisted shape we see in some animals, while in others whose bodily exertions and arterial functions are less laboured than those of the horse (the sheep's for example) it is more round: and, indeed, the heart of this last mentioned animal is, from this circumstance, usually termed, in the London district, a "round heart." Out of the small chamber on the left side is the blood driven through a valve, as before, into the large chamber of the same side; which in its turn contracts, with much force, upon its contents, expelling the same into the great artery. This periodical rushing of the blood into the arteries, imparts to these vessels a motion we term pulsation, whereby the blood is propelled forwards, to the remotest parts of the body. Lessening in size, and increasing in number, the branch-arteries, which receive this blood, become more sensitive as they are farther removed from the heart, and afford us, at certain places, the means of ascertaining by the touch the degree of heat at which may at any time be the state of the animal's blood; the contractions of the heart being regulated, in quickness and force, by the degree of stimulation the heat of the blood may occasion it. This is termed the irritability of the heart, and the medical test of which I speak, is called "feeling his pulse." Even in the tail may this contractile power of the arteries, and its effects, be seen to advantage; when a colt is being docked, the blood squirts forth with frequent gushes, answering in a tolerable accurate manner to the state of his pulse, although the artery at that remote part is very small. But the blood that is so changed in the lungs, as I have just said, is still found to contain certain particles or properties, which would render it unfit for the purposes to which it is to be applied, or those particles are required to effect certain purposes elsewhere in the system. Accordingly, the major part passes into the liver, there to be refined of its bitterness; whilst a portion is attracted to the kidneys, where it leaves its saline qualities, that pass off by staling, as the former is voided by dunging. See "Liver," farther onwards.

41. These purposes, however, are not always effected alike regularly, from several causes. Frequently, the blood comes to those organs in a state too vitiated for their utmost activity to cleanse; sometimes a diseased state of the blood, at others, languor of the parts, indispose them to the performance of their functions, and the blood is suffered to circulate, filled with humours that war with the constitution, and form what is termed "constitutional disease," or predispose him to receive disorders of varied malignity, according to circumstances. I shall come to advert to this point when, shortly, the liver and kidneys claim our attention; but, the chiefest cause of disorders incident to the animal by reason of defective liver, is the great exertions he is put to, and the consequent rapidity of the circulation, whereby the blood is propelled through

it with tremendous velocity: it then acquires inflammation, and becomes ul cerated (see sect. 39); a disposition which must be increased whenever the blood has been formed imperfectly, either as regards the lungs or the lacteals So much is this the case, that the blood drawn from a very aged patient of mine lately, that was incapable of grinding his corn properly, showed evident proofs that particles of ill-digested food had entered into the circulation, and is an answer to those writers who aver that the lacteals reject the bile and suck up only the fine parts of the chyle; whereas, all inquiring persons must know, that they take up the lighter parts of all substances whatever that pass through the intestines. If this were not so, how is it that the slow poisons just spoken of enter the blood? or how would it come to pass that so much good is performed by alterative medicines, that, in like manner, insensibly in- troduce themselves into the blood, and produce invisibly those permanently good effects we so much admire? The opposition of a few is no obstacle to this commendation of an obscure but safe and certain class of medicines,—to say nothing of an alterative regimen, which is more safe still, and certainly more natural, though slower and less positive in its effects than active medicine. How much longer is the horse to be treated with nothing else but violence?

42. BLOOD-VESSELS, or tubes, I have already observed, pervade every part of the body, and are of two kinds, whose office is directly the reverse of each other. Arteries, it will be recollected, convey the vital fluid to all parts of the body, and the construction of these, it may easily be conceived, from what has been said, is simply that of a tube with great contractile powers: they are large near the heart, but soon branch out of a lesser size, until, entering the smaller organization of the solids, they become very minute, infinitely nume- rous, and more sensitive, thus affording the means of renovation, or growing to the flesh, bones, skin, hair, hoofs, &c. The large artery communicates with minor branches, soon after leaving the heart, by two rows of openings, like perforations, in its lower sides, at two or three inches asunder. Some one has likened the arrangement of these vessels to the stem, branches, and twigs of a currant-bush, and so might the veins that run nearly parallel to the ar- teries, through every part of the animal, but are so constructed and arranged, as to take up and reconvey the blood (which the arteries constantly deposit) back again to the heart; and the similarity will still further hold good, if we extend it to the leaves of the bush, and compare these to the glands (see Sec- tians 27, 28), in every one of which an artery terminates, and deposits its nu- tritious contents, and where every small vein begins the absorption of what the artery has so left behind. This absorption, when obstructed, lax, or other- wise imperfect on the surface, may be restored by stimulating the parts with spirits; when it is too high, and labouring greatly in consequence of the vis- cidity of the blood, local inflammation is engendered, and one or other of those diseases I shall hereafter treat under "abscess," is the consequence of this constitutional derangement. Blood that is thick, heavy, or viscid—call it which we like—causes heat, which being general, is fever; the rapidity of the circulation increases violently, and the blood becomes more fluid than when the animal was in good health. Arteries may be distinguished from veins in the dead subject, by the property they possess, of retaining their tubular shape after the blood is discharged; whereas, the veins collapse, when empty.

43. If the arteries are plain tubes, lying for the most part concealed, the veins, on the contrary, are more frequently found exposed to the sight under the skin, next to the muscle. But, more delicate and more numerous, the veins perform their part of the circulation by a totally different means than is found in the propulsion and contraction of the arteries. One of the means of effecting this purpose is by the obvious and simple movement of the body and limbs, as well as by the act of breathing, whereby the blood is pressed out

of some one part or other of the veins into the adjacent part; but these vessels being furnished with innumerable valves, within an inch, or less, of each other, the doors whereof open only towards the heart, the blood must necessarily force itself out that way, and no other. If we suppose that any given portion of the vein is hereby emptied; what follows from the circumstance, but that the next-door valve, which kept back the blood contained in the adjoining portion below, while it was pressed upon from above, being thus relieved from the pressure, will now open? Then, in rushes the blood, and the space is again filled, but only to be emptied in a similar manner. It is with a view to accelerate the circulation through the veins, and to keep the blood warm in winter, that mankind betake themselves to forced exercises, as hunting, running, or beating the arms athwart the chest, and that beasts with the same propensity gambol and frisk about, or rub themselves: both promote the same ends, by brea hing short, so as not to cool the lungs too much, or by drawing in the same warm air over again, in sheltered situations.

44. By the process just described, the blood being once admitted into the minuter veins, finds its way to the larger ones, which convey it to the heart; near to which, as the quantity becomes great in the large vein, it receives a powerful auxiliary in the filling and emptying of the lungs, and the working forward and backward of the midriff, as described before. Passing through it near the back bone, and, consequently, as its more muscular part, the great vein must at every inspiration receive from the midriff considerable aid in pouring forth its contents—to say nothing additional of the pump-like action of the heart itself. At this part the vein acquires the appearance of a double tube, the outside thickened and muscular, the inner one membranous and collapsed, as if too big for the space in which it is placed. **But the curious fact—** how the blood which had been sent into circulation through the arteries, gets into the veins at first, deserves consideration; as this must be effected laboriously, when the fluidity of the blood is lessened, or else accelerated with frightful rapidity, whenever it so happens that fever prevails: local inflammation, by the same rule, must cause an unusual flow of blood to the part affected; and, as most of the impurities of the system will then be drawn towards it, at that place must they leave the occasion of the most direful effects. In proof of this doctrine, it happens frequently, that when an animal is attacked with inflammatory complaints at two places at once, the greater evil of the two increases, while the lesser one "runs off," as it is called, or gets cured of itself. So, on the contrary, when a diseased horse (glanderous, for example) is well fed and well kept, he not unfrequently overcomes his disorder without medicine, to the great wonder of the unknowing; the solution whereof is, that the constant supply of new blood has quite changed the nature of the animal, his disease has been "taken up," or absorbed, by the animal system, and ultimately carried off in the common evacuations. In short, absorption and effusion are the great internal secrets of animal life; although we can not say with certainty how they are carried on, it is no less a positive law of nature, that the veins at their commencement in the glands absorb or suck up, the blood that is deposited there by the arteries, and separated by the lymphatics; equally well known is it, that effusion takes place, of arterial blood, into much larger veins, as well as from one part of the system into another, internally. These points are deducible from a multitude of facts and well-known operations, some proofs of which the reader will find more in detail in the course of this treatise. In some parts, a positive connexion of the capillaries, or smallest blood-vesssels, is found to exist; in such a manner, that the section of one or more of these, and the consequent obstruction of the means of life to all the parts below it, which some would naturally expect, is hereby remedied.

45. THE PROCESS OF DIGESTION, whereby the food is prepared to be converted into blood, is no less curious than "the circulation" itself, is equally conducive to the support of life, and being obstructed, is also the harbinger of disease. If, on the two other great functions of animal life, depend his immediate existence, no less does the horse's capabilities, his present health, and the engendering of future obstinate, incurable, and often mortal diseases reach his vitals by way of his stomach and intestines, that comprise the organs of digestion.

Every one knows that oats and hay are the chief sustenance of the horse in a domesticated state, and these, together with one or two other similar productions, and water, are given to him for nutriment alone; it therefore follows, that whatever substances are taken into his stomach, which act not to the same end, must operate injuriously, even though containing nothing hurtful in themselves: these must obstruct, if they do no more. But, when matters obnoxious in their nature, and possessing strong powers and effects, are being administered to the horse, it seems but fair to examine whether such things are likely to agree with his common nutriment, with the blood that is made from it, or with his constitution, his habits, or the tasks he has to perform. Or whether, on the other hand, they do not prove destructive of the food itself, of the vessels that contain it, and of those which draw up its finer particles that are to be converted into blood.

Let him who practises by violent means consider a moment the natural structure, formation, and functions of the organs on which his medicines are destined to act, and compare their altered state, that has been brought about by reason of the treatment the animal has received at his hands. Some protection, doubtless, is afforded by nature in the secretion of the fluid so often mentioned, which defends the several organs against ordinary injuries; but these are often found insufficient in quantity (notwithstanding the supply which may be drawn by effusion from other parts of the body), and the frequent repetition of the monstrous mixture of the doctor's skil-less art, ultimately effects its purpose in destroying the tone of the fine vessels just alluded to (the lacteals), and then the symptomatic disease of the skin, called hide-bound, follows, as a natural consequence.

46. THE STOMACH is a bag, or pouch, with two holes in it; the one receiving the end of the gullet, the other opening into the small gut. See it; plate I—L, 26—29. It lies behind the midriff, inclining a little to the left, having the lungs contiguous on the other side that membrane, and the liver next behind or under it. This main organ of digestion is but small in the horse, as compared to that of any other animal, being so designed to assist his fleetness, out which would be impeded by his receiving large quantities of food at a time, and points out the reason why he requires to be fed and watered frequently: he larger and more distended the stomach of any animal may be, the more sluggish and vicious he is. As one proof of this position, I formerly obtained the stomach of a man which was larger than that of any other subject I ever saw or read of, and its possessor in life had a heavy, slow and sordid manner, together with one or two other bad qualities which brought him prematurely to his end.

At the upper orifice of the stomach, a membrane, nearly insensible, coming from the gullet, enters it loosely, and spreading along its lower part, lines about one half of it, thus defending it from acrimonious or poisonous substances, whilst the coat of the other parts thereof consists of striated muscular fibres, very sensible and given to contraction, and running transversely to those of the insensible coat, a circumstance which has given rise to the fanciful notion, that digestion proceeds in the horse's stomach by what is termed "trituration," or pounding; but, upon reference to the figure which illustrates another organ, (at section 56), the reader will be able to form a more accurate notion, near enough for his purpose, how the stomach contracts its sensible part upon its contents. This sensible coat secretes a juice which, from its strong gastric nature, not only digests the food, but would also corrode the stomach itself (insensible though it be), were it not supplied with another fluid for its protection, in the saliva or spittle that descends the gullet along with the food. This saliva is also secreted, in some measure, when the animal may not be feeding: but, whenever this supply fails, the gastric juice predominates so much as to cause a galling pain in the stomach, and occasion in the animal a ravenous desire for filling it, if not with food, at least with some substance that may keep it distended, and perhaps carry off the painful superabundance. Horses so circumstanced, when in harness, gnaw the pole or shaft, or bite at each other, and soon learn to become crib-biters, gnawing any thing they can come near, as well as the manger; litter, bits of old wall, and dirt, at length, are found by them agreeable to their palate. See further at section 49.

The insensible membrane I spoke of, by its loose folds, forms, at the entrance of the stomach, a kind of valve, which prevents regurgitation of the food, like that of ruminating animals; and a similar contrivance at its termination in the lower or right orifice occasions a short obstruction until the pulp is mixed; for, when the stomach is filled, the relative position of the two orifices alters in a great degree. From these premises it seems apparent that any substance entering an empty stomach does not act upon the sensible part of it, but being soon mixed up with the gastric juice, it proceeds into the intestines, there to communicate its effects—whatever these may be. Whether nutritive or medicinal, poisonous or beneficial, the intestines receive all with but little alteration.* But when it so happens that the food does not pass readily out of the stomach, a fermentation commences, and the sensible part thereof being then distended, the ill effects ascend the gullet, reach the head, and cause vertigo, staggers, &c. At times, a specific inflammation takes place, and communicates itself in four or five days to the whole of that surface, taking its course downwards or upwards, according to the orifice that may be most affected; this being all the way down through the intestines, blocking up the influx of gall (as described sect. 48), and causing yellowness of the eyes, until its appearance at the anus; or, in the other case, it ascends up to the nostrils, making its appearance first about the head, and communicates either way to the skin and its coat.

47. Of the intestines, guts as they are usually called, it is important to keep in mind, that, notwithstanding the appearance of great tenacity they assume, they are, nevertheless, extremely irritable, being composed of two coats of fine muscular fibres that cross each other, the one circularly, the other lengthwise; and having a lining which secretes a fluid for its protection, they admit in their intervals an innumerable quantity of absorbent vessels, that are constantly sucking up the finer particles of their contents. This sort of conformation

---

* At this place, for the information of those who would practise the veterinary art by comparison, it may be useful to observe, that in the human stomach is digestion principally performed, in the horse's very little; in both, the small intestines appear to mix the food with the bile and other digestive juices; but man having no cæcum, or blind gut, like the horse, to receive the heavier parts of the food as they escape from the small intestines, his lacteals begin higher up than those of the horse, which lie wholly on the large intestines. It follows that, whatever is received into the stomach of man is felt through the system immediately; with the horse this does not take place until it has reached the intestines. One other dissimilarity in the mode of digestion is worthy of notice: in man, the work of digestion is nearly finished when the bile is mixed with the food—say at an average of twelve hours from its being taken, whilst the horse passes his feed into the intestines in about two hours, before it has well assumed an homogeneous appearance, which the bile seems to effect for him. With us liquid remains in the stomach; the horse passes water immediately into the cæcum.

renders the intestinal canal obnoxious to repeated strong drastic purges, particularly aloes of the Barbadoes kind, that heat and irritate the parts by their coarseness. Inflammation is most likely to succeed such irritation, in summer-time especially, and the animal is usually destroyed by the pretended remedies of the farriers; or, being pressed forward in his work during the attack, goes until he drops down and dies. At the fundament may be seen the earliest indications of this species of over-physicking, in the disgusting protrusion of the inner coat whilst expelling the contents thereof; an ordinary effort of nature to get rid of what is offensive to it, which, considering the horizontal position of the horse, might appear wonderful to us bipeds, but for the well-known double operation of the coats of the intestines. From the top to the bottom of the canal a spiral motion is kept up by the alternate contraction of the two coats thereof, the one in circumference, the other lengthwise, resembling that of a worm, and appearing as if a corkscrew agitated its inside. By this means the mass is pressed backwards, and as it increases in quantity and becomes less and less clearable, the offended nerves excite the guts to renewed efforts for its expulsion, in which the lower part of the belly, with its covering, from the cœcum to the sphincter, concurs with all its powers of contraction. Partial retention of the breath, and consequent pressure upon the midriff, and parts behind it, contribute to lessen the longitude of the intestine at every effort. This kind of excitement, if repeated too often, it is plainly to be seen, must keep up the irritation of the parts concerned in it, and dispose them to contract inflammatory complaints.

48. In length about thirty yards, the intestinal canal has in its course two or three different offices to perform towards digestion, whereof the smallest gut nearest the stomach is for receiving the gall, or bile that has been formed in the liver for that purpose. At the termination of that small gut, at the end of twenty yards, an immensely large one occurs, called the sac (cœcum), or blind gut, where the contents are prevented from issuing too soon, by reason of the internal coat of the small gut getting into folds, as it were. We may as well consider this as another valve; and that it was provided by the Author of Nature to correct the animal's propensity for transgressing his laws against repletion, as well as to prevent the contents of the cœcum from returning upwards, when this latter is compressing the large intestines backwards, in the act of dunging. But inflammation sometimes, obstructions oftener, produce at this place more tedious affections than is generally imagined. When it so happens that the stimulus of the bile is insufficient (as in diseased liver), and acrimonious particles are left behind, or the half-masticated food inflicts injuries on the very sensible surface of this passage, then the noisome effluvia reascends to the stomach; the bile, too, enters it soon after, by reason of the intestines having lost their power of compression and elongation, when the corkscrew motion downwards is changed to an upwards motion, and all becomes disorder in that region. Loss of appetite, fever and dullness, with drooping as if in pain, and a staring coat, follow each other in succession; for the secretion of bile, which I shall come presently to describe, as affecting the skin, is thereby vitiated. These appearances it has been a fashion to consider "symptoms of the worms," or of "debility" (another term for low fever); and the practice of administering bitter medicines, that are supposed to kill the worms, is only successful on account of their restoring the tone of the stomach, and by supplying to the intestines a congenial stimulus in the place of bile. This was the case with Mr. White's statement, in vol. i. p. 170, where he says, "I have sometimes succeeded in destroying worms by giving aloes, one dram and a half, every morning until purging was produced." That is to say, "the horse became well;" but whether he had any worms to be destroyed is another question; and then, if a dram and a half would succeed *sometimes*, {

should apprehend a larger dose (as eight drams, his favourite quantity) would more inevitably have poisoned all the worms his horses may have had, of whichsoever kind they might be; but this mode, as will be perceived, though more destructive of worms, would not have acted as a *tonic* restorative on the stomach and intestines, like small repeated doses. I, however, who am a man of no fashion, generally have found those kind of attacks accompany a repetition of irregular feeding; that is to say, very little one day, very much another; now all, now none; the attack varying in degree, and changing from simple obstruction to the inflammatory, as the animal may or may not have been allowed water with his food.

49. To supply this deficiency, in some measure, does the *cæcum*, or blind gut, seem to have been placed at the termination of the small intestines. In this second cavity digestion is supposed to be completed, much liquid being found therein; and we know that here, in a corner, termed its "appendix," are frequently deposited hard matters, as earth, stones, and other substances, little compatible with the purposes of nutrition (as noticed at sect. 46); but whether these ever pass off by stool remains in doubt, and we are left to conclude that it is much less sensible than the other intestines. In size it may be about thrice that of the stomach of the individual; and it is placed near the surface of the belly, lying on the left, about midway between the fore and hind near leg. Here it is exposed to damagement from a variety of causes: the groom, while dressing him, often hits the horse here with the curry-comb; the dealer tries whether his new purchase is a roarer by striking him hard with his lash whip, whilst he holds up his head short; and I have frequently seen one of the most noted jobbers and breakers in London terrify his "restive customers" into obedience, by a kick of his foot skilfully placed on this part. The facts are notorious; what is worse, they long remained uncorrected. The consequence of all this hard usage is, that the *cæcum** loses its functions, more or less, certain heavy particles are not expelled as they ought, but, remaining behind, attach to their sides some earthy particles of the food which would otherwise pass into the colon, and the heat of the animal's body causes them to become stones of great magnitude. Six or eight pounds in weight, and nearly as many inches in diameter, are quoted as by no means uncommon sizes, in certain parts of the country, where humanity is at so low an ebb, and the police equally unmindful of their duty, as in the district of London before alluded to.

50. The colon, or large gut, commences at the only orifice of the last-mentioned sac; then, turning underneath the small intestines, and proceeding forward to near the stomach and liver, it turns about, and, in its course backward, makes a great number of zig-zag turnings, by means of two ligaments that run along its whole length, and coil it up. Such a shape, or rather *noshape* disposition of its folds, would inevitably obstruct the progress of its contents, but for the fore-mentioned double motion of its muscular coats, with which it is furnished, as well as the smaller intestines. Being heavy, it is suspended the whole length of the horse's hinder part, by a strong half-transparent membrane (called mesentery), which being fastened to the bones of the back, and hanging down in folds, or plaits, admits of the gut's filling up the same from side to side, in the semi-globular manner we may perceive when the animal is opened. But where the mesentery embraces the intestines the tightest, as if to prevent the too ready escape of the food, there is placed along the whole length of the depression, between the folds of the gut, a white vessel

---

* The muscles of the cæcum being stronger than any other part of the intestines, are compelled by the blow to contract forcibly, as do the intercostal muscles of the lower ribs, whereby the air in the lungs is suddenly expelled, and, if he be affected, the horse groans as the air passes the upper part of the wind-pipe.

having numerous branches to the right and left, full of nutritious juice, making its way towards the fore-part of the animal.

This is the lacteal duct, which, from its situation between the folds, formed by the mesentery, is by some termed "the mesenteric canal," and by and by, (in Gibson) "the mesenteric artery." *Mesocolon* and *mesorectum* being the names of parts which usually merge in the general term "mesentery," for the whole, I have made no distinction. But all this does not signify so much as the manner in which this duct gets filled at first by the lacteals, how it constantly flows in health, or is obstructed by disease, and what is the mode and the effect of discharging its contents near the heart, as before alluded to in sect. 37, second paragraph, as well as just below in sect. 51. Herein may be found much matter for pleasing reflection and study, by him who aspires after obtaining a more accurate knowledge of the curative art than is generally possessed ; and to attain to perfection wherein, he must study the thing itself by inspection, since nothing that I can find room to set down here can give him any thing like an adequate notion of its importance ; nor, indeed, was it ever my intention to employ strict anatomical description, or to enter into learned definitions, any farther than should be found necessary to illustrate what I have to teach, respecting diseases in general, and some long standing errors of respectable veterinary surgeons in particular. On no other point, throughout my present labour, do I so much desire to be rightly understood, as on this one of the absorbents, and absorption altogether ;* for it is only when this function takes place with regularity that health can be preserved ; when it is disordered, our business is to restore it, too much or too little being equally productive of a disposition to diseases, though opposite ones. An indolent or an impoverished absorption requires our care no less than a too rapid or feverish performance of this function: the fleam and cathartic medicines reduce the latter kind of symptoms ; a generous mash, tonic alteratives, and good grooming, are the best restoratives of a languid system. Pulsation is the test of either state of derangement ; and he who is the cleverest at discovering, by this prognostic, what is going on in the system, will always make the most humane, as well as the most successful, horse-doctor.

51. Towards its termination, the colon makes a short turn, as if to prevent the too easy escape of the dung into the rectum, or straight-gut, without an effort of nature to straighten the curve at that place ; as we see it performed when the animal strains the part, while contracting the lower muscles of the belly, together with the cœcum, in order to produce a stool—the whole transaction being most intelligibly termed "a motion." Several such impediments occur in the course of the intestinal canal, and some of them are so abrupt, as no after-art is ever capable of reducing to a straight line: the reason for which kind of contrivance is, that its contents still possess some nourishment, which it is desirable should be extracted, and they are thus detained that nothing might be lost : to say nothing of the existing opinion, that the food which has thus lain some time in the animal must impart a juice differing considerably in its properties from that which was but recently received into the stomach. No operation in the system is more beautiful than this one of drawing from the food, now properly mixed and softened, what becomes the milky fluid called chyle, first, and blood immediately afterwards ; the first mentioned being performed by innumerable transparent vessels, whose fine mouths open every where on the inner surface of the intestines. From the word *lacta* (milk), these vessels are termed *lacteals*, their function being ab

* Generally termed "the absorbent system," and until lately, wholly unattended to in veterinary practice: Gibson, in his lengthy particularities respecting the horse, not having once mentioned the lacteals (as if they existed not), and contenting himself with just loosely naming 'lymphatics' at page 55 of his first volume.

sorption (like the lymphatics); tne largest whereof lying along the mesentery (as I said before), sends out smaller branches, and these again more minute ones, to encircle and penetrate the gut; in this their mouths do incessantly suck up, or absorb, and convey to the larger vessels the material for replenishing the system with new blood. Passing along the spine, the large tube, filled with this milky fluid, at length reaches the fore part of the animal, and acquires the name of the *thoracic duct;* here it mixes with a portion of lymph, and is conveyed immediately by a large vein to the heart. Ascending the pulmonary artery, as described in a former section (37), the air entering the lungs, changes its colour to a fine healthy scarlet, and at the next pulsation it is driven into the circulation to mingle with the mass, to impart its newly acquired properties, and to return again and again, wasting away, until at length it becomes used up and extinct, its place being supplied with other new matter by the continued process of digestion. A change of substance this, which is said to take place with the whole body of the horse in the course of every year and a half, or two years : so that at no time has he a particle of flesh, bone, hair, hoof, or other matter which formed his body two years before, and affording a fit subject for reflection and admiration in us, as it invites those who have the care of providing for his health to take advantage of the well known circumstance, the more securely to effect those changes by gradual means, which too frequently are attempted by violence, and fail. Nature will not be forced, rather seek her in her recesses, and humour her ways. Those who act differently, generally induce some lasting disorder to appear upon the surface, which they treat as if local; when, alas! these are seated in the very vitals of the animal, sometimes in the most delicate parts of the mesentery. Heating, or cordial medicines, as well as those other untoward mixtures, which corrode, or blunt, the mouths of the finer lacteals, thereby dispose them to receive materials improper for the making of good blood; the consequence whereof is, that tubercles frequently are found, which fill up the cavities of the lacteal duct. The matter of these tubercles is usually hardened, and resembles the yolk of a hard boiled egg. The disease appears in a staring coat, is commonly considered to be the worms, and treated as such, with more heating or drastic medicines, which but increase the evil. Other obstructions are thereby formed, and if a solitary worm or two are found on dissection (as frequently happens), they have been generated in the obstructed part, but have not caused any disorder.

Of the mesenteric canal, it may be useful, as well as curious, to remark, that I have always found its state of health or disease to correspond with the appearance of his coat; when this is smooth, the former is full and free from obstructions; when rough, the contrary. In hide-bound, this canal is yellow; in farcy, red, as well as the bowels; these appear bluish, when the horse dies in consequence of being worn out, though at the same time, the flaccid lacteals still preserve their healthy white, if no other cause to the contrary prevails. But, upon such further particulars as are connected with the study of digestion, as influencing or influenced by respiration and circulation, I shall take occasion to say more hereafter; adding thereto a few cursory remarks, that were not absolutely necessary for the present illustration of the animal system, but will be found more in place in the succeeding section.

The LIVER with its sweetbread, the Kidneys, and the Bladder, being liable to certain diseases peculiar to each, besides the property of affecting one another readily, as well as being at all times mainly instrumental in maintaining and restoring health to the other parts of the whole system, now claim our undivided attention. I shall, therefore, proceed at once to a brief description of the uses and functions of each, and accompany the same with a few gene ral, but pertinent remarks on the present received mode of treating the disorders incident to the several parts that impede those functions, reserving particulars regarding the causes, sympsoms, and method of cure, to a subsequent part of the volume. Consult the Index.

52. The Liver is a very important and immensely large glandular body of a dusky red colour, almost divided, like the lungs, into two lobes, having two smaller subdivisions; and is attended by its pancreas or sweetbread, a small flat part thereof, which has the property of secreting a sweet kind of saliva. This secretion was noticed before, as entering the gut near the stomach, along with the bile from the liver: both are therefore conveniently situated underneath the stomach and behind the midriff, to the skirt of which the upper part of the liver is attached; but the exact functions of this pancreas, or its diseases, are no farther known to us, except that it partakes a good deal the appearance of its joint neighbour, and that it is indeed sweet to the palate.

Before he proceeds farther, the reader had better consult the place of a skeleton as to the situation and extent of this important organ (important in a curative point of view), as relates to the midriff, stomach, and kidneys, where it will be seen included between the squares marked K—N as intersected by the lines numbered 21—27. The side view therein presented is necessarily the left or near side, but the other lobe or right is of greater length and more substance, it touches the right kidney, and its upper surface is contiguous to the diaphragm, which presses upon it at each inspiration of the lungs. This tendency of the liver to the right side seems to have been designed by nature to counterbalance the leftward position of the heart, and of the lower part of the stomach; the pyloric orifice of which is seen at the intersection of the lines K and 26. In a former page (sec. 27.), I took occasion to describe the minute glands with which the extremities are furnished, and to advert to the secretory glands, all which are formed by arteries that deposit their contents, and which is again taken up into the veins; but the liver, the largest of all glands, and a secretory organ, differs from the others in one great and signal respect: it is formed of an assemblage of veins only. Its structure, in other respects, is much the same as that of the smaller glands.

Into the liver is brought the blood which has been sent from the heart to circulate and nourish the whole system (except a portion which the kidneys attract); a service that is performed by means of a great blood-vessel they call *vena porta*, that passes along the right side of the spine. In size very large, and always filled in health, a sight of this vessel shows how busily employed the liver must be, in separating from so great a quantity of blood the bitter qualities it has obtained by having passed through the animal's system, and imbibed whatever might there lurk of the offensive, the diseased, or the infectious. It proves, also, that any disease with which it may be attacked, must be proportionably violent in its progress, and tedious to cure, inasmuch as both will depend upon the state every other viscus may be in, through which the blood happens to have passed. Are the kidneys, or either of them, inflamed? the blood which has recently passed through them comes to the liver to get rid of its noisomeness, in the form of bile. Is an abscess to be dispersed, and the acrid matter driven from the part, to be taken up by the lymphatics (see sect. 29), at the liver it is strained off, and here must be imparted a portion of its baleful qualities. It follows of course, that whatever medicine is directed towards the liver must go thither by means of the circulation, i. e. through the absorbents: for schirrous liver this is best accomplished by the lymphatics; for inflamed liver by means of the lacteals; in other words, these are the internal and the external modes of exhibition, and the preparations of mercury are here mostly kept in view.

The secretion of too much bile, and the consequent inability of the vessels

6 *

to carry it off, it may easily be foreseen, would be the harbinger of jaundice; and its approach may be discerned by the yellowness of the eyes, by the increased number and thinness of the animal's dungings, and the constant emptiness of its belly, which both feels and looks loose and flabby. On the contrary, too little bile, must leave the intestines without the requisite stimulus to expel their contents, which, soon getting dry and hard, a constipation usually follows, that defies the remedy by purgatives: nor is the operation of backraking with clysters always of effectual service. Whichever extreme affects the liver, the patient becomes weak: but in case of deficiency, though he may look more brisk for a few days, stretching his hind legs out when unemployed, he afterwards becomes feverish, hot under the tongue, sluggish and dull in the eyes. Schirrous liver—a corrosion or rustiness of its fine surface, accompanies this deficiency of bile, and when it recurs often, the disease becomes permanent; but whether caused by, or causing the same, I am unable to ascertain. Ulcerated liver is occasioned by a too great heat in this organ: if occurring upon its thin extremity, the disorder cures itself by a natural operation, i. e. by adhesion to the gut, and passing off by stool; but when seated higher up, it terminates fatally, by wholly debilitating the system, and sooner or later destroys the patient.

The preparations of mercury, before alluded to, act variously upon the system, according to the mode of exhibition the practitioner may adopt: in the form of calomel it assists the liver to discharge its functions by lowering its tone; the blue pill (*pilul. hydrarg.*) is finely adapted to solve the crudities of stomach and bowels in carnivorous animals, but has never been extensively tried on the horse. For any disease of the whole system, or "bad habit of body," as Richard Lawrence properly calls that predisposed state of it which ultimately produceth tumours, grease, fistula, farcy—mercury, in all its various shapes, is the only specific.

Too great a secretion of the bile, although it pass off, produces a roughish meagre coat first about the belly; the patient becomes languid, especially after being compelled to any great exertion, when he perspires too readily on the carcass, his manner is uneasy, and after a while, partial hide-bound commences under the chest. Should the bile be of a less acrimonious nature, those symptoms are then perceptible lower down (i. e. farther back), and when his eyes appear yellowish, it is then a confirmed jaundice; but in very bad cases, producing death, people vulgarly call it "broken hearted," because commonly brought about by bad usage*.

Inflammation of the liver generally accompanies those appearances; but we can not be certain, though it is to be presumed, that inflammation is always consequent upon an over-quantity of secretion. After much procrastination, medicinal remedies are of little avail when tried on the most extensive scale, although no disorder to which the horse is liable is easier of cure, if it be taken in time: the patient requires only a treatment directly the reverse of that which brought on his ailments, and he gets well, almost of course. Regular work, moderate feeding, and tolerable behaviour comprise these natural remedies: they are usually found efficacious in the earlier stages of the disorder, and then only. But those natural remedies being neglected, and alteratives (the assistants of nature) never thought of, languor of the whole system prevails sooner or later, and the best of medicines fail to act by reason of that languor. The absorbents are then accused of not performing their

---

* Since writing the above, I have ascertained upon the view, that a horse, rankling under the effects of maltreatment, absolutely broke the cells of his heart through high-spirited chagrin. The case is described much at large in the "Annals of Sporting," for July 1822, a paper which I was induced to draw up at the instance of my friend John Bee, Esq., who was present at the death and the dissection

function properly; or, if they do so, then the liver and the kidneys fail in re-
fining the blood sufficiently, so that, at its getting to the extremities once
more, those particles which ought to have been carried off are there deposited,
and form the nidus of those external maladies that are mistakenly considered
local diseases, and treated as such, instead of correcting the foul habit of body
which is thus plainly indicated. Of the whole series of tumours or abscess,
grease is the only one which people in general think of taking up into the
system; the matter that proceeds from the pustules that form grease is so pal-
pably composed of *urea*, or the principle of urine, which ought to have been
attracted to the kidneys, that every body who would cure the grease, very
properly, as if by instinct, administers diuretics; and when this means of
cure is adopted early, always with a proportionate degree of success. But of
these things more in the following sections.

53. THE KIDNEYS, although the seat of only one disorder (inflammation),
yet are they so intimately connected with the cure of other diseases, which
are constitutional, that a right knowledge of their functions can not but prove
highly serviceable in the judicious administration of the universally approved
method of cure, by the urinary passage. Diuretics, or urine balls, are so constantly in the hands of grooms and others, that I would admonish them thus
early to reflect a little on the consequences of going on from day to day in
urging these fine glands to over-exertion, whereby they are kept in a constant
state of irritation, are rendered incapable of acting their part, or literally become rotten. They are situated, one on each side of the spine, close to the
last two ribs (see plate G, H, as intersected by figures 28—30), where they
are attached as well by the blood-vessels which belong to them, as by stout
cellular membranes which cover them underneath. With this exception the
kidneys of horses seldom have the covering of fat, termed suet, which we find
in other animals, owing, no doubt, to the very great action of the parts. Mr.
Richard Lawrence must have been thinking on the ox or sheep's kidneys,
when he wrote his 289th page. For my part, so little of this fat on the kidneys has been noticed by me, that this book was already at press before I was
convinced they were ever covered; and yet I have assisted in opening
and noting the state of as many horses, I believe, as any man in England
who ever wrote a line on this subject: in France, I have reason to conclude,
they are more industrious in this respect. The left kidney lies close to the
ribs; the right one farther forward, is loose, and is connected with the right
lobe of the liver; which being much longer than its left lobe, seems to extend
itself backward for that purpose. Excitement, no doubt, is the mutual intent
of this connexion; and that deviation from her true system, which nature allows in the effusion from one part to another, takes place, when either the
one or the other may be diseased, obstructed, injured, or destroyed. On no
other grounds can we account how it is brute animals so long survive the total
destruction of some vital part, as we frequently find.* One consequence of
this loose situation of the right kidney is, that inflammation generally makes
its appearance upon it earlier than on the left, a circumstance which is partly
derived from its proximity to the liver; it also imparts some of its own feeling to that organ, when inflamed; two facts these which ought to be well
kept in mind, when we wish to excite unusual secretion in either. In shape,
the left kidney approaches the angular more than the right one; from which
I infer that, although the functions of the two must be so nearly the same,
in affections they differ; at least a gall or slight blow will affect the left much
sooner than the right kidney.

* Latterly, Mr. Travers has given the public the results of many curious experiments on
this subject.

54 The section of a kidney, which should be performed lengthwise, wiłl show in the centre its pelvis, in which the tube (or ureter) that carries off the water to the bladder takes its rise: in this pelvis stone is sometimes formed, that often finds its way to the bladder, unless it remains in the ureter, or comes away entirely.* The ureters communicate immediately with the bladder, and the water they convey is formed by the outermost red part of the organ draw ing the blood into it, and through which it is filtered by the vascular or whitish part which lies next withinside; here numerous little tubes convey it to the centre one, or ureter, that enters the cavity of the pelvis at H I, 33, 34, of the plate of a skeleton.

The blood, which has been so filtered of its water, is absorbed by a vein, which is plainly visible in the section of the kidney; and the whole function shows how rapidly circuitous any medicine must act, which being poured into the stomach is found, in so short a space as two or three hours, to have work-ed its passage through the bowels into the lacteals, thence through the heart and arteries into the kidneys, filled the bladder, and caused a staling of the noxious water, which is to carry off disorders of one sort or other. Here it is worthy of remark, that the operation of internal medicines is much more cer-tain in the horse, when directed against the absorbing vessels and the kidneys, than when intended to act chiefly on the stomach; for, as hath been observed, his stomach being one half of it insensible to stimulants, we are not certain of producing upon it any effect whatever. In all swellings of the legs, the good properties of diuretic medicines may be discerned almost immediately, by reason of the connexion which subsists between the functions of lymphatics and of the kidneys; so likewise, *diaphoretic* medicines no sooner excite the lacteals to a performance of their function, than the skin shows evident signs of its good effects. But both means of cure may be abused, as I shall show more particularly in the sequel: the first, being administered too often, wears out the functions of the kidneys; the second, being carried on too long, at length refuseth to act upon the skin.

55. An idea respecting the deposition of water in the membranes was thrown out in the twentieth section; and another, as to variation in the pro-portions of urine and perspiration in summer and winter, at the bottom of section the twenty-second, to which the reader may refer. On this topic a foolish notion having got abroad as to the small quantity of acrid matter con-tained in the urine of the horse, induced Dr. Thomson† to submit a portion

---

* I was called in to examine a horse, whose diseases had baffled the skill of many clever farriers. He had been long declared to have "a complication;" that is to say, none knew his disease, tor he occasionally voided blood with his urine, in great pain; they had therefore given him diuretics to such an excess, that he could not bear the hand's passing along his back over the kidneys: his sheath showed signs of œdematous swelling, and upon that region being pressed he became unruly. I, however, saw enough to ascertain, by the heat and tension of the part, that it was inflammatory, and as his pulse was high, his tongue hot and dry, I pro-posed to bleed him, and to foment the part; the operation, however, was scarcely performed when its owner resolved to take no further trouble, and the horse was slain. On examination I found his kidneys were rotten, and as pervious as dough: ulcers appeared upon both lobes of the liver, and the neck of the bladder was inflamed a little. The sheath preserved its size; and on the top of the penis a small shapeless stone, the cause of all this mischief, lay buried under the cuticle; and would, I should apprehend, have come away in the course of a day or two spontaneously. How it got there is most inscrutable.

† Of Edinburgh, in his Annals of Philosophy, for August, 1820. By the way, on this sub-ject it is worthy of remark, that for seven or eight years past, the French and Italian doctors have made a great fuss about this *l'uree* (urea), or proportion of the principle of urine, calling it "a discovery;" whereas our own people, in every branch of medicine, have been acting up-on the same doctrine for better than forty years, to my certain knowledge. Some have regu-lated their practice (human) by the appearance of the water, with various success; and I have a great notion, that this test of the state of the horse's health may be added to those other symp-toms by which we endeavour to ascertain the ailments of an animal which nature has forbidden

of it to chemical analysis in order to decide that point. "The result was, that it contains an unusually large proportion of that principle, so that without being concentrated by evaporation, it yielded crystals of nitrate of urea, very readily on the addition of nitric acid." This fact being thus satisfactorily ascertained, accounts for the strong ammoniacal vapour of stables that affects the eyes of the attendants, and being inhaled (as said in sect. 39.), is clearly the harbinger of several diseases in the horses confined in them—glanders among the rest.

56. THE BLADDER, or receptacle for the redundant water of the whole system, as it is separated from the blood by the kidneys, is situated within the hollow of the pelvis, at the intersection of H I with 33, 34 on the plate of a skeleton, with its outlet or neck turned towards the place of exit, varying a little according to the sex. It consists of three coats or layers, the outer two being muscular, and having their fibres crossing each other—(as may be seen upon splitting asunder a stale bladder), the better to enable it to contract upon and expel its contents. The inner coat is membraneous, sensible on distention, and secreting a mucous fluid to protect itself against the effects of the urine. When, however, the bladder becomes full, the secretion is insufficient for its protection, and irritation commences in order to induce the muscular coats to concur in the expulsion of the urine. This desire must be very great in the horse, for the reason assigned at the close of the preceding section, and shows the necessity of permitting him to void his urine upon his first intimating an inclination thereto. The shape of some horses' bladders differs a good deal from that of others,—particularly about the neck, those of the female being considerably wider, and shorter, than those of the male, a circumstance to be remembered when I come to treat of the disorders incident thereto; since in inflammation of its neck, for example, in one sex we are obliged to have recourse to instruments, in the other the urine may be discharged by the fingers. But it so happens that horses are more liable to the disorder just named than mares. My reader will also please to note, that the thin membrane which defends the whole intestine against the friction of the surface, (termed peritonæum,) reaches backward to only half way over the bladder; so that it offers no obstruction to our operations upon its neck in cases of disease.

57. To recur once more to the subject of a preceding section (the 55th)—the principle (of *urea*) that resides in any given quantity of urine evacuated by the horse, it may be here observed, that when the animal, on a journey, has been pushed onward, and thus prevented from staling for a considerable time, he at length produces it of a deeper colour and less in quantity than usual, a change which has been effected by the great heat of his body having taken it up again, by the absorption and effusion which nature has provided, of aqueous particles from one part of the system to another. The principle, or urea, however, remains in the bladder, and produces one of two evils; either the inner or sensible coat becomes inflamed, and loses, after an attack of diabetes, some part of its function of secreting the mucous fluid for its defence, if it does not terminate fatally; or, being less severe, but often repeated, a de-

to complain. Whatever practitioner should undertake to judge of the horse's diseases by its urine, must prepare himself to undergo a good deal of ridicule, and may expect some calumny; he would not, however, be far from the right path towards making a proper estimate of the quantity or violence of its ailment, though he might not so readily ascertain the precise nature of the disorder. The terms "nephrin," and "uric acid," the oldest and the newest for this principle of this evacuation, show the assiduity of which it has been deemed worthy, in that practice where it is confessedly of less importance than it is in ours.

position of earthy particles takes place, which is generally converted into stone or gravel.

Palsy of the bladder is induced from frequent repetitions of thus neglect ing the calls of nature, as well as from injuries of the spine; in both which cases the nerves having lost their sensibility, the coats do not contract suffici ently, and some water is always left behind. In all diseases of the bladder, a disposition to fill speedily manifests itself: and in palsy, this is the leading symptom. When this evil takes place, the horse, while staling, seems un willing, or is incapable of discharging the last drops of each voidance; and, if the usual practice of giving diuretics be adopted, the animal is ruined, if he does not burst the fundus of the bladder and die immediately: rather, the contrary method of discharging, instead of filling the bladder, should be sought, and the readiest way to effect this is to introduce the hand into the funda ment, which having emptied, the bladder may be felt much distended. In this case, we are told, "too much pressure might terminate fatally;" but by smoothing the bladder gently with the tops of the fingers, from its neck for wards, is usually successful. In fact, I never should have thought of its fail ing, but for what is said in one of the books on diseases of horses (White, vol. i. p. 121), where we find a good number of pages bestowed upon "suppres sion and retention of urine," which are not diseases in themselves, but the effects of disease; the first arising in defective secretion of the kidneys, the second in the bladder, or its neck. At all times a good deal of sympathy exists between this organ and the kidneys, and the kidneys with the liver; inflam mation of either being soon communicated to all three, in a degree propor tioned to the animal's general state of bodily health previously to the attack.

CHAPTER III.

General Observations on the Animal System of the Horse, with Reference to the Origin of Constitutional Diseases; Recapitulation and further Development of Veterinary Practice, upon the principles before laid down.

SEEING that a recapitulation of the preceding chapters, and a few general observations arising therefrom, would be necessary, before we examine into the particular diseases to which they have reference, I shall here add the no tice of such minor parts of the horse, as may seem to have been overlooked; and then draw such conclusions from the whole, as to the principles upon which veterinary medicine may be most successfully conducted, as appear to me best adapted to your acquiring those just notions of the theory as lead to favourable results in practice.

The animal system* (which has been so often mentioned) whereby life is continued and strength renewed, diseases are contracted, and the disposition to throw them off is constantly manifested, and by which the ordinary wear and waste of the various component parts of the body is unceasingly supplied with new and healthy matter, has been shown, in the foregoing brief account of the separate parts that contribute, by their united actions, to make up this system. A system that, although apparently complex and infirm, is, in re ality, simple, magnificent, and robust. It is we (mankind) who derange the its action of those parts, by our vanity, our wants, and self-will; or, by our

---

*A system is a course of action, according to some known rule or law of nature; and the term has been applied to some of man's contrivances also, not very happily.

our ignorance, put the whole system out of repair, when we endeavour to control nature, instead of humbly following her track, and working after her fashion; and every mechanic knows, that a system, or a machine, being once put out of order in its minutest part, incurs the danger of complete disorganization in those that are more material to the performance of its functions as a whole: an observation that applies as well to a watch or steam-engine, as to a worm, to man, or the horse; but which, of course, I intend should be applied to the last mentioned animal particularly.

Our Creator, however, as if prescient of the barbarities his *image* would fall into, in the exercise and abuse of the power he gave us over the living things of the earth, hath, in his goodness, conferred on brutes the means of supplying from one part of the system the losses which accident may occasion in another part: a subject well worthy our patient scrutiny, as furnishing the means of effecting cures in desperate cases, and not to be disregarded in first attacks of malignant diseases.

But "the animal system," as a term, or in fact, may be taken to imply as well that of all animals as particular kind of animals—descending sometimes (not improperly) to individuals of those kinds. Some persons, however, descend still lower, and the term "system" has been sadly misapplied, and bandied about from one thing to another, until it is brought to describe particular parts or portions only of the individual's system. The dog kind, the horse kind, and mankind, are good and proper distinctions, for the system of animal life differ in all three: they are not in every case moved in a similar manner by the same class of medicines; whereby we first perceive that their systems differ, and we examine the dead subject of either kind (as in the preceding chapter), to find out how this takes place, and in what degree, and we regulate our practice conformably to the discoveries so made. The several individuals, too, of the same kind, have particularities in their respective systems, arising from habit, from country or climate, or from *crosses*,[*] that demand our serious analytical reasoning, in the application of similar remedies, and adapting their proportions to the removal of similar symptoms. So, a sensible difference is known to exist between the constitution of a cart-horse and a blood-horse, between a galloway and a hunter; each requiring accurate discrimination in ascertaining the state of disease,[†] and this consideration ought to inspire us with carefulness in applying the remedies, since that which restores the one might be injurious to the other. Among those four breeds, we frequently find individuals variously affected from the same causes according to their built, shape, or make (see pages 2, 6, and 18), according to the constitution and co-adaptation of the dam and sire; as age may come on, accidents have taken place, or chiefly as the individual may have been mistreated by his unworthy master, the sordid farrier, or unfeeling ostler. To all which important distinctions in the state of his patient's particularities, I beg to call the studious reader's most serious attention, while examining his case, in order to apply the remedy most appropriate to the degree of attack.

In the two preceding chapters of this treatise, more of the animal might undoubtedly have been described, or the same subjects considerably enlarged upon, and more parade of learning might have been displayed, but the reader would not have benefited one jot by that course of proceeding: he might, probably, have bewildered himself (as many do) in the mazes which would then surround him; whilst the description of those parts of the animal, which

---

[*] The system of the same individual, also, may undergo changes by time; so that a medicine may operate differently now from what it formerly did.

[†] The surest barometer of health, the pulse, would indicate an approach towards fever in one individual, which might be the certain standard of health in another. See 'The Pulse,' at page 60.

contribute but inferiorly to the system I had in view to illustrate and explain, might have led him to look upon these in a light, too important for the functions they perform—as regards my purpose.

The eyes, the tongue, the ears, the skin and hair, the tail, the genitals, and the hoof, or foot, though each deserving our most sedate attention, for many good reasons, yet, as they do not originate disease, I then purposely avoided taking particular notice of them.*   Nevertheless, I do not mean to deny, that they all, according to each its functions, accurately indicate the existence of disease, as they do of health, and the degree of both is marked on them with wonderful precision.   Hence it was easy to conclude, even though we did not know the fact to a demonstration, that they are subject to some deplorable maladies that are peculiar to each, arising out of constitutional defectiveness, to say nothing of accidents, nor of the fancied improvements man presumes to make upon the works of his Maker.

Under this last reproach lie all those farriers and others, who give pain unnecessarily to the animal in the indispensable operations.   Among these, I class that of docking, notwithstanding the gibes of our continental neighbours (the French) conveyed to us in something like the following couplets, about the period of king James's abdication.

> Proud Englishmen avaunt, barbarians as ye be,
> Who cut your monarchs' heads off—oil horses take the *queue!*
> We Frenchmen, better bred, who reverence the law,
> Never meddle with our kings' heads, and let our horse-tails grow.

Although of no moment in themselves, these verses show the then French customs, and mark the period when docking and nicking came up among us in England, to be in the early part of the seventeenth century.   But I put it to the reason of any, the most strenuous advocate for this custom, whether he ever contemplated the probability of a horse being subjected to this operation three or four several times; yet it is no less true, that at a market dinner-table, in the town of Watford, in May, 1820, I heard of a horse which had been so served five several times, from no other authority than that of the last owner of the unfortunate creature.   I took occasion to show, in a preceding page, that in all great exertions of the animal powers, the tail and head had a share.

Firing is another of those barbarous practices that are much oftener resorted to than is necessary or proper.   In fact, we may observe that this and similar painful operations are adopted in an exact ratio that the operator's education may have been neglected.

Of the foot, I have already, in the first chapter, noticed some general faults, arising from constitutional defects in the form of the whole limb; and I shall thence be led to enter into further consideration thereof, with more particulars, under the article "Shoeing," as well as when I come to treat of the several disorders incident to this important part of the frame.   Meantime, I am induced thus early to reprobate one other species of that busy intermeddling in the affairs of nature I took occasion to advert to higher up.   This consists in the baneful practice of cutting away, unmercifully, the horny part of the sole, that lines and defends the sensible sole, whereby injuries upon the road become more frequent, and lameness from unknown causes is incurred; but if not so, canker, rottenness, corns, are sure to follow, or the hoof contracts, and fever of the feet and founder succeed each other.

The skin and coat received some attention under the article Secretions,

---

* The diseases of the foot, I consider as those of accident or infliction, and with a brief anatomical description, will form a separate chapter.

and elsewhere, as the reader must recollect, or refer to; but he must never forget, that the first mentioned may be safely and powerfully stimulated as the outlet for many constitutional affections of the system, the proper time for their use being indicated (as I said before) by the appearance of the coat.

The tongue always partakes of the general state of the system: in the horse, it does not afford to the sight so sure a prognostic of the state of the stomach as in the human subject; but, to the feel, it communicates to us the state of the blood with so much accuracy as demands our assiduous attention, to the acquiring, by practice, the most intimate acquaintance with its monitions. This member of the body, in conjunction with the coat, I have always considered the health-gage of my patients. See observations on the "Pulse" at page 60.

The eye is a most material organ of sense, and is much studied by those who would render themselves good judges of the general soundness or unsoundness of the horse's constitution. It beams bright and steadily in health, projects most fiery when the animal is most vigorous; in lassitude it sinks, it blears with a cold, and under extreme circumstances is extinguished. After a heat, horses full of blood, with foul stomachs, certainly alter in their vision, shy and become troublesome; and, so sure is the eye the barometer of vigour that horses got by old sires have the eye more sunken than others, with a hollowness over it.

The ears, by their movements, show the apprehensions of the horse, if not his disposition. When he fears the lash, he turns their cavities backwards. Is he disposed to be resentful, they are laid flat on his poll. Following his companions, or the hounds, or going homewards, the cavity of the ear turns sharply forward: asleep, as well as under other circumstances of easy watchfulness, one ear turns forward, the other backward; but, when roused suddenly, they alternately change position. Who, then, would destroy those useful appendages of the horse's organ of hearing? Who would singe off the hairs, which, passing from side to side of the cavity, catch the sounds and convey to his rider the first notice of danger from wild beasts, as well as pleasure from the cry of the hounds? The Arab knows, by his horse's ears, of the approach of enemies; but the Englishman relies too securely upon his own comparatively imperfect hearing, and cuts off those better intelligencers of distant occurrences; or, he more assiduously abridges their utility, by clipping away the inside lining; or, worse still, by applying flame to the part, he renders the horse skittish ever after. Those are the only disorders of the ears of horses; if, for want of this hairy defence, premature dulness of hearing, occasioned by rain, dust, and other substances entering these organs, be not another.

That the genitals draw off from the system and store up a noble secretion, for the purpose of continuing the kind, is certain; but I shall pursue the matter no further than to notice the change to which the coats of geldings are subject as to colour, compared to those of perfect horses; and all the inference I mean to draw from that fact is, the still further corroboration of my previously maintained opinion, as to the seat of perspirable matter residing in the lacteal part of the system.

As it is the blood which by its deposite forms all those parts, so by means of the blood must we endeavour to correct any derangement of the system of animal life, whether of quadrupede or bipede; for the working of the system in making of new blood and cleansing the old is the same in all, though differing in degree, whilst mainly agreeing in the process. Would any one demand how it comes to pass, that quadrupeds draw so much substantial nourishment from herbaceous vegetables, whilst man can only extract a watery juice, devoid of all nutritious qualities? let him be answered, that all depends on the

7

digestive powers, these being greater in the brute, than in man. It even appears plainly to me, that the animal food taken by man is the same as the herbaceous taken by quadrupeds, only that it has meantime undergone the process of digestion, sanguification and deposition in the solids, &c. and hence arises the difference in the practice of the curative art as applied to the one animal and the other. Every disease is in fact a compound, varying in different constitutions, and the composition of the remedy should be adapted to every variation thereof, even of the same attack.

BOOK II.

THE CAUSES AND SYMPTOMS OF VARIOUS BODILY DISEASES INCIDENT TO
THE HORSE; WITH THE MOST APPROVED REMEDIES IN EVERY CASE.

CHAPTER I.

Of Internal Diseases.

INFLAMMATORY DISORDERS, GENERALLY.—FEVER.—From all the infor
mation the reader may have collected together in his mind, respecting the
"circulation of the blood," as described with instructive minuteness at pages 33
to 42, he will naturally conclude that the horse is ever most liable to contract
one or the other of those disorders we term inflammatory. The great heat
of his blood, combined with his bulk, and the amazing exertions he is compelled
to make, all together constantly predispose him to incur fever of the whole system or inflammation of particular parts, according to concurring circumstances.
Nor is the matter changed one whit, when we reflect that fever sometimes
terminates in local inflammation, which we term "critical," as being the crisis
and cure of the disorder; and that the inflammation of one part or organ (the
liver in particular) frequently devolves into fever of the whole animal system,
by means of the rapid circulation of the blood through the diseased organ.

Let us proceed to discuss the subject generally at first, and to pursue each
in detail afterwards; simply premising, that all the disorders incurred by the
horse are referable, more or less, to this over-heated or inflammatory state of
his blood, and its consequent unfitness for the purposes of promoting animal
life, health, and vigour. For, the more heat, the more viscidity or thickness
there will be in the blood, and less will it be found capable of circulating the
longer such unnatural heat continues, up to a certain point of the disease:
when the animal is so far affected as to lose its appetite, and consequently no
fresh blood can be formed by the digestive powers, the blood then becomes
thinner every day, because its more solid particles are constantly being deposited in the cellular membrane, to supply the waste that is unceasingly going on there. The reader would do well to read over again what is said concerning this process of the animal system at page 48, with the references there
made to page 37, to page 23, and, in fact, to the whole tenor of the second
chapter. But this supply soon fails, as necessarily it must, when it is not re-
plenished at the source, and wasting of the solids succeeds of course, unless
nature is assisted by our art judiciously;—the right application of this art is
what we are now in search of.

One of the immediate consequences of the horse being hard worked, or high
fed and physicked with stimulants, is the constant heating or feverish state
of the blood. Increased action of the heart and arteries accompany and keep
up this state of irritation, which may be further accelerated by the animal's
being allowed to take cold whilst in that state, whereby the perspiration is
checked of a sudden, and the blood which may then fill the smaller vessels is
detained there, to the further annoyance of the larger ones: he then contracts

inflammation of all the solids and organs of life, or, more properly, *fever*.
But when only a certain part of the system, or a single organ is thus checked, we consider the affair under the name of *inflammation* of that part, as of
the lungs, the kidneys, &c.; always keeping in mind, that, by continuance,
these extend their baleful affections to other organs, with which a certain
sympathy is known to exist.   In like manner, when external muscular parts
swell and secrete matter, this is in like manner an inflammation of that particular part, or tumour, or abscess, with a great variety of names, according to
the place where it may be seated : poll-evil and fistula are among those external
complaints to which I allude.

The latter, or local kind of inflammation, is the effect of the former or continued internal fever, and whenever such a tumour or abscess makes its appearance near the surface, the general inflammation or fever subsides; when
it discharges offensive matter, the fever is cured.   If such a tumour appear
without previous general fever of the system, we repel it, so that it may disperse and pass off by stool.   It may usually, however, be considered as an
effort of nature to relieve itself of offensive viscid matter that lurks in the system; and in this case only, when well ascertained, would that reduction of
the system which I shall shortly insist upon as proper in all inflammatory attacks, be least advisable, as nature would then require aid to assist her in her
efforts, rather than subtraction from her powers, by the bleeding, purging,
&c. so recommended.

But whenever a cold is caught, whereby the trunk is affected, one of two
evils is experienced, that are quite contrary in their effects : 1st, Either the
bowels lose the power of retaining their contents, and of contributing their aid
to the purposes of digestion, chylification, and sanguification, i. e. the making
of fresh blood, and diarrhœa ensues; or, 2d, The extreme heat of the body
causes the dung to harden, and if the obstruction be not speedily removed,
the most distressing consequences usually happen.   Either extreme may come
on gradually and imperceptibly ; but as the latter (termed constipation) is of
most frequent recurrence, is a disorder of over repletion, producing vertigo,
staggers, apoplexy, megrims, or fits, I have considered it under a separate
head, as "costiveness;" seeing that it sometimes supervenes without previous
fever, though always accompanied by it.   One or other species of affection
of the bowels is also produced by catarrhal inflammation, or fever of the organs
of respiration, when this is violent or of long continuance.

Respiration of confined or noxious air in close stables, as described at page
39, also produces quicker circulation of the blood ; with perspiration and temporary fever, which may be confirmed by sudden exposure to the open air,
and the consequent detention of blood in the small vessels which we term
capillary.   Sudden immersion in cold water whilst sweating and respiring
with difficulty after a run, wading through a river, or standing in a current
of cold air, are all prolific sources of inflammatory disorders.   Indeed fever
and inflammation are so closely allied to each other, that we run little risk of
creating confusion of terms by considering them as derived from the same
origin, and none whatever in treating of both in the same chapter.   For most
stablemen and farriers, as well as many veterinary writers, do speak of the
one and the other promiscuously, as if they were the same, when describing
the symptoms of either; nor do I see any good cause for my deviating from
this practice upon the present occasion, after the slight distinction just drawn.

One other general observation may be aptly made in this place, which may
stand instead of much discussion hereafter.   As fever is a necessary consequence of any inflammation whatever, so without fever there would be no inflammation.   Every run you give a horse heats or inflames his blood, quick-
ens his pulse, and he sustains temporary fever.   Whilst in this state, if any

viscus, or organ, that constitutes a vital part of his system, receive such a check or damper as I have described, obstruction of the finer blood-vessels ensues—as, of the lungs, by their drinking cold water, or mere affusion of it on the chest, and inflammation is the name: if the whole body of an animal or its entire surface be so affected, the evil consequences are similar, and fever is the name by which we designate it. Horses out of condition, or already in a low state, though feverish, with quickened pulse, do not require further reduction; since this is evidently "low fever," which I have treated of under a separate head; as I have also "Typhus fever," or the affection of the whole system which arises from a vitiated or corrupt state of the blood. But, in all cases, the best guides to the practitioner for his prescriptions, and indeed all his operations, are the causes, the symptoms, general health and peculiarity of constitution of the animal; when it so happens that such particulars can be extracted from those about him; as will be the case in all studs of a superior cast. If the feverish affection arises from inactive kidneys, the diuretics recommended lower down will be all the treatment that is requisite in such a case; if a dull heavy pulse and the state of his dungings show that the bowels only are at fault, purgatives alone will restore health. So of any other visceral obstruction, when these give pain fever ensues, and is best removed by the exhibition of mercury; if the internal irritation continues, rowelling is the remedy most appropriate to such cases, and the state of the pulse will tell the doctor when and why he should bleed. This will bring us to an early consideration of "the pulse," its indications and general rules. In all cases of inflammation, whether of the whole system, or fever, or of particular organs, let bleeding be resorted to immediately, in quantity proportioned to the amount of heat, which is ascertained by the temper of the pulse. "Open the *prima riæ*," also, is a good maxim of a late respected lecturer on those subjects, meaning thereby—purge the bowels or chief canal, and keep them open. Copious clysters of warm water-gruel assist the latter materially, particularly if a solution of salts be added, according to the nature of the case; but rather than delay the clyster through want of the ingredient being at hand, use simple warm water only. Very often, in slight attacks, the animal requires no other treatment, if resorted to in time; but delay is dangerous, for with every hour the symptoms increase in a three-fold ratio, and the animal becomes weaker and weaker every moment, and therefore less able to bear up against the attack. In all cases, be quick, for ruin is going on with rapid strides, whenever the animal shows signs of great internal pain. Fresh air, diluting liquids, and clysters, in all cases of inflammation whatsoever, are found of as much service in the restoration of health, as the best active medicines that can be administered; the first mentioned most positively so, unless the animal perspire greatly at the time, or it suffers under a fit of shivering. Danger is to be apprehended in the latter case, and the fresh air need not then be admitted; but if shivering is succeeded by sweating, or even a small degree of moist heat, it may be considered as the crisis of the disorder, when something has taken place that is favourable to the cure—of which more particulars in the proper place. Continued shivering, by the way, denotes the termination of all inflammatory diseases—in death; cordials then may do good, but more frequently accelerate the catastrophe, whilst the diluting liquid —water-gruel, will afford relief in some measure, but can do no harm. But fresh air, that issues not in streams, is of all other restoratives that upon which I place the most reliance; even removal to a fresh stall, or up and down the stable, effects great changes in the animal's spirits, that can not fail to strike the eye of an attentive observer, and bespeak, more than words can convey, the vital necessity of a cool atmosphere.

The PULSE—Being the chiefest criterion for judging of the state of the cir-

7 *

culation of the blood, and as I have sat down with the notion that my book will be read straight-an-end at first, let the reader attend a moment while I say a word or two on this preliminary topic.   Without an accurate knowledge of this touch-stone of the main spring of life, no one can form a judgment fit to be acted upon as to when it is necessary to bleed or of the quantity to be taken : thus, in cases of fever, the groom begins very properly by bleeding; but he almost invariably takes too little, or in case of increased action of the pulse, through over exertion of the animal's powers, he bleeds when such a course is detrimental, and almost always administers cordials, thus reducing with one hand, and increasing the action with the other.—See pages 33, et seq.

When in health, the pulsations or strokes are from thirty-six to forty in a minute; those of large heavy horses being slower than of the smaller; and of old ones, they are also slower than of young animals.   When either may be just off a quick pace, the strokes increase in number; as they do if he be alarmed, or terrified, or hear the hounds' familiar cry.   Fever, of the simple or common kind, usually increases the pulsations to double the healthy number; hence the propriety of ascertaining the state of this index of health, while the animal is still free from disease, goes to prove over again the propriety of my plan of teaching the curative art in animals by closely examining the indications of health, and setting down in one's mind every deviation therefrom as the approach of illness, that ought to be met and combated at the threshold.

In this view of its utility, why might not the attendant groom, or horse-keeper in more humble establishments, keep a register of the state of every horse's pulse, when it comes first under his care, and renew the same examination at intervals of a week or ten days?   This practice alone would render him expert in all cases of imminent danger; to say nothing of those other indications, the dungings and the water voided.   On this latter point the reader will turn back to what is said of " Urine" in page 52, 53.

As the fever increases in violence, likewise, when the animal is in great pain from inflammation of the intestines, &c., the pulse beats still higher, and reaches to 100 in a minute, or more.   The danger is then great, and less than three or four quarts, drawn from a large orifice, would do harm rather than good, by increasing the action of the blood, and the hardness of the artery would also be increased.   To ascertain either state, the attendant should apply the points of his fingers gently to the artery which lies nearest the surface.   Some prefer consulting the temporal artery, which is situated about an inch and a half backward from the corner of the eye.   Others again, and they are the greater number, think it best to feel it underneath the edge of the jaw-bone, where the facial artery passes on under the skin only to the side of the face.   In either case, too great pressure would stop the pulsation altogether, though by so trying the artery against the jaw bone, will prove whether it be in such a rigid state of excitement as attends high fever; or elastic and springy, slipping readily from under the finger, as it does when health prevails and the strokes follow each other regularly.

The presence of high fever is further indicated by a kind of twang, or vibration, given by the pulse against the finger points, resembling much such as would be felt were we to take hold of a distended whipcord or wire between the fingers, and cause it to vibrate like a fiddle-string, sharply; whereas, in health, a swell is felt in the vibration, as if the string were made of soft materials, and less straightened;—facts these which owners would do well to ascertain by practising upon the pulse of their own horses.   Languid or slow pulse, and scarcely perceptible in some of the beats or strokes, indicate lowness of spirits, debility, or being used up: if this languor is felt at intervals only, a few strokes being very quick, and then again a few very slow, this indicates low fever, in which bleeding would do harm. Quickness, however, is the chief indication of the whole class of inflammatory fever, and this being my principle object at present, I shall postpone further consideration of the pulse until I come to treat of "blood-letting."

FEVER.

There are two kinds of well-marked fever, simply so called—first, that which arises from the pain an animal may be put to by the derangement of some main organ of life, by misusage, hard riding, wounds, &c.; and secondly, that which consists in a general inflammation of the blood arising from a cold, a chill, or sudden check, as before described. The ancient vulgar name given to this alarming disease conveys to the common observer a better idea of its force and danger, than those which are settled by consent of the faculty of horse medicine; and the phrase "inflammation of the blood" may be taken as more plainly indicative of the cause of fever than aught the moderns have substituted in its place. Had our plain-speaking ancestors termed it "inflammation of the blood-vessels," they would have been still more accurate, probably: but no mistake is more common throughout life, than to speak of the thing contained for the thing containing it, and *vice versa*. When the symptoms come on quick or acutely, the most prompt measures must be taken: a mild attack may be easily reduced if taken in time, but, if neglected, it assumes the most alarming symptoms. Evacuations and diluting drinks are the proper means of reducing the patient; but before purgatives are administered, see what is said a few pages onward respecting "Costiveness;" for it not unfrequently happens, that this is all that ails the animal, except his being worked too hard while costiveness is upon him.

In either case of accelerated pulse from those causes, bleeding should presently be had recourse to, and let the quantity taken be regulated by the force and quickness of the circulation of the blood: for this is what constitutes the fever. If the pulsation advance to above 60, two quarts should be drawn; if above 70 in a minute, three quarts of blood would not be too much to take away at once. If the number of beats be much more, ascending rapidly, with the rigid feel of the artery above described, four quarts at least must be drawn, and that from a large orifice. Should this rigidity, or hardness of the artery continue, notwithstanding the bleeding, a quantity that shall cause faintness or tottering might be taken, or rather a repetition take place of the same operation in lesser quantities, until that hardness of the artery is no longer felt. Some skill, derived from practice, is required in watching for this last mentioned symptom; but whatever is to be done, let there be no delay in the first operation: twelve hours should intervene between the two bleedings.

Immediately hereupon, let a mild purgative be administered, adapting this, as well as the amount of bleeding, to the size of the horse, if he belong to either extreme of exceeding large or very small. For one of the moderate coach-horse kind give the following

Purgative Ball.

Aloes, 7 drachms.
Castile soap, 4 drachms.
Oil of caraways, 6 drops.
With mucilage sufficient to form the ball for one dose.

In all cases of fever arising from accidents, hard runs, &c. which may be considered as temporary excitements only, the above treatment in its mildest form will be found sufficient completely to reduce the symptoms; but in the fever simply so called (arising from inflammation of the solids as before described), a repetition of the purgative becomes necessary, with mashes, a quiet stable, and an attentive groom. When the fever arises from indigestion, or any derangement of the stomach or bowels, its immediate cause will be found in hardened fæces; and in addition to the forementioned remedies, give a

Purgative Clyster.

Water gruel, 6 to 7 quarts.
Table salt, an ounce to each quart.

Let it be applied assiduously, and some assistance be given to bring away the first hard fæces that appear: the remainder of the hardened dung will come away, naturally, in good time. See further under the head "Costiveness."

Castor oil, in the quantity of a pint or more, will open the canal partially only, passing by the main evil in the cœcum and great gut,* and producing but a small quantity of the offensive cause of disease. But help must be afforded in this respect; and if the bowels yield not to the purgative ball, other means must be resorted to, though I should never think of having recourse to oil in the first instances. Although the constipation or obstruction be obstinate, yet very strong diuretic purgatives are ineligible, as they might kill the animal, or at least injure the intestines materially, by reason of that very circumstance.

Distinctions have been drawn by some writers between "symptomatic and simple fever;" that is to say, whether the excitement, called fever, originate in a check of the circulation received externally or internally; but as the treatment in both cases is so nearly the same, I shall make no such distinction. The internal attacks alluded to, when confined to a single organ, and not extending to the whole frame, are more properly termed inflammation of that viscus or organ, and therefore will be treated of hereafter, under the following heads, viz.

Inflammation of the Lungs,
Inflammation of the Stomach and Intestines,
Diseases of the Liver—Inflammation, &c.
——————— Kidneys and Bladder.

All these produce fever throughout the whole system, when either exists but in a slight degree; for those parts are all of them vital, and communicate their feeling to the solids by means of the circulation. It is not, however, until these attacks are well marked, that they deserve separate consideration; for some horses suffer under the one or other during life, with more or less malignity according to exciting circumstances, the lungs being the most general sufferer, the bowels the seldomest attacked of either, but usually prove the most fatal of this whole class.

The symptoms, in all cases, are heat and acceleration of the pulse, as before described, and which in fact, brought me to the consideration of this portion of my subject before the others. A hot mouth soon comes on; shivering takes place early, and the animal evinces signs of internal pain by looking at his flanks or chest. The fever is then likely to fix on the lungs if not speedily reduced. Loss of appetite follows; but too gradually to be waited for, as a

---

* The practical reader, whilst waiting the progress of the disease, will not waste his time by turning back to the first book, at p. 46, and see what is said of the conformation of those large guts, and the difficulty of escape that must attend their offensive contents at the turns or sinuses (which I have there considered as so many valves), when inflammation or fever has once begun.

a criterion for judging and acting promptly. He will evince languor and dulness, with half closed eyes, and a small discharge from them, as if tears es
caped; sometimes, this last will happen in cases of mere debility or starvation
also, when it is not too much to suppose the animal may be deploring his hard
fate. Consulting the pulse, however, will settle any doubt as to which ailment the animal labours under; for this main characteristic of health will, in
the latter case, partake of his debility, and strike now hard and then soft, a
few beats each : in this case a feed of corn or water gruel, would probably restore a more healthful even pulse, whereas bleeding would go to destroy the
patient. It has been termed low fever, though not very properly; and *lentor*
or more justly *lenteur* (slowness, dulness, heaviness) by the French veterinarians; yet, having no better name for it than "low fever," under that head, I
shall shortly bestow a few lines on this species of systematic debility.

The dung and urine are always good indications of the state of the body;
if the former fail, fever is the cause, it subtracts also from the quantity of urine;
and if he stale small quantities at short intervals, some internal inflammation
has taken place. See Inflammation of the Kidneys. In fever, the mouth and
tongue become drier than ordinary; and if any saliva be secreted, it is tough
and ropy. If the animal be in condition, upon lifting the eyelid an uncommon
redness appears; if he be out of condition, or in a low state, this does not always happen; so this indication may be reckened among the uncertain
symptoms.

If the remedy and the symptoms of fever are thus pressed forward together
upon the reader's notice, as exemplifying the assiduity he should display in
repelling the attack, let him know that his work is but half completed when
he finds the heat and acceleration of the pulse reduced by his endeavours to
the ordinary standard. The tone of the patient's stomach and the whole digestive process require restoration, and this with a careful hand, that the bowels may not again get overloaded; because why, a second attack of this sort
would be more difficult to surmount than at first; for the bowels have partly
lost their function of expelling their contents, through the violence of the disease, if not by the harsh action of the remedies employed. Hardy working
horses, of course, recover their appetite as soon as the fever abates; and no
further care is required for such than an occasional laxative or purgative, according to the amount of obstruction. The ball prescribed at page 63 may be
given at intervals with the fever powders; and subsequently, the fever drink
prescribed below for all other descriptions of the horse recovering from fever.

Fever Powder. No. 1.

Powdered nitre, 1 ounce.
Emetic tartar, 2 drachms.

Mix for one dose.

No. 2.

Powdered nitre, 6 drachms.
Camphor, 2 drachms.
Calx of antimony, $1\frac{1}{2}$ drachms.

If either be deemed more desirable in the form of a ball, this may be effected
by mixing the powder with mucilage and meal; but in the form of powder
mixed with his corn is most eligible, as the medicine then acts earlier, whereas the ball presently descends into the great gut.

Fever Drink.

Cream of tartar, 1 ounce.
Turmeric, 1 ounce.
Diapente, 1 ounce.

Mix in powder, and add to a pint of warm gruel, to be given once or twice a day. This is a good cool stomachic, and restores the appetite, at the same time that the disposition to the return of fever is kept down: if found of marked service, the doses may be repeated to three or four times a day for a week.

LOW FEVER.

Together with TYPHUS, or putrid fever, and RHEUMATIC fever are diseases incident to the horse, though attempts were long made to deny the application of these terms to any of his numerous afflictions, by those who dread, inordinately, the falling into analogies with the human practice; a fear that may be carried too far, notwithstanding all our care should be employed in separating this from the veterinary practice.

Cause.—Of low fever, under the idea of debility, a few words fell on the preceding pages: and truly, if "high fever" may be produced in a subject that is full of blood or condition, by over-exercise, and the other causes thereof set down above (pages 59, 60, see also Book I. at page 42, &c.), these same causes, operating upon a horse out of condition, or which has not sufficient blood in his frame to receive inflammation, necessarily occasion that languor which attends debility of the entire system. The reader will, perhaps, oblige me by turning to book I. at page 40, and reading over again what is there said as to some causes of low fever. But the respective terms we give to the various kinds of attack would signify much less than they deserve, were it not for the danger we should otherwise fall into of treating one disorder for another, when the symptoms (some of them) so much resemble each other. This danger is more likely to come upon us in cattle medicine than in the other, since we are under the necessity of finding out what is the matter with our patients, whilst the human doctor receives the information at once, in words.

As inflammatory fever is more prevalent in the spring and summer, owing to the high condition of most horses when first attacked, so does low fever, or irritation of the animal system of a horse in low condition, mostly prevail in autumn and winter. We owe this latter in great measure to the debility or weakness brought on by the shedding of his summer coat, when the autumnal equinox sets in. Being then much exhausted by the heat of the season just gone by, he sweats profusely on the least exercise; then his coat becomes dry and husky when at rest, and his skin sticks tight to his ribs, slightly resembling hidebound. The animal having lost much of his natural covering and no care being taken to palliate this loss, he is more liable to catch cold if exposed and still pushed in his work. If not relieved from its severity, coachhorses in particular become unserviceable in great numbers, to an alarming degree, resembling much the distemper of the spring season. Too often it happens, such knocked-up horses are considered as done for, and the owner sells off; whereas experience tells us, that a nourishing regimen would restore them to their wonted vigour; for the serious or watery part of the blood (chap. 2, sect. 20, 21) having been drained off by the violent perspiration they were exposed to by their summer work, the muscular fibres become too rigid, and the blood too thick for circulating in the finer vessels; it therefore remains rioting in the larger ones, distending their capacity and increasing the irritation. Working horses are then usually deprived of their corn, because they can not work; this only adds to the irritation of the vascular system and solids which constitutes the low fever we are now considering.

Symptoms.—Parallels, or distinctive characteristics, of such diseases as somewhat resemble each other, are therefore very proper, inasmuch as they prevent those dangerous mistakes in practice that happen oftener (even in the human practice!) than suits me even to hint at in this place. They are most particularly serviceable to veterinarians: for this reason it is I recommend the reader to compare what is said of the symptoms of high fever, just above, with the present page, as regards the symptoms of low fever. They are placed near together for that purpose, as I then said (at page 64). The pulse in this case never mounts high during an entire minute, but beats quick a few strokes, and then slow, and so low as scarcely to be perceptible; this denotes, that though fever be present, there is not strength sufficient to bring it to a crisis. The artery feels rigid, at intervals only, and again becomes supple, if not elastic, to the touch; his flanks are agitated more than usual, and his hind quarters and ears become cool if not cold. As in high fever, his eyes are dull and heavy, and water will occasionally fall from them. Though in the former species of fever he evince considerable pain, in this no such symptom appears, but despondency assumes its place.

Remedy.—Unless his body be already too open, give the laxative draught, as under: and as he will still feed, diuretic powders may be mixed occasionally with his feed, consisting of nitre and rosin, of each about one ounce Should the urine appear turbid, or come off with difficulty, in small quantities, the diuretic ball is indispensable; and these, with good gruel and care, accompanied by tonics, will restore to the animal a comparative portion of health. Time and moderate usage will accomplish the remainder.

Laxative Draught.

Aloes and carbonate of potash, of each 2 drachms,
Mint water, 4 ounces.

This will correct the urine also, and its laxative quality may be increased by adding to the quantity of aloes.

A Diuretic Ball.

Turpentine and soap, of each 4 drachms, with mucilage to form the ball.

A good restorative for lowness, occasioned by the moulting fever of autumn, is recommended by J. Clark, of Edinburgh: he says, "the end of autumn proves very severe to those horses whose flesh and strength are exhausted by hard labour. In this low and spiritless state the moulting season comes on, and carries off numbers that good nursing and feeding, with rich boiled food, at this season might have preserved. Carrots and potatoes recover some horses surprisingly; it renews their flesh and the fluids generally, and promotes the secretions; it operates upon them nearly in the same manner as spring grass, and its effects are presently visible on their coats." Many stable men give oatmeal mixed up into bergue, or crowdie, for horses that evince signs of languor and lowness of spirits, after fatiguing work in winter: if made into stiff gruel, i. e. boiled, the restorative effect is found still more desirable, and a smaller quantity of oats then sufficeti.. A gradual return to hard food does all for the horse's working condition which can be desired.

Fever is brought on, in some degree, whenever it comes to pass that either of the vital organs may be deranged in its functions. Not unfrequently it happens that a diuretic is all the patient requires, which may be judged of by the state of his pulse after the medicine has operated. When this is the case, the feverish symptoms owe their origin to suppression of urine, and the reabsorption of the contents of the bladder into the system. See Bladder and "Suppression of urine;" and, after treating the attack simply as such, a cordial ball should then take place of all further treatment. as the immediate tall

ing down of his pulse to a healthy standard will show.   Too free use of urine
balls, however, in the hands of horse-keepers, spoils the action of the bladder.
See chap. ii. page 51.   The French give a bottle of their routine wine made
warm, and most of our farriers administer a quart of ale with the same view.
Those are mostly wagon-horses, full of flesh, that so absorb the aqueous particles of the urine, and ultimately the principle thereof (termed *urea*), and
light up the fever anew.   Another cause of feverish attacks, generally of the
slighter kind, but liable to prove fatal, if neglected, is the retention of his
dung, or constipation, which means costiveness.

COSTIVENESS

May be considered an original disease, and as one producing as well as being
produced by fever.   That is to say, hardness of the fæces generally attends a
fever, and is frequently the chief cause of it : like the preceding ailment, we
have only to remove the cause, and the effect ceases.   See also " Diseases of
the Liver."

Causes.—Want of the necessary or usual evacuation by stool, that is sometimes occasioned by the bowels having lost the power of expelling their contents, as described in the second chapter, page 45.   Simply speaking, the individual having been a long time dosed with purgatives, any neglect hereof
causes the dung to harden and obstruct the contractile functions of the intestines : heat ensues, and re-absorption takes place, as in case of retenti o of
urine, until the dung loses all moisture and becomes as hard as baked clay,
forming in the rectum (or straight gut), small round lumps.

The same kind of big fleshy horses as are liable to suppression of urine, are
also principal sufferers by constipation or costiveness.   Hard food and hard
work in warm weather is very productive of this malady, which is often mistaken for inflammation of the bowels, the means of prevention, therefore, are
obviously the direct contrary mode of feeding, and also keeping a good watch
on the dunging of each horse in the team.

Symptoms.—When constipation attends general fever, it is then but a corresponding symptom of that disorder, and the reader is referred back a few
pages to what is there said on this head.   But, when the pulse is not so high
as to warrant us in pronouncing it fever, and the dung is ascertained to be
hard, there is no difficulty in treating it as simple costiveness.   It may be distinguished from colic and from inflammation of the intestines, by the quiet
state of the animal when he is down, which is not the case with either of those
disorders, in which pain of the bowels is most evident ; whereas, these do not
appear to suffer from the costiveness, though the brain and the whole of the
nervous system, become more or less affected from sympathy with the stomach,
and ultimately producing delirium and frenzy.   His eyes offer the earliest
symptoms by their duiness, contraction, and expansion, succeeded by sleepi-
ness ; he refuses his food, he will not work, the mouth becomes hot and dry,
the ears cold, and the breathing difficult or nearly imperceptible on account
of the pressure of the stomach and bowels upon the midriff.   See page 34.
The pulsation usually increases, if he be in a tolerably good condition ; but
this increase is ever inconsiderable until fever comes on, and marks the period
when blood-letting would be necessary.   A dull heavy pulse is more common,
until the paroxysms of madness may render this symptom a little sharper and
quicker for a short period.   At length he tumbles down, regardless of the
situation, and the action of the head shows how greatly this part is affected,
until stupor and death ensue, if the sufferer be not relieved.

Remedy.—Purgatives are not always the most eligible medicines even in
the earliest stages of the disorder ; for, if the constipation has lasted a con-

siderable time, great injury would be done to the intestines by forcing a passage, whereby a commotion might be raised in the stomach, but would act in efficiently where the evil chiefly lies, viz. in the large intestines and rectum. As soon as it is ascertained that the animal has not dunged for some days—when he seems uneasy, a fulness is perceptible towards the flank, the fundament, &c. and unusual dryness and tightness is discovered at this latter part, the operation of back-raking should be resorted to. Castor oil, one pint, would indeed find a passage in the first stages of the attack, but good part of the evil usually remains behind; in the more advanced stages, especially when the patient drops, nothing else will relieve him but back-raking. Let the operator strip his arm bare, and having well anointed it with soft soap, lard, or butter, (the first being the most eligible,) he will bring his fingers to a point, and gently introduce the hand and wrist, when he will feel and draw forth a portion of the indurated fæces he will there meet with, in lumps hard and dry. This he may repeat three or four (or more) times, and leave the animal to himself a little, whilst a drench is preparing. Trivial as the relief may seem which has been thus afforded to the patient, he will immediately evince proofs of its benefits, by a more sane conduct, by licking forth his tongue, opening the half shut eyes, by looking about him, and sometimes by getting upon his legs. In this latter case, plain water gruel, as warm as a person might take it without inconvenience, may be administered in the quantity of two or three quarts, if he will take so much; but if the animal be exhausted, and does not get up without difficulty, or without help, one half the drench may be ale or porter. Although he will seem recovered, and may produce a stool, his bowels must next be emptied. In order to this, give a

Laxative Draught or Drench.

Castor oil, half a pint.
Aloes, 2 drachms,
Prepared kali, 2 drachms.
Water gruel, 1 pint.

Repeat this next day, leaving out the oil, and doubling the quantity of aloes; or, after an interval, give the usual purgative ball, containing seven or eight drachms of aloes, as prescribed at page 63.

INFLAMMATORY DISEASES of every sort leave behind them a good share of weakness, which full feeding will not always amend. We must therefore restore the tone of the digestive powers by the aid of medicine, that may be repeated according to circumstances, and the returning strength of the convalescent animal.

Tonic Ball. No. 1.

Jesuit's bark, 7 drachms.
Prepared kali, 2 drachms.
Mucilage sufficient to form the ball for one dose.

In ordinary cases, one of these per day for a week will be found to have done as much for the animal as could be desired. But should the coat still appear rough and staring, give the following:

8

Tonic Ball. No. 2.

Salt of steel, or sulphate of iron,
Columbo root, and
Bark, of each 3 drachms.
With mucilage to form the ball.

Great precaution is necessary to prevent a relapse, which would render the patient's case more dangerous than at first; the animal being less capable of bearing up against a fresh attack, by reason of the reductions he has been subjected to. Soft or sodden oats, fine hay, clover, a few vetches, carrots, grass cut fresh from a sloping ground, may succeed each other in small quantities, until he may be returned to oats and hay as usual. If the heat return at intervals, as usually happens towards nightfall, give him

A Cooling Decoction.

Linseed, 2 quarts.
Coarse sugar, 2 ounces.
Water boiling hot, 6 quarts poured upon the seed.

Let it simmer three or four hours, and pour off the liquid for use when nearly cold. The linseed will bear another water, less in quantity; but some horses will take the seeds also, which may be permitted. Give the whole in the course of the day, at two or three intervals, and repeat the same decoction once or twice more.

TYPHUS, OR PUTRID FEVER,

Is caused by long-continued debility, or slow fever, as much as by the injudicious use of medicines administered for the cure thereof. Of these, the most common error consists of cordial medicines, diapente, wines, &c.; which, as they give a short-lived vigour to the animal, are supposed to have done some good, and are therefore persisted in, until the digestive and secreting parts of the system are spoiled.—See chap. 2, page 22, &c.

Symptoms, the same as those in slow fever, mark typhus fever, only the pulse is accelerated upon taking the medicines just alluded to: its irregularity is also greater, until, by continuance of the disease, it ceases to denote any particular state of the body long together. Hence, the supply of new blood carries with it similar effects: the vitals lose their tone, and the muscular part of the system wastes and becomes rotten on the bones, and if the same stimulating treatment has been kept up until the animal dies, its flesh will be found on dissection to have acquired an uncommonly bright purple colour, not only on the surface, but wherever incision is made. Putrescence, in a high degree, has already taken place ere that catastrophe seals the sufferer's fate!

I mention these minor circumstances to prove (so far as I can do so) the real existence of this main type of putrid fever. Another symptom of typhus goes to the same proof, namely, delirium, which follows a continuance of the stupidity discoverable in slow fever. A well-marked case is reported in the Annals of Sporting, for Nov. 1824, to which work I have since been some months attached; and, although I was precluded by absence from examining the subject, I have reason to rely on the report afterwards made to me by Mr. Ford, that its flesh was putrid in an extremely offensive degree, and wholly unfit even to be cast to the dogs.

From the very unaffected and detailed account of the narrator, it appears plain that unskilful persons might be led to apprehend such paroxisms denoted hydrophobia; but a short inquiry into the habits of the horse previous to its last delirium, would go a good way to relieve the anxiety usually instilled into a neighbourhood by such events. None can say, however, until the experiment be tried, whether animals fed on such meat might not acquire rabies thereby.

The *mad staggers*, as the term is, which has never been satisfactorily accounted for, can be no other than this delirium of the typhus fever, brought on by pushing the animal in his work although labouring under slew fever. None but common or ordinary cart-horses are lost in the staggers; whilst none but a very ordinary owner would so force his cattle to the last extremity during illness. As the above is all I shall find it necessary to say of staggers, I must here remark on the singular impropriety of Mr. Richard Lawrence's considering this as an attack of apoplexy! Since one pang alone denotes the death so to be named.

Rheumatic Fever is one of those disorders in the horse, upon the existence of which doctors disagree; but doubtless the vicissitudes of heat and cold to which the horse is subjected, whereby the whole system is checked so as to occasion general fever, is equally likely to check the circulation in one or two limbs only. And the pain the animal would thus labour under in the performance of its duties would constitute one of the causes assigned higher up for simple fever. Little good, however, would ensue by my considering it separately; so I shall content myself with referring the reader to the head of simple *rheumatism*.

Epidemic fevers—Distemper.

Cause.—When these appear, from time to time, they may fairly be ascribed to the season; for the kind of attack is not of a nature to become communicative, unless by continuance putridity follows: then, indeed, infection may begin, as it would also happen in any of the preceding species of fever. A rainy spring after a mild winter produces an epidemic catarrh, as well as sudden chill, among horses that are out at pasture whilst shedding their coats, and the most delicate receive this influence earliest. We may as well consider, that whatever may give one horse a cold, or affect his lungs, singly applied to him, would, if applied to many, in like manner affect the whole: this constitutes *epidemy*, or the distemper. Cloudy weather and cold easterly night winds, when the weather is warm or murky by day, is more likely to check the action of the lungs or of the whole system, than when a colder season has prepared the animals to withstand the influence thereof. An epidemic prevails sometimes in autumn; but, happen when it may, horses at grass acquire it less often than those which are kept in, upon hard food.

Symptoms.—As just intimated, a cold, that harbinger of so many other evils, is what marks the epidemic in every case; in addition to this, the animal will labour under the other symptoms of fever before described, according to its actual state of body at the time of attack. Thus, if the horse be in full flesh and vigour, his veins quickly fill with the stream of life, inflammation of the blood will ensue, or rather, to speak more accurately, of the vessels which contain it; hence, simple fever, or fever of the whole system follows, as before described, pp. 62, 63, but, be he poor, with little blood to receive inflammation, low fever is that particular affection which accompanies the original cold or catarrh.

Hence, I feel no hesitation in classing the epidemic—at least all those which have happened in my time, with one or the other of those diseases, and recommend treating it accordingly. At its earliest stage, of course, as it assumes the shape of a catarrh or cold (which in the more malignant cases becomes "Inflammation of the lungs,") I should treat it as such; but if not called in until this attack had extended to the animal's whole system, and catarrh had subsided into general inflammation, no reason exists why we should consider it a different disorder, merely because the patients may be more numerous than ordinary! The reader had therefore best proceed on to the next head of information, for the details as to the sufferings and cure of a single animal, which I apprehend will instruct him how to treat the many; for, neither the name nor the character of the disorder can be changed by this circumstance, however alarming its extent.

INFLAMMATION OF THE LUNGS.

Causes.—Like all other of its class of disorders, inflammation of the lungs is occasioned by a sudden check being given to the circulation, by cold when the animal is heated, either by exercise, food, or close stabling, as before described. How it happens that this organ of animal life is much more frequently deranged than any other, the reader who has well studied the second chapter, pp. 31, 32, will be at no loss to account for; adhesion of the pleura, or of the lungs, to the ribs, &c. as described at section 32, being very common: the labour of action, not to call it pain, is greatly increased thereby, and a certain degree of fever is thus engendered and kept up. The animal is in this manner always predisposed to acquire cold or catarrh; and ultimately inflammation of the lungs comes on, if the cold be neglected. Excessive exposure to the rougher elements, added to the changes in our humid atmospheric temperature, accounts for the prevalence of affection of the lungs. Out of the same causes arise several minor evils, to be considered hereafter; as,

> Simple cold, or catarrh.
> Broken wind, of three kinds.
> Roaring.
> Chronic cough. ·

The symptoms of inflamed lungs rapidly follow each other; shivering, difficulty of breathing, loss of appetite and sluggishness, with drooping of the head, become visible in quick succession. In a few hours, if the animal be in good keep, longer, if out of condition, those symptoms increase, with unusually quick action of the flanks, accompanied by hot mouth and hectic cough. Its ears and legs become cold, and he cares not to lie down, or being down, he rises languidly, as if mourning his fate. Sometimes the progress of this monstrous disease is accelerated by its previous habits, if the animal's constitution be predisposed towards inflammation.

The cure is sometimes mainly effected by the effusion of water in the chest, which frequently takes place upon bleeding the patient; the practitioner has little more to do than place himself in the situation of the handmaid of nature, and all will go on well towards perfect restoration. How this effusion is performed, none can know. Suffice it for our purpose, that such is the case, as I have shown in the second chapter, where I undertook to investigate the animal functions separately, and imagine I can not be misunderstood: See sections 19, 20, 21, in particular, at pages 22, 23. We may ascertain when this effusion has taken place, by an evident remission of the desponding symptoms just set down: the flanks cease to heave so much as hitherto, the animal looks up more cheerfully, he tries to eat a bit, the cough almost ceases, and the warmth of the ears returns, all in a partial degree: but the roughness of his coat, which always accompanies inflammation, does not so soon return to its original suppleness but assumes the first symptoms of hide-bound. When those favourable symptoms appear, much assiduity in the minor helps to recovery should be kept up, though further bleeding will be evidently unnecessary.

I have presumed that the patient has been already blooded in this as in all other inflammatory attacks, and that to an amount commensurate to the virulence of the attack, even to the amount of five or six quarts, if the animal is of full habit Of this proportionate degree, or quantity, let the reader more precisely inform himself by turning back to what is said on this head and the pulse, under general inflammation, or fever, at pages 63 and 64. The operator will of course follow up the bleeding with the purgative ball prescribed at page 63, in the case of general inflammatory disorders. Were I to repeat over again such general instructions, however diversified in language, I should add no new information. In every case of bleeding a laxative should follow, as before directed, and clysters or water-gruel be administered in aid of both, at intervals of three or four hours. Neglect not tolerably warm clothing; and by good hand rubbing, beginning [gently, for 'tis sore] at the neck and chest, and so proceeding towards the hind quarters, endeavour to obtain external heat, if not perspiration. When these appear, it is a sign that effusion has taken place, in a greater or less degree, according to the quantity of perspiration. This may be assisted in some degree, after the laxative and clyster have well subsided, by administering a

Sweating Ball.

Take tarter emetic and asafœtida, of each one drachm.
Liquorice powder and syrup, enough to form the ball for one dose.

Repeat the same in twelve hours, unless much perspiration has supervened in the meantime, when there will be no necessity for repetition. Thin water-gruel will assist the expected perspiration; or, if the animal be a fleshy one, a bran mash may supply its place: either must be given blood-warm.

The heat of the lungs, which is the immediate cause of the disorder, is visibly reduced by every inspiration of fresh air the animal takes. Naturally, then, this air should be fit for its purpose, or pure; at least not the confined air of an over-filled stable, replete with noxious effluvia; nor, on the other hand, a current of air that issues by doors and windows to the right and to the left, particularly in cold weather, or even in warmer weather whilst the animal is yet sweating with the diaphoretics just now recommended. As in most other affairs of life, the best will be found the medium course; for the noxious stable air having irritated and so predisposed the lungs to receive the blighting influence of the cold air, it follows that either extreme of stimulating, or bracing overmuch, must do harm one way or the other. A full and free inquiry into the best means of employing this main auxiliary in the restoration of health in inflammatory disorders would be well worth the labour of any veterinarian competent to the task; but as regards myself at present, such a course would ill suit my views in writing this too brief treatise. I shall, therefore, content myself with observing here, that since it is to this want of ventilation in stables, and crowding many horses together, that we owe all pulmonary complaints and most fevers, the subject is worthy consideration as a preventive as well as a remedy.

Formerly, the general practice was to clothe the animal almost to suffocation, and to close up every aperture by which air might enter the stable; the consequence of which mistaken notion was a severe attack of the lungs that

8 *

usually proved fatal, wherever these addenda to stable management could be
employed in supposed perfection. Not so the poor man's or the dealer's
horses under inflammation of the lungs, or the more dreaded "epidemic
distemper;" his stables being more or less pervious, and his horse clothing
without the nap, it was no uncommon thing to find these had recovered,
whilst the more pampered and more valuable animals fell victims to every species of inflammatory diseases. These results were known to many, in various
circles, about the time of the establishment of the Veterinary College; and
the mutual communications that thence resulted, proved the impolicy of the
old plan of adding heat to heat, and increasing the disposition to acquire disease,
of the lungs in particular. A revolution which had recently taken place in the
human practice regarding the treatment of inflammatory and febrile disorders,
also contributed to open the eyes of our veterinary practitioners in this respect,
and they adopted the direct contrary practice in its greatest extremity. Mr.
Colman advised turning the horse into a loose box, leaving open the apertures,
without clothing or paying any regard to the seasons. Nought, however,
could be more absurd than to suppose that a disease which is produced by
cold should have the continuance of cold prescribed for its cure.

My practice has been to afford the animal as much fresh air to breathe as
could possibly be allowed consistently with keeping out a draught or current;
taking care also that none whatever should be directed towards his body, nor
any enter the stable from the windward in stormy or cold seasons. With these
precautions, in a loose box and well covered up about the chest, but not tightly, he would ever be found turning round to that side where the most air was
.o be obtained, as if by instinct, knowing whence the readiest natural relief
from his sufferings was to be found. In one case, of an aperture being made
into an adjoining shed, the patient was frequently discovered inhaling the little
air which was to be drawn thence, though the orifice was no other than a displaced knot of the wood partition.

In general, the disease bends before the remedies prescribed; the hand-
rubbing must be continued, particularly of the legs, which in the worst period
of the disease are uncommonly fine, but should it last him some time they
swell, and in either case prove they are the barometer of the disorder, as well
as the necessity of rubbing them. On the other hand, should the pulsation
increase after bleeding, and no favourable symptoms appear (as indeed they
can not be then expected), this necessary operation must be repeated to the same
amount as at first, or up to a state of tottering as recommended before, at page
63: this necessity will occur but seldom, and that always with patients in
previously high condition. Therefore, no danger can be apprehended from
this copious discharge; for, at the end of twelve hours or less, which is the
period at which I should again resort to the fleam, the blood would have returned to its former courses in every respect; the continuance of fever up to the
same original height of the pulse, shows that the particular animal then under
treatment, possessed an uncommon quantity of blood, and therefore that an
unusual quantity should be taken away in order to alleviate the heat that is destroying it, and will destroy it, if the heat be not subdued at this second
bleeding; for, should this fail, I expect little good from further attempts,
though it is desirable to try what I always consider as the forlorn hope.

Should those remedies fail, suppuration takes place usually in six and thirty hours, and the animal is lost. Occasionally, however, it happens with low-
priced animals, that the inflammation fixes itself and terminates with destruction of one lobe only of the lungs, generally that on the right side, the
other performing all the functions, but how perfectly, or for how long time, I
had no means of ascertaining. At this point of his inquiries, the studious
reader had better consult over again what I thought it necessary to say upon the

dissection of the lungs, in chap. 2, pages 31 to 34; but he will please to remember, I am not at present prepared to maintain, that the real cause of a destroyed lobe, which I have just suggested, is more correct than that ventured by me at page 33.

Bleeding, though highly beneficial at first, when the animal system is in full vigour, is extremely dangerous after the inflammation has continued some time. When (the fever continuing) weakness is indicated by swelling of the legs, or nature seeks to relieve itself by a running at the nose, then bleeding will be harmful; this latter was considered a most favourable indication of crisis in the epidemic fevers of my youth; but I sincerely hope that the groundless fears the rumour of such a plague engenders, never more will visit us with affright: the idea of infection, in such cases, is too ridiculous to admit of refutation.

Weakness follows, of course, every attack of so vital an organ as the lungs, and is a necessary consequence of the great evacuations of each sort his extreme danger has rendered indispensable. But cordial balls, or indeed, stimulants of any sort, are very improper, and might occasion a partial relapse, if given before the animal is quite recovered. Good grooming, diet, and exercise, constitute the means of restoring his strength. Let him be well rubbed down, daily, and his nostrils sponged out clean and often, when the discharge takes place, which most commonly attends the cure; the same offensive matter must be cleaned away from the stall and manger, and he may be led forth daily whilst this business is going on. Hand-rubbing the legs should continue, so as to promote warmth, and they may be subsequently wrapped up, especially if the weather be chilly, with hay-bands, &c. Exercise may be gradually increased as the patient gains strength and appetite. At first, good stout oat-meal gruel, sweetened with coarse sugar or treacle, alternating this with wheat-meal, in order to coax his appetite; then oats which have been steeped in boiling water may be given, and next put him to hay of fine odour, in small quantities at a time. If grass or green vetches can be procured, a little, and not too much, may be cut for the now convalescent horse, in order to keep open his body; on the contrary, should he appear low spirited, a little malt occasionally will give him more vigour before leading him to the field every day, or leaving him there in clothing, whilst the sun may be out, if it shine at all.

BLISTERING and ROWELLING are recommended by most veterinarians, as tending to divert inflammatory heat from the more vital part of the surface. The theory is good; in the practice of human medicine I believe blistering is universally adopted; and this is one reason why I ever looked upon this means of cure with suspicion, even before I ascertained that the general heat or fever is always increased by the employment of either blister or rowel. Both are of the same nature; and the practitioner may learn how either operates on the system by ascertaining the state of the pulse previously to the application, and comparing it with the increased action of that barometer whilst the remedy is taking effect. Subsequently, however, it must be confessed, the agitation of the pulse will subside; and although I seldom find occasion for employing either blister or rowel, yet I am free to allow, that the manner in which inflammation of these organs sometimes terminates (namely in abscess, or soft tumour under the skin), seems to invite an early adoption of artificial means to bring about the same ends. The hand-rubbing just recommended effects this to a certain extent; and if it has been neglected, or lazily performed, then will blistering become necessary to prevent suppuration within.

As this tumour usually makes its appearance and marks the crisis of acute attacks, the practitioner may form an estimate of the probable beneficial ef-

fects of blistering in any case, by comparing and noting the earliest symptoms
of any two cases, in one of which the crisis has been subsequently attended
with such a superficial tumour, and in the other not so. He will then employ blistering with more reliance on its efficacy than I have found necessary after the hand-rubbing.

When this remedy is adopted for inflammation of the lungs, employ blistering ointment composed of cantharides and sweet oil, or hog's lard, or all
three—or the following

Blistering Ointment.

Cantharides, powdered, 5 drachms.
Hog's lard, 4 ounces.
Oil of turpentine, 1 ounce.

Mix, for one extensive application over each side of the chest; which is a
neater and more expeditious method of attaining the desired end than rowelling. When the latter method is adopted, let the tow used for the rowel be
dipped in a mixture of sweet oil and oil of turpentine; and the skin of the
breast or belly,—if more than one such seton is employed,—be separated only
just sufficient to admit the rowel, in order to increase the irritation, but if the
surrounding parts swell to an inordinate size, change the tow for some which
has been sodden in digestive ointment.

PLEURISY, or inflammation of the pleura, a membrane covering the two
lobes of the lungs (see chap. ii. p. 42)—has been described by Lawrence as a
separate disease; but, as the treatment is the same as the preceding, I can see
no propriety in making the distinction he does, especially as we can not know
the difference until after death discloses all imperfections.

A COLD OR CATARRH.

Causes.—If I sought much nicety of arrangement, the disorder termed "a
cold," would have preceded the similar but more malignant attack I have described under "Inflammation of the Lungs." Both are occasioned by cold
applied to the animal's organs of respiration at a time that he is most susceptible of its influence, differing only in the part which may suffer. Thus, when
the canal through which the air passes receives the check (before described),
which is the immediate cause of inflammation, every one agrees in its being
merely "a cold," though in most cases no attack is more replete with danger
if neglected.

But the origin and progress of such a check upon the functions of the
membrane that lines this canal, having been already fully described in the second chapter, pages 33, the studious reader must turn back to that part,
if he would trace causes to their effects, and does not presently recollect all
that is there said on this topic.

One prolific source of the disorder termed a cold, is found in the shedding
of the coat in spring and in autumn, a process of nature always attended with
a certain degree of debility or general weakness. Hence it is that the animal
sweats profusely upon the least exertion; and being in this state suffered to
stand (harnessed perhaps) in the open air to cool, the sweating is too suddenly
stopped, and he gets a cold at least. That the lungs should suffer the soonest
of any other organ is not at all astonishing: the very great exertions made by
the lungs in the business of progression, is much increased by adhesions and
other obstructions to the action of its several parts; and this, added to their
exposure externally, and the constant inhalation of fresh, cold damp air

altogether, the prevalence of pulmonary affections in every varied stage ought no longer to be matter of surprise to any person, however casually he may look at the matter.

The horse is subject to cold or catarrh at every season of the year, and some animals retain chronic cough all the year round, and some during their natural lives. But the cold which is contracted in the spring differs materially from that of the autumn. The former attacks the animal when he is full of hard meat and gross feeding—"full of humours," according to a homely but intelligent phrase, and a malignant sore throat or an inflammation of the lungs is the ultimate consequence, however slight the cold may have been at first. Sometimes access of all those symptoms of diseased lungs, which I have already or may hereafter take occasion to describe, will be found in the same animal, and he usually bends before the complication of evils and dies, unless speedily relieved by bleeding, &c. From its prevalence at some seasons, we then agree to call it "epidemic," and to recommend a treatment corresponding with the prevailing symptoms, if these be mild, as a simple cold; which form the epidemic fever or distemper always assumes in its earliest stages. On the other hand, the cold or catarrh which the moulting animal acquires in autumn, finds his system reduced by the heat and labour of summer; his blood, in quality or quantity, is scarcely capable of being excited to inflammation, and the first attacks are more easily subdued. Neglect, however, increases the evil at all times, especially with the more valuable well-conditioned animals, some of which are so tenderly managed, that they scarcely can stand the opening of a door or shutter after dark, without catching cold. Neither autumn or winter is the season for remedying this defect in stable management,—if ever it can be got over at all.

Symptoms.—According to the precise part attacked, these vary not only as to appearances, but as to virulence or malignity, always increasing as the complaint descends lower and lower down towards the seat of vitality; the danger being also greatly augmented when the animal is pre-disposed to acquire catarrh in its worst forms by some previous misfortune—as adhesion, &c. A simple cold consists in slight inflammation of the membrane which lines the nose, windpipe. &c. the functions of which membrane in health are described in the 34th section, chap. ii. together with the manner in which the disease is engendered. As we find in all other inflammatory disorders, variations in the symptoms occur, according to the previous constitution or evils of the individual, and its actual condition—much more than is attributable to an adverse season, or the immediate cause of disease. For example, if two equal animals be exposed to a chilly night air, that horse which had performed a journey previously to turning out, would catch a cold for certain,—the other most probably would escape; but, if both had performed the same journey, let us suppose, and one of them laboured under the constitutional defect of "adhesion of the pleura," (see page 32), he would acquire the more malignant cold, known as "inflammation of the lungs,"—his less unhappy mate a simple cold. What horrid symptoms denote the former, I have attempted to describe; the simple cold, at its first appearance, is too well known to require minute description.

If the cold extend no farther than a check upon the mucous secretion of the membrane that lines the nose, a purulent discharge is first observed in the morning, its eyes become dull and a little bleared; and, in twenty-four hours, a short cough denotes that the inflammation is creeping onwards, and has reached the epiglottis. The attack upon this point of conjunction between the throat and mouth, will be greatly accelerated by the injury most horses sustain which have been subjected to the brutal operation of being "coughed" by the dealers;—an injury that thus produces latent effects, though the pain were originally little, and that little long ago departed.

We hear this kind of first attack termed "a cold in the head," the second symptom is "a cough," and feel no disposition to quarrel with either term.

In proportion as the attack may be more severe, the symptoms increase, as does the danger. Passing the hand down over the windpipe, at the epiglottis, the animal will shrink if it be sore within, and he will soon evince difficulty of swallowing, and refuse his food: inflammation has begun. When these are not preceded by a discharge from the nose, this symptom does not appear until the inflammation is lowered by bleeding and other remedies: the discharge is then an indication that the inflammation, or heat, has subsided and no longer demands the adjacent secretions. See page 33, for a more minute description how this demand takes place.

With those symptoms of sore throat others become apparent, and the whole assume a malignant tendency proportioned to the severity of the attack and previous state of the suffering animal. As happens in all other inflammatory complaints, the pulse tells of the existence of fever, in its degree: accompanied by languid eyes, breathing quick and laboriously, and general heat of the skin without perspiration. In some cases the sore throat is substituted, in some measure, by enlargement of the glands underneath the jowl, which are also attended by soreness more or less; and as this species of attack is occasioned by the humidity of a cold spring or wet autumn acting upon moulting horses, great numbers feel its influence at once, and gives reason for veterinary writers to consider this general distemper as "the influenza," and an "epidemic." Enough has already been said under the latter head of information, therefore let us proceed to treat of the thing as it regards the individual patient.

Remedy.—When the glands swell, as just mentioned, and there is no reason to doubt, according to the corresponding symptoms, that it is the effect of a cold—which may further be ascertained by their heat and tension, let some discutient application be used—as camphorated spirits of wine: but if the inflammation be to a great degree, bran poultice may be applied to advantage. If those enlarged glands already contain matter. the tendency to irritation will thus be reduced; if merely sordid tumours, either application will effect relief, by reducing the size and tenderness of the part, so as the animal may take his medicines with less difficulty. Steaming the head for an hour, or applying hot flannels that have been steeped in boiling water, will be found serviceable, taking care to dry-rub the coat immediately after, which also assists to reduce the swelling. If this symptom does not give way before those applications, and the throat is ascertained to be sore, blistering may be resorted to, taking care to extend it over the whole of the parts affected. See page 76.

As in all other inflammatory diseases, bleeding to an amount proportioned to the violence of the attack, with purgatives and clysters, should accompany the foregoing external applications: and these, with plenty of bran mashes, sodden oats, and the fever powders prescribed at page 65, will reduce the symptoms. Similarly to those also will be the precariousness of his complete recovery, and so should be the care that the relapse, to which he is for a time daily liable, should not reach to a great height. I need not repeat the general precautions which are set down at page 70.

Unwilling to leave the reader in a dilemma as to the mode of applying the bran poultice just recommended and upon the efficacy whereof I mainly rely, I have taken the pains to sketch a bandage proper for that purpose, with its fastenings, the ingenious contrivance of some Frenchman, whose name I love to have been BOURGELAT.

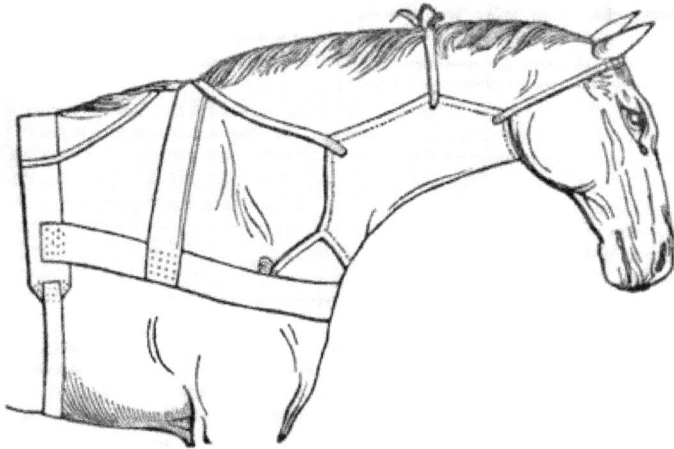

It will be seen, that unless the remedy proposed is practically applicaole, the preparation thereof would be wholly unprofitable; therefore, when the poultice, the steaming, or the blistering, be found necessary, we should endeavour to secure it in the best possible manner; and as most persons are out poor horse milliners, I have undertaken in this instance, as well as in cases of Strangles, Poll-evil, and Vives, to exhibit the best means of retaining the remedies in their proper places.

The cloth to be employed should be of stout but supple linen, as Russia duck: or hempen sail-cloth; or in failure hereof, a fresh sheep-skin, or a piece of Shamoy leather might be substituted.

Some recommend steeping the cloth in a solution of gummy substances, to render it water-tight; but such contrivances only add to its unconquerable stiffness, and I should prefer oiled silk, such as used for umbrellas, if readily procurable, and not too dear for the pockets of those more immediately concerned.

When spread abroad, the cloth will be of an irregular octagon shape, at each corner whereof it is to be strongly sewed on a piece of broad tape for the purpose of fastening to the girth, or round the neck, or to a breasting of broad web, which is supported by another piece, that passes over the withers, and which two should be previously fastened together by stitching the cross-piece ends upon the breasting. The two extremes of the bandage will be the fillet across the forehead and the fastening at the girth; therefore measure should be previously taken of the whole length proper for the individual patient, lest the tie, which would otherwise be necessary at the ears, might discommode the animal, and occasion skittishness; or on the other hand, the application would not be kept in its place properly. A single glance, however, at the cut will instruct a tolerably expert workman, or work-woman, how to manufacture such a bandage as would answer every purpose.

THE COUGH

Which accompanies this disorder will frequently remain after the other symptoms have abated; in some cases a cough is the only symptom of catarrhal inflammation that the animal suffers under, and in both we should apply ourselves to reduce the inflammation of the wind-pipe, &c. which occasions the cough; for if not cured at once, it baffles all our efforts for a long while, and ultimately becomes what is denominated (from the length of time it has lasted) a CHRONIC COUGH. But no absolute necessity exists for considering these as separate or distinct diseases, the one being but a prolongation or fastening of the other on the system, as described at page 85 below: therefore should our attention to the first attack be unremitted, and the remedies applied in turn to each variation of the symptoms. If these are accompanied by the swellings and soreness of the throat and glands, just spoken of, the cough will generally cease, when these symptoms are removed; but if not, the cough must be considered as a simple disease, and be treated accordingly. By the way, seeing that after all our care and anxious examination, we can but imperfectly distinguish between some cases of ill-cured catarrh, or the chronic cough, and the incipient cough, or a fresh cold, the practitioner would do well, in cases of doubt, when he finds one of those remedies fail to afford the expected relief, to try another, and another, for example.

When the cough continues, and there is reason to apprehend, from the frequent and violent efforts of the animal to expel the mucous secretion, that this is thick or viscid, and does not come away, though the animal evidently sneezes for that purpose,—the lungs must be relieved by softening the agglutination; otherwise termed "cutting the phlegm." Venesection always effects this end; but, when blood-letting is not rendered otherwise necessary, the drenches Nos. 1 and 2 will afford relief. As the cough always becomes more and more troublesome as the discharge lessens of itself, or ceases altogether, we may conclude some lurking virus that has fixed upon the lungs is the immediate cause of the cough. In order to enable the lungs to throw off this cause by a more copious discharge, give the

Expectorant Ball.—No. 1.

Sulphur, half an ounce.
Asafœtida, 1 ounce.
Liquorice powder, 1 ounce.
Venice turpentine, 1 ounce.

Mix for four doses, and give one on each of four succeeding nights. See his exercise be moderate, and allow him the cooling regimen before referred to (page 67), as proper for convalescent horses after inflammatory attacks.

Expectorant Ball.—No. 2.

Powdered squills, 2 drachms.
Gum ammoniacum, 4 drachms.
Powdered ipecacuanha, 4 drachms.
Opium, 4 drachms.
Ginger and allspice, of each 1 ounce.
Balsam of sulphur, 4 ounces.

Mix, for six balls, with Castile soap, 2 ounces, beaten up with mucilage; treacle, or syrup: to be given once or twice a day.

If this regimen can not be followed by reason of want of attendants, his bowels at least should be kept in a proper open state by mild laxatives; or, if costiveness prevailed when the cough first came on, simply opening the bowels will then procure ease, if it do not effect a cure. This may be attained by giving, for three or four days,

The Laxative Ball.

Aloes, one and a half drachms.
Ipecacuanha, one and a half drachms.

Mix with liquorice powder and mucilage for one dose.

These medicines, and every modification of them, which the experienced chemist can suggest, it is desirable should be tried in succession, as the seat of the disorder is so very various and uncertain, that the partial good which one may effect, will frequently be aided by another. To this end the following ball and drenches have been prescribed and used with success—

Diuretic Ball.

Yellow resin, 2 ounces.
Turpentine, 4 ounces.
Soap, 3 ounces.
Salad oil, 1 ounce.
Oil of aniseed, half an ounce.
Powdered ginger, 2 ounces.

Rub the two last together in a mortar, with a little linseed powder. Melt the first three articles over a slow fire, and then mix in the powders. Divide the mass into eight balls, and give one a day until the water is affected.

Drench.—No. 1

Vinegar, 8 ounces.
Squills, 2 ounces.
Treacle, 6 ounces.

Bruise the squills and pour on the vinegar boiling hot; simmer these near the fire two or three hours, then strain off and add the treacle. Divide into three or four parts, and give a portion two or three times in the course of the day.

Drench.—No. 2.

Bruised garlic, 4 ounces.
Vinegar, 12 ounces.

Pour on the vinegar boiling hot; let it simmer four or five hours, strain off and add six ounces of honey. Divide into three parts or four, and give in the course of the day at intervals.

But no ultimate cure can be effected unless the diet and regimen is properly followed up; nor, if the animal be pushed in his work whilst the disorder is virulent; and, after all our care, if the cough does not abate, but becomes worse by reason of a new cold, it fixes upon the lungs, and the animal drags out a miserable existence. This has been usually treated of as consumption, by reason of its resemblance to the same disorder in human medicine, from the wasting away, or consumption of the animal system, which accompanies a diseased state of the pulmonary arteries. Of the importance of this part of the system to animal life, to existence and health, the attentive reader can not fail to be sufficiently aware who has well perused that part of the second chapter of this little manual, in which the functions of the organs of respiration are described with requisite care—page 31 to 35. The hopelessness of bringing about a cure, after the ruin has proceeded so far as we have just contemplated, must likewise be most apparent to him : I will not therefore, pursue farther in detail the last wastings of this vitally essential organ of the animal system, but proceed shortly to notice some other effects of an ill-cured cold or protracted cough.

BROKEN WIND

Is already so minutely described, as to its causes and symptoms, in the second chapter, that I apprehend repetition in this place would prove worse than use-less. The reader will therefore turn to page 34, and the recapitulation of my treatise on the organs of respiration which immediately follows, at page 35. Generally speaking, broken wind is brought on by inflammation of the organs of respiration, and acquires a different name, though requiring but little variety of treatment, according to the part which may be the more im-mediate seat of disease ; for it must be clear, that although this may lie in the uppermost part or larynx, in the lowermost part or midriff, or more centrally —the communicable nature of inflammation is such, that the whole must par-take in some degree of each and every partial derangement. And this de-gree will be proportioned to the excitability of the individual's organs of res-piration that may be the subject of attack : if the animal contract cold or cough in the vigour of age and health, he will experience its effects in the most frightful shapes ; it proceeds to encroach on and obstruct the right func-tions of the lungs with rapid strides, and if the symptoms do not abate, he dies. But, being partially removed, it becomes a chronic disorder* to the end of his days ; and, agreeably to the part which may experience the attack, has it been the practice to denominate chronic cough either roaring—broken wind —thick in the wind—or asthma. Hereupon, however, the doctors disagree.

How this difference arises may be worth a moment's investigation here, al-though so large a portion of the second chapter has been already devoted to the subject, and the reader must absolutely turn back to it. At page 34, the thickening of the midriff, in consequence of inflammation attacking the ad-jacent viscera, was minutely described: this thickening of the membrane also extends to every other part of the lungs, wind-pipe, &c. whenever cold or inflammation prevails ; and in the event of its continuance, the thickening of the membrane remains long after the virulence of the disorder may be sub-dued. If this state of the organs of respiration extend over them generally, the patient may very justly be said to be " broken winded;" when this ex-tends to the thickening of the pleura only, he would then be thick winded, or short in the wind, as he would also in case of adhesions of the midriff, as described in page 34, already referred to. Neither affection, however, can fairly be set down for broken wind ; though both those membranes being af-fected might properly enough be considered " a broken manner of drawing in and expelling the wind," for the inspirations and expirations are in this case extremely irregular, broken, or variable ; whereas, when the air-cells are really broken, or burst into each other through great exertion, then the air escapes with difficulty, and the expirations are now slower than the inspira-tions (as before observed), and both together constitute irregular respiration, or true broken wind.

But of controversies there is no end. J. White and R. Lawrence were for some years at issue on these points ; White having taken up Lawrence rather sharply, and somewhat unjustly, if he meant to impute error to the

* Chronic disorders are those which, having lasted a long time, become almost second nature, and plague the organs of respiration more than any other viscus : thus, a tickling cough may stick by an animal for years, but it becomes worse upon any great exertion, or on catching fresh cold.

atter, as regards the symptom of respiration just spoken of, for each writer was right in his separate position : as they disagree as to what constitutes broken wind, so they could not of course agree as to the symptoms. See pages 159, 160, of White's first volume. This author also disorders his own positions at the same place, in two other instances, which I should not have noticed, but for his tart rebuke of R. Lawrence for attributing the term broken wind to the thickening of the membranes. In this view of the case, it will be seen, I certainly can not agree with this very clear-headed veterinarian ; but I do not therefore, harshly refute a gentleman and scholar for not agreeing with me upon a simple term of science : it was this unamiable attachment to trifles that so long impeded the progress of chemical knowledge, until the plain-speaking Davy, Nicholson, Park, and Paris, came into vogue, and drove Lavoisier from his prostrate coterie,—Dickson was put to silence, and Fourcroy's reveries were laid in the dust of oblivion.

White says, " The lungs of broken winded horses that I have examined have generally been unusually large, with numerous air-bladders on the surface." p. 160. Yet, in the next page, he opens a broken winded subject, and says, " The lungs were lighter [meaning less] than usual, and without the air-bladders, contrary to the state Mr. Lawrence describes." What Lawrence had said was this: "The most common appearance of the lungs in broken winded horses is a general thickening of their substance, by which their elasticity is in a great measure destroyed, and their weight (i. e. size) specifically increased. On this account air is received into the lungs with difficulty, but its expulsion is not so difficult. Thus, in broken winded horses, inspiration is very slow, expiration sudden and rapid, as may be seen by the flanks returning with a jerk." (p. 123, octavo edition.) And he is correct as to these two motions accompanying the thickened membrane or substance of the lungs; only I should have termed the disorder thick wind, and not broken wind, when all would have coincided with White's statement, barring his own self-contradiction as to the size of the lungs, which Lawrence had mistaken for weight, and which had met with the counter assertion of being "specifically lighter." On this point of their dispute, however, neither the one nor the other could possibly *know* aught with requisite certainty; and I, for my part, am inclined to believe, that the lungs of high-bred horses are specifically lighter than those of the cart breed, saving that the whole organs of respiration are much less muscular in the first kind than in those of the latter, the skirt or border of the midriff in particular. On the other hand, the hearts of blood horses invariably run of a larger size than those of the common English horse. Vide page 37. One cause of broken wind, or rather that mainly predisposes the animal to contract this disorder, is voracious feeding, which distends the stomach inordinately, and for a while gives to the animal a short-lived vigour and healthy appearance. This induces its proprietor to put him upon his mettle, and try the extent of his powers at progression ; and as he will best perform those feats upon a plentiful feed, the action of the midriff and lungs thereby becomes laboured, and the proper expansion of the lungs is impeded. Heat and tension are the immediate consequence, and broken wind of one or the other species is the remote consequence. Horses that eat their litter, and what other hard substances they can come near, are similarly predisposed to broken wind; namely, by the great distension of the stomach, and inability of inspiring a sufficient quantity of air to fill the lungs, whence the inert cells, or the portion not distended, fill up, contract, and become useless, or, upon sudden action and over distension, they burst at once.

Cure there is none for broken wind, and therefore all that can be done by way of alleviating its symptoms must be effected by management, or as it is more generally termed, by

Regimen. Of course, any person would avoid exposing the animal to fresh cold, and not push him too hard on a full stomach; nor indeed, give away a chance of increasing the malady by the same means as I have just said originally brought it on. He will, on the contrary, follow an opposite course of treatment, and as much as possible regulate his feeding and exercise upon moderate principles, for the stomach and bowels are always affected by broken wind. Hence it is, that flatulency accompanies broken wind of every kind, so that the animal in his endeavours to cough, usually breaks wind after an effort or two. Much medicine is not requisite, and, in slight cases, far from desirable; tonics, bracing air, and regular hard meat feeding, broken or sodden, and given in small quantities, will do more for the horse than physic of any sort. For the first, Peruvian bark, or cascarilla in small doses, may be given occasionally adapting the quantity to the bulk of the animal.

Tonic Ball.

Cascarilla, ⎫
Gentian root, ⎬   1 to 2 drachms of each,
Oil of Carraways, ten drops; with

Mucilage enough to form the ball. If irritation of the bowels is indicated by a certain protrusion of the anus, add of opium 10 to 12 grains.

When the cough is particularly troublesome, or the animal seems to labour much in respiration, give the following

Ball.

Dried squills, powdered, 1 drachm.
Gum ammoniacum, 3 drachms.
Opium, 10 drachms;

With mucilage sufficient to form the ball.

If there is reason to apprehend the horse swallows his corn without grinding it, as commonly happens, bruised or sodden oats should be given, and the bowels discharged by purgatives, when alteratives may not be deemed equal to the urgency of the case. Those prescribed at pages 86, 87, are applicable in this case also; inasmuch as the two disorders bear very near resemblance to each other in this respect. Give green food, succulent roots, and bran-mashes, as there recommended. Let the water be soft, not too cold, and given in small quantities at a time, and frequently.

As broken wind produces disordered bowels, and is re-produced by it, the connexion or sympathy between the two, thus plainly demonstrated, should be employed in the alleviation of the former in all its stages, when it has been of long standing. The means of attaining this object has been shown; and when the animal under treatment is equal to the care and expense, he frequently recovers so much of his former powers of free respiration, that his cure will seem for a short time fully effected. These appearances, however, are completely illusive; upon the least extra work he relapses into his former difficulties of continuing it, and the cough, the roaring, wheezing, or labouring of the flanks and chest, return as bad as ever. If the work be very hard, as always happens when the horse has been sold deceptiously, and the new master would try his utmost powers, the relapse is then worse than before; he hereupon becomes a confirmed roarer, by the wind and lymph being driven inside the membrane that lines the wind-pipe, and causes inflammation of the very fine blood-vessels that traverse it. Hence the number of lawsuits that are instituted to recover the valuable consideration paid for broken winded horses that are returned upon the hands of the sellers as roarers, that never were known to either groom or stable-boy for roarers, before the day of action or trial. Hence, too, let us charitably suppose, the contradictory evidence often given, and the flat, downright cross-swearing that usually takes place on such occasions. For the horse having been partially made up for the purpose of sale, i. e. nursed, patched up, and to all appearance "right in all his parts," the fact of his going in pain comes out by way of his skin at first, and the new purchaser being generally desirous of trying all he can do, the ruin is effected, by pushing him too much, of driving the wind inside the membrane, as before described.

Hereditary Roarers. Early in the present century, a question arose among breeders, whether the gift or the curse of roaring descended from parents to their progeny. The decision was looked for with unusual anxiety among the breeders of farm-horses in Norfolk and Suffolk, where a famous well-built horse in every other respect was much sought after, even subsequently to his being denounced a roarer prepense. Would his stock take after him ? was a problem very desirable to be set at rest, when Mr. Wilson, of Bildestone, late Sir T. C. Bunbury's, propounded the question to Mr. Cline, an eminent surgeon and anatomist in London. In reply, Mr. Cline said, "The disorder in a horse which constituted a roarer, was caused by a membranous projection in a part of the wind-pipe, and was a consequence of that part having been inflamed from a cold,[*] and injudiciously treated. A roarer was not therefore a diseased horse, for his lungs and every other part might be perfectly sound ; but when a horse was in strong action, his breathing became proportionably quickened, and the air, in passing rapidly through the wind-pipe was in some degree interrupted in its course, and thus the roaring noise was produced. The existence of this in a stallion could not be of any consequence. It could not be propagated any more than a broken bone, or any other accident."[†]

Unfortunately, however, for this opinion, and not exactly in accordance with my own, several of that horse's get became roarers, but we are left to guess whether hereditarily or acquired. An account of the horse in question appeared in the Annals of Sporting for 1823 ; but the colouring given by an evident partisan of the stallion-master induces one to lament the absence of that candour, from which alone useful truths are to be drawn ; for, we are deterred from indulging in pathological investigation where the grounds of inquiry are so impalpably sandy as were those adduced upon the occasion.

CHRONIC COUGH

Is already defined to be the remains of an ill-cured cold, which may or may not have been a cough originally. It bears close analogy to simple broken wind that is seated in the wind-pipe or its branches, of which it may be considered a continuation, or the natural consequence of neglect, with more inveteracy. How this effect would so accrue was described at page 80 ; and the analogy is still further corroborated by the fact, that the treatment for cough of long continuance is precisely that which is found serviceable for broken wind, the situation of the two disorders making the only difference in either respect. Again, the symptoms of both may, by long and careful treatment, be so reduced as to seem cured, for a longer or shorter period, and both will return in the shape of roaring, upon the animal being put to sudden hard

* Not always so, Mr. Cline.
† Our human anatomist is very nearly right as to an accident not being descendable ; but seeing that roaring did descend to the first generation, we must infer that this was "an accident of birth," and not a contracted one, which might possibly go no farther.

G *

work, as mentioned in the last pages. The corresponding symptoms of both are also so nearly alike, that I merely comply with custom while I recapitulate these for the use of readers who might not choose to consider that horse broken winded, which to all appearance is only affected with "an old cold in the wind-pipe." But, let the first term appear to an owner ever so formidable in sound, the latter is no less dangerous in effect, and both are alike liable to terminate in roaring.

The symptoms which indicate chronic cough are nevertheless so slight, that it is too often considered as but a small remains of the more alarming catarrh, which its owner vainly imagines will go off in time, as the other disagreeable symptoms have done. In this hope he is invariably disappointed, if the means of reducing it be deferred. After the more violent symptoms of catarrh have subsided, and the cure may reasonably be considered as complete, the horse returns to his usual feeding, and, as in the former case, eats voraciously; he is denied water oftener than twice a day, perhaps not so much; but, when at length he does drink, he gulps it up as if famished. This is commonly the cause, and the first indication of the cough which follows immediately after, but is often mistakenly attributed to his improving too fast after his long illness, and it is considered only fair that "he should be allowed to recover himself completely." Precaution is thus lulled in fancied security, and unless prompt relief be afforded before the damp season of autumn returns, the symptom increases to obstinate confirmation; until time renders cure hopeless, alleviation or abatement of the coughing being all that lies within the power of medicine or stable management to effect for it—the aid of the former being then of little avail. Very few small proprietors of horses use timely precautions in this respect, and the disorder goes on: large owners having more experience, adopt early measures, and if pursued with proper vigour, these usually prevail in lowering the symptoms.

An occasional cough is also brought on by high feeding, which, as it arises from the rapid production of fresh blood, is termed *plethoric cough,* by way of distinction. Of this symptom it would be needless to tell the better informed, perhaps, at an interval of six days, that we have but to take away the cause, and the effect ceases of course. This, however, does not always follow; for the cough sometimes remains after the gross feeding has been reduced in quantity and quality. In this case, it must be considered as chronic cough, and treated as such, by emptying the bowels, &c. as above directed.

Remedy. As in the case of broken wind of every other kind, the horse eats every substance he can come near, chronic cough being sometimes produced by over feeding, as well as always producing that symptom. Therefore, when a horse has a cough, occasionally, for two or three days, his appetite being good, we had best conclude he is too full and must be emptied by an alterative or purgative, according to the emergency of the case: if he be of gross habit, or has failed in the proper evacuations; if his heels swell of a morning, or his coat stare like hide-bound, the cough will vanish before the following

Purgative Ball.

Barbadoes aloes, 8 drachms,
Castile soap, 2 drachms,
Ginger, 1 drachm,
With mucilage sufficient to form the ball.

Failing to stale properly, the patient's heels will swell, in addition to the cough, and both may be got rid of by a diuretic ball or two at farthest. If

the evacuation by the skin be at fault, through cold or otherwise, accompanied by cough, the perspiration will be restored, and cough depart, by the exhibition of Emetic Tartar, one or two grains, every day twice in powder, until
its effects are perceptible on the skin, and the cough then diminishes.   This,
however, is a very slow remedy, though sure, and is sometimes given in much
larger quantities.   The preparation is very simple when given in the form
of a ball, being made up of liquorice powder and mucilage only, of a sufficient
consistence to retain that form.   If much heat of body is perceptible, though
the pulse may not indicate inflammatory disorder, add to the foregoing ball,
nitre, 4 or 5 drachms.

But whatever course is pursued, if symptoms of a bad habit of body are
discoverable, it will be advisable to administer the foregoing pargative ball once
or twice previous to adopting any other means of cure.   If those symptoms
of a bad habit of body do not appear, then the purgative should be of a milder
nature and given at the same interval.

Mild Purgative.

Aloes, 4 or 5 drachms,
Castile soap, 3 drachms,
Calomel, 1 drachm,
Ginger, 2 drachms,
Oil of Carroways, 10 drops;
Mucilage enough to form the ball for one dose.

Some horses are more delicate than others, and being then irritable about
the throat and chest, are liable to contract a periodical cough, which becomes
chronic without due care.   Such animals should be exposed as little as possible to any violent weather, or sudden change of the temperature : these are
the kind of animals that will benefit greatly, or suffer the most, by a summer
run at grass, according to the heat, the dampness, or dryness of the season,
and the precautions used previously to, and at the turning out. Neither should
such tender animals, under circumstances of chronic cough, which generally
affects their coats also, about the chest in particular, be treated with a purgative, even of the mildest form but with alteratives instead.

Alterative Ball.

Aloes, } 12 drachms each,
Hard soap, }
Emetic tartar, 1-2 drachm,
Ginger, 1-2 oz.
Oil of carraways, 1 drachm;

With mucilage enough to form the balls into six doses. Give one every morning until a loose stool is produced, which may happen on the third or fourth
morning or longer, as the animal may be more or less relaxed.

Even with this moderate employment of laxative medicine, the kind of animal for which it is most desirable will be very unfit to turn out to grass of a
sudden; as, on account of its delicacy, it will in that case be more likely to acquire a small hectic cough, which no one attends to because of its triviality,
until time renders it chronic, with all its attendant consequences.   Roaring,
broken wind, are among these evils, and have already received as much attention here as they separately require.

Frequently it happens that a horse has a constitutional cough, or one which comes on only upon high feeding, or a disposition to plethora will produce the same kind of cough, and, in either case, it seems but an effort of nature to relieve itself. In this case, the rapid repletion of blood drives it into the smaller vessels that line the windpipe, &c. and there causes the titillation which after two or three efforts ends in cough, and so on repeatedly. None but those which are in some slight degree or other already afflicted with chronic cough are ever so attacked, I apprehend; indeed I have frequently remarked how excellent a test of " bad in the wind" was good feed, or a large feed, with work upon it. In this case, the administering of nitre and resin will thin the blood, and give immediate relief.

Drench.

Nitre,   } of each half an ounce,
Yellow resin,   }
Oil of aniseed, 20 drops.

The oil should be first well mixed with the resin, and the whole given in a quart of water-gruel. Recurrence of the same affection may be prevented in some measure by giving the same in another form, which is in general reckoned more convenient—namely, as a cough powder, substituting aniseeds, 1 ounce, for the oil, and pounding the whole together; mix with the corn.

INFLAMMATION CF THE STOMACH AND INTESTINES.

Whenever one of these organs is affected, with inflammation particularly, the other soon feels the effects of the attack. This arises from the proximity of the two; or the continuity of the digestive faculty, which is mostly carried on in the intestines, as the reader of tolerable recollection well knows was so described in Book 1. page 44, &c. Corrosive poisons, indeed, carry on their work of destruction upon the internal or villous coat of the stomach until the ruin is complete; but, although horrid inflammation accompanies its ravages, I would not class such a species of accident under any other head than " Poison :" to call it by its symptom would be delusive. Neither is the inflammation caused by worms, proper to be taken into consideration here, though in this case both organs are affected at the same time; but the bott question involves other considerations, besides the best means of destroying them, of preventing the access of this irritating insect, or of alleviating the effects of its bite and adhesion to the villous coat, alike of stomach and intestine.

With those exceptions, there is no greater difference in the causes, symptoms, or means of cure of inflammation in the stomach and intestines, than exists between those of the great and the small gut. Inflammatory pain in the smaller parts of the alimentary canal will ever be more acute than those which attack the larger ones; thus, when the stomach is the seat of disorder, the pains will be duller, the paroxysms less distinctly marked, and the pulse but little altered; but, when by continuance it reaches the small gut at the lower orifice of the stomach, then will the pain and anxiety of the animal increase greatly, and the symptoms thereof, visible in his manner (to be described shortly), will become more distinct, rapid, and vehement. The pulse increases in number, in sharpness of vibration, and irregularity. Such is the difference also that is discernible between attacks upon the colon or great gut, and on the smaller guts. But all this refers to the first attack; for after a while, if the means adopted are insufficient to check its career, the ruin goes on to affect the whole abdomen, and the animal dies in excruciating torments.

Causes.—Much the same as those which occasion fever in all ordinary cases; that is to say, a sudden check given by cold to the action of the parts, while these may be in a state of excitement, or through over action, hard work, excessive heat of the weather, the operation of cordials, &c. By this latter means stallions and brood mares are sometimes destroyed prematurely, even without catching any cold, or this part of the system receiving any check whatever; in these cases, excitement has been carried to the utmost pitch by high feeding, and stimulating the male, until nature gives way, or rather, I might say, catches fire almost, and if not speedily arrested, the heat soon destroys the functions of all the abdominal organs of life.

To stage-horses, inflammatory complaints usually prove fatal, from the same immediate cause; the animal being fed high, and pressed forward to the accomplishment of his daily task, regardless of the first indication of this disease; and in summer time, we witness numbers of such dropping down in harness, sometimes whilst going along, seldom giving warning of approaching dissolution. But, whatever be the previous state of the animal's bodily health, he can rarely stand the maltreatment he receives from his driver:—viz. that of being driven through ponds and large rivulets, while he is yet perspiring greatly through fatigue and the heat of the weather. Long rests in currents of air, or unsaddling horses under similar circumstances, are alike productive of inflammation of those or some other part of the animal's inside, if it do not bring on fever of the whole system—as before observed, p. 59. The kidneys or the liver are sometimes alone affected by this species of culpable neglect; but in either case the effects are not immediately perceptible, and the disorder creeps on unheeded, or seizes the animal violently, so that it dies at the next going out.

Neglect of the necessary evacuations, or the discontinuance of those which have been customary, even though injudicious, will occasion an accumulation of dung in the intestines when they are least capable of bearing it: upon this, pressing the horse in his work will bring on inflammation, as it will sometimes after a heavy feed and water, which some injudiciously give on account of a hard day's work lying before him. The same happens to horses that are inordinately fat, when hard worked; the dung that is then eliminated bears with it a portion of the slime or mucus that lines the intestines, and this appearance has obtained for this species of inflammation the term molten grease. I postpone, for a few pages, the consideration hereof, in compliance with custom rather than in obedience to propriety.

Adhesion of the gut sometimes takes place, so as to cause partial obstruction to the passage of aliment; at others, tubercles are formed on the mesentery that holds the bowels in position; and in either case the secret is disclosed by a staring coat, which some mistake for the worms. Both those affections are the effect rather than the first cause of inflammation of the part, and may be distinguished from "the worms" by the state of the pulse, by the heat, tension, and soreness evinced by the patient on passing the hand over the belly. See page 46, book I. The reader will also perceive, upon turning back to page 22—24, in what manner this adhesion is effected, by the exhaustion of the moisture that is designed by nature to lubricate the parts.

Colic of long continuance, if the animal is worked while this is on him, is another prolific source of inflammation of the intestines; as is the drinking cold water copiously, while in a state of perspiration, or after a trying journey, which is always attended with spasmodic colic of the stomach and bowels, at first, and of inflammation sooner or later, according to the temperature of the individual. The necessity of getting rid of the lesser attack before it acquires a permanent and dangerous aspect must be obvious; and as the treatment proper for either, is at total variance with the other, the one requiring warmth and stimulation, the other a cooling and reducing treatment, our first duty is to ascertain precisely the exact nature of the attack ; for a mistake on this point would, and does frequently, prove fatal—ay, in human as well as in horse medicine.   Therefore it is, that I have judged it expedient to set down here a table of the symptoms that will enable the practitioner to distinguish between the two kinds of attack.

For this mode of setting before the eye in parallel columns the discriminating symptoms of two such apparently similar disorders, I am indebted to Mr. Ryding, who inserted it in his "Veterinary Pathology," 1801, pages 86, 87 ; and it was copied by White into his "Compendium," 1803, with a few alterations, by no means for the better, I have adhered chiefly to Ryding, with but one slight alteration.

SYMPTOMS.

A *table for distinguishing between the* Colic *or* Gripes, *and Inflammation of the Bowels, by the symptoms that mark the character of each.*

Spasmodic or *Flatulent Colic.*	Inflammation of the Bowels.
1. Pulse natural, though sometimes a little lower.	1. Pulse very quick and small.
2. The horse lies down, and rolls upon his back.	2. He lies down and suddenly rises up again, seldom rolling upon his back.
3. The legs and ears generally warm.	3. Legs and ears generally cold.
4. Attacks suddenly, is never preceded, and seldom accompanied by any symptoms of fever.	4. In general, attacks gradually, is commonly preceded, and always accompanied by symptoms of fever.
5. There are frequently short intermissions.	5. No intermissions can be observed.

Whilst marking these distinctions, which ought to be kept in mind while prescribing for disorders so nearly alike at first view, but differing so widely in effect, the reader is earnestly requested to turn to the *Index*, and there find the page at which I have thought proper to treat pretty much at large of "Inflammation of the Kidneys," "Diseases of the Urinary Organs," &c.   He will there perceive how fatally these affections have been mistaken for "Colic;" he will learn that this unhappy error is likely to happen more frequently than would at the first glance be imagined; and he will observe that the symptoms correspond in many respects with those in the second column above—therefore require an equal correspondent course of treatment, but that the deposite of the stone in the kidney is an incurable disorder that admits of no remedy.   Furthermore, the reader will observe, that the whole of the article alluded to, on "Calculus, or Stone," requires his strict attention : and also bear in mind what is there said as to calculous substances which are deposited in the cæcum or blind gut, producing symptoms so much like spasmodic colic, that much care is necessary in applying the appropriate remedy in each case, lest he hastens the patient's end.

Of those symptoms the state of the pulse is the surest indication of the approach of an inflammatory attack of the bowels, or any other viscus ; and the particular part which is then suffering must be gathered from other circumstances.   If he has long suffered colic without relief, doubtless inflammation has taken place, and gangrene is likely to follow : this is the harbinger of death.   Adhesion of the gut sometimes baffles the best treatment for colic.

and soon devolves into inflammation. In either case, the remedies proper for colic must be abandoned, and others more adapted to the change of circumstances be employed instead.

Whenever the cause of inflammation of the bowels may fairly be ascribed to the quantity or quality of their contents—without adding thereto by any extraordinary exertion, its approach will be very slow, and denoted by sluggishness and the refusal of food at first. As they are mostly working cattle that are thus attacked, the evacuations are not sufficiently minded, or the attendant neglects to make mention how these have discontinued in a great degree, or changed their appearance—the dung being then hard and the urine high coloured; hereupon the pulse increases, and the outrageous symptoms described in the second column of the table of symptoms go on to a frightful degree, endangering the lives of bystanders. Even in this stage, the progress of the disease may be arrested by prompt and vigorous measures, adapted to the kind of animal that may be the subject of attack, and the circumstances under which the present alarming symptoms may have been brought on. If a heavy lumbering wagon-horse, that owes his disease to alimentary indulgence, we shall find no higher operation necessary than emptying the overcharged canal by force of arms, i. e. back-raking; but the high-couraged stagecoach horse, which falls under the exercise of the lash, and the influence of a vertical sun, has seldom aught within him of that kind to part with, and requires the introduction of some substance or liquid that shall cherish the afflicted stomach and bowels, and alleviate the burning heat that, ascending to the head, causes his delirium. Presence of mind, however, or the adroitness which much practice teaches, is frequen y wanting for the first mentioned remedy; and the means of applying the second is so seldom at hand, that the animals are too often left to their fate and are lost. But I anticipate the remedies. See also pages 62, 68.

Remedy.—From the rapid progress made by this disorder, when left to itself, and its usually disastrous termination, the duty of attending to the pulse of his animals as before insisted upon (at page 62), will strike every intelligent horse proprietor, as the very best means of guarding against the fatal consequences of inflammatory attacks. He will by this means be apprised of the earliest approach of the disease, and thus enable himself to meet it in its mildest form: he will compare this certain indication of heat—whether fever of the whole system, or inflammation of a particular part, with the state of the patient's urine, which will then be high coloured, and the dunging defective. The rectum will be dry, hard, and hot; the belly, on passing the hand over it towards the sheath, will have the same feel; the animal will shrink from the touch, his eyes appear languid, or partly shut; as the disorder proceeds they assume unusual redness, or what has been termed bloodshot.

Up to this stage of the disorder, the first remedy will be clystering and bleeding the animal freely, if he be not very aged or of spare habit, immediately after giving the following

Laxative Drench.

Powdered aloes, 2 drachms,
Subcarbonate of potass, 2 drachms,
Water gruel, 1 pint,
Castor oil, half a pint. Mix.

If delay is to be apprehended in procuring the above drench, give castor oil, one pint, or in default hereof, salad oil, two pints, whilst the drench is preparing.

In ordinary cases, a voluntary stool will be produced at or soon after bleed
ing, occasioned by relaxation of the tenesmus that constitutes the disease.   If
the dung comes forth in small quantity and small hard knobs, the anus must
be cleared by the hand, according to the directions given at a preceding page,
69.   And, when the constipation has endured for a long time, the hardened
dung will not come away at all without this manual operation of back-raking,
which must be performed the more assiduously as the difficulty may be greater and the dung harder.   Let a warm clyster be thrown up that is copious
enough to fill the emptied gut, at the least.

Clyster.

Water gruel, from four to six quarts,
Epsom salts, 4 or 5 ounces,

Inject warm, with a large syringe, or ox-bladder and long pipe : perform this
operation effectually.
    A second and third should follow, a little warmer than the first, and after
an evacuation has taken place, the next clyster may be made without salt, and
a small degree thicker than at first.   Its effect will be to remain and nourish
the parts nearly in the same manner as a poultice does an external inflamed
wound.
    Too often, however, those early indications are entirely neglected; the ani-
mal is harnessed in to his day's work, and the consequences are both dreadful
and dangerous to behold.   If he be a stage-coach horse, or destined to take his
turn at a posting-house, his sluggishness and refusal of food is usually attributed to "a little overwork;" and the much abused cordial is commonly administered; which brightens him up for the renewal of his daily task, and ac-
celerates his fate, unless rescued as by a miracle that is very seldom wrought.
In these cases, the first symptom perceptible to the driver is the horse's leaning against its next horse; but, upon being touched up, it makes fresh exertions according to its quantity of courage, until it falls down with closed eyes,
in excruciating torments, lashes out behind, and beats about on the ground,
seldom having the strength to get upon its legs again.   Bundles of straw should
be placed for the afflicted animal to roll upon, and his head pressed down with
the hand whilst the severest paroxysms expend their force.   When at length
he gets up—which may be considered a favourable sign, that proves his
strength is not wholly subdued—he may be supported into a stable.   Meantime, however, an examination of the rectum must take place, and the manual operation of emptying it be employed—if need be; that is to say, if hardened dung should be accumulated there.   At any rate, water gruel in large
quantities must be prepared, as well for administering by way of clyster as of
drench; in both, giving it now without the addition of salt, and in the latter
manner nearly cold.   By these means, the alarming symptoms will diminish
greatly; but if there is still reason to apprehend that obstruction may prevail
in the larger intestines, this must be got rid of by means of the oily laxative
prescribed at page 91, and the repetition of clysters in quantities, and administered with a vigour sufficient to reach the evil.
    Bleeding, of course, would be adopted to the amount of four, five, or six
quarts, according to the exigency of the case and the size of the animal.   If
the blood become buffed, as it is called by some, or sizy on the surface, a
second blood-letting is necessary to complete the cure.   Low, but nourishing
diet, should follow; as bran-mashes, stiff gruel, and afterwards sodden oats;
the return to hay provender being made gradually, and then of good quality.
    In very bad cases, the return to full health and vigour will be slow, and a

relapse is to be dreaded, as a fresh attack would prove much more obstinate than the first. The dung, by its quantity, consistence, regularity, and general appearance, will afford the best means of judging when the bowels are completely cleared of their offensive contents; for it not unfrequently happens that several tolerable stools may be procured by the help of medicine, and yet some lumps, replete with danger, remain behind. The pulse, that great criterion of health or disease, by dint of low living, may have regained its natural state, and so remain steadily for a tolerably long period: but watching the dung for a day or two will corroborate that main indication of health, or by its irregularity dispel an ill-founded reliance on the completeness of the cure. Yet will the administering of purgatives, or even alteratives, of aloes in particular, be found full of danger, as tending to irritate the bowels anew. The same may be said of all stimulants whatever, whether applied externally or given in the form of cordials, notwithstanding the animal may evince signs of returning pain, and these be ascertained by the corresponding symptoms of low pulse, warm legs and ears, to arise from spasmodic or flatulent colic only. For these returning pains are usually occasioned by the soft kind of regimen just recommended; to which the patient may have been subjected during this illness for the first time since it was a foal. I have known a small feed of corn or two effect relief from lowness, in the case of horses which had been long time previously used to hard food: if these be devoured voraciously, this will tend to prove 1st, that the change is desirable, and 2dly, that the next feed should consist of broken oats—or a new disease will be engendered. Adopt the tonic system, recommended generally, at page 69.

MOLTEN GREASE

Is but a variety of inflammation of the intestines when the subject of attack happens to be very fat, and little accustomed to exercise; when marked by costiveness, it may be treated as such; or, if attended by a looseness, may rather be considered as a spasmodic effort of nature to relieve itself of an unnatural load. The vulgar name given to this affection of the intestines is farther supported by the popular notion that the fat, or grease, which the individual possessed in a superlative degree, had melted (or was molten) and passed into the guts, whence it was expelled with the faeces. This, however, is physiologically impossible, notwithstanding the support such a notion has received from some revered authors; the appearance of slimy unctuous matter along with the dung, more particularly when this is much hardened, being no other than the mucous secretion described at pages 22, 23, as designed by nature to defend the surface of the intestines from the injurious action of hard substances that might be taken into the stomach. Indeed, this intention of nature in providing such a defence is demonstrable in the fact, that the harder the knobs of dung may be that the animal presseth forth, the greater is the quantity of this greasy, unctuous, or mucous secretion that is eliminated along with it, and which gives name to the disorder. Probably, the secretion of this grease may then proceed with more celerity; its access may be greater, the more it is thus required by nature to defend the alimentary passage. This supposition is drawn from the fact just stated; but, whether the well-founded conjecture be too hastily hazarded, is for the more minute inquirer to conclude upon, or investigate farther, as may seem good to him.

At any rate, the doctrine of effusion, or the passing of those secretions, whether mucous or aqueous, from one part of the system to another, as nature or accident may require the supply, is tolerably evident from another circumstance that is often recurring in cases of molten grease. [The subject is more fully treated of at the page just referred to.] The perspiration of the two se cretions in succession, here referred to, is pretty well recognised, and is easily proveable, in the manner there set down ; the unctuous, mucous, or greasy secretion (call it which we like) of the external surfaces following that of the more liquid, or watery kind, after any great exertion. Horses that contract molten grease are ever those which have been highly fed, without exercise sufficient to excite visible perspiration thereby ; and the feverish heat of the body occasioned by high living and indolence, in time exhausts the whole supply of the aqueous secretion. So much is this the case, that the animal's discharge of urine becomes less and less as its seclusion is continued, until the decided access of fever takes place, and we notice its colour is higher and higher as its quantity decreases. [Look again at sec. 55, page 52.] As before explained, the secretion of mucous matter takes place within the guts, &c., or that surface which is next to the food ; on the other side, and every other part of the animal system, the watery secretion destined to lubricate the parts, to keep them supple and to prevent adhesion, takes place. On that side (which is popularly considered the outer surface!) good quantities of fat accrue, all along the whole length of the intestines, which is usually scraped from slain beasts, and preserved as tallow. From this source is derived that access of grease, which, as I have said, is greatest as the inflammatory symptoms may be higher. When this has long been the case, and stools are at length procured, a long thin wormlike portion of this fat comes away with the dung ; which would be of itself a sufficiently alarming appearance, though wanting animation, but for the well known, but inexplicable, doctrine of effusion, or communication through the gut : this appearance, then, of a long tenacious fatty portion of thin membrane, which usually accompanies molten grease, should be considered as little more than denoting the crisis of the disorder.

Let the system be reduced according as the state of the pulse may dictate —for which consult again page 62, as to bleeding, and page 68, 69, as to treating him for "costiveness" simply. If heat and irritation be perceivable to the touch and sight about the anus, without high pulse, the first symptom may be reduced by administering

The Sedative Clyster.

Camphor, 4 drachms,
Spirits of wine, 3 or 4 drops,
to promote the solution, and add
Sweet oil, 2 ounces.
Mix well, and then add thin warm water gruel, 2 or 3 quarts.

As before intimated, molten grease is rather an effect than a cause of disease, and partakes of colic in one of its forms and of inflammation in the other ; the symptoms that enable us to distinguish when the one or the other prevails being precisely those set down at page 90. Allowing somewhat for the feverish symptoms that always prevail with such fat and bloated animals as are subject to this disorder, the practitioner can not commit himself to the guidance of a better test than that just referred to, nor more safely adopt a treatment that is more likely to reinstate his patient in health. For the treatment which is proper in case of spasmodic colic affecting fat animals, the reader is referred to the next head of information.

THE COLIC, GRIPES, or FRET.

This disorder has been frequently referred to, under the preceding head of Inflammation of the Intestines, to which it bears great affinity in some of its points—as already stated, the cause of both being nearly the same in most cases, and long continued colic always ending in inflammation, if not effectually checked in time. Much of the difference that exists between the two kinds of attack depends on the previous state of the animal attacked : if it be a high fed and hard-worked animal whose digestive organs receive a sudden check, he contracts inflammation in the first instance; but one that is lower kept, and therefore not so irritable in any part of its system, is soon troubled with spasmodic affection of the intestines, which receives the name of gripes, or fret in different counties, as it does that of flatulent colic in most of the books that treat of animal medicine. Colic, however, is the general name given by most stable people to every pain of the inside (of man and horse) that occasions writhing, or other demonstrations of that pain, which few can discriminate in their own persons ; much less in their horses. To this undiscriminating manner of naming disorders that require such very different treatment at our hands, is to be attributed the loss of many lives annually. Into this anomalous manner of treating those disorders it is painful to notice one of the most scientific veterinary writers of our time has fallen. We do not find in Mr. Richard Lawrence's "Complete Farrier" any reference whatever to inflammation of the intestines : though under the head of "Colic or Gripes," he proceeds to describe the symptoms of inflammation in such a manner as might mislead ignorant or half-taught persons to treat both alike, and thus destroy their horses.

A violent cold, or a slight one, will also determine the disorder one way or the other, when the individual's system may be of no decisive character at the time of contracting it. That a low state of the animal system is favourable to engendering spasms of the intestines, is inferred from the circumstance, that subsequently to a horse afflicted with inflammation undergoing the copious evacuations recommended in the preceding pages for the cure of that disorder, he is frequently visited with spasmodic affections that require sedatives and tonics to restore the patient to complete health.

Causes. Next to drinking cold water, and catching cold by exposure to air or water whilst heated, the eating of bad, ill-got, or rank hay, is a prolific source of spasmodic cholic. If it lie in the intestines chilly and comfortless, and thus predispose the animal to acquire cold, the cause of epidemical colic is plainly attributable to such bad hay; for it then prevails usually over certain districts, and mostly among country cattle. Pushing a horse in his work when large lumps of undigested matters distend parts of the gut, will bring on spasms, torpor, and inflammation in succession. Horses that gormandize much, being worked hard, and the stomach becoming empty, occasions the fret, and inordinate action of the intestines expels the muceus secretion that is designed for their defence. This constitutes molten grease, of which I have treated largely just above, and am decidedly of opinion that the expulsion of offensive matters in all cases where the animal evinceth but small sensations of pain, is but an effort of nature to relieve itself, and ought rather to be assisted than abated by hot or "cordial medicines."

Strong astringent purgatives, oft repeated, or neglect during the operation, are frequently succeeded by flatulent colic, that soon becomes inflammatory if the internal commotion be not judiciously arrested by sedatives. Cordial balls and drenches, as they impart a short-lived vigour, so when their stimulating effects die away, they leave behind a debility that is more excessive as those factitious effects have been most intense ; in this respect, the cause and its consequences assimilate closely with those which succeed the disease of inflammation, and the debility which follows the cure thereof, with spasmodic colic. Diapente, and other provocatives, that are given to stallions in the season, leave behind them the same species of debilitating effects after cover-ng, and would devolve into colic first, and inflammation afterwards, but that those horses' evacuations are well looked after, and the system of stimulants is kept up by repetition. This treatment, however, can not always succeed, so we frequently find that stallions die suddenly of inflammation in the intestines, in the spermatic cords, or other parts of generation. Cases of death, *in actus coitu*, from the same causes, are upon record. I mentioned this before, at page 18.

All horses that have been pampered in the above manner, or by being kept in close stables, or having their water chilled, when they come to be subjected to common usage, are most likely to suffer by colic in its worst forms. Horses that are made up for sale by dealers and cunning breeders, in order to give their coats a sleek appearance, upon passing into the hands of new owners, commonly undergo attacks, more or less acute, of spasmodic colic, if they do not at once fall ill of inflammation of bowels, kidneys, or bladder. With animals so circumstanced, mere flatulency or looseness may be considered a favourable termination of the making-up system before alluded to.

Symptoms. These, as contra-distinguished from those which denote inflammation of the intestines, will be found in the table of comparative symptoms at page 90. In addition thereto, other symptoms, that mark the degree of spasmodic attack, require equal discrimination, seeing that treatment which may be highly proper in the more virulent attacks, would be injurious if employed upon every slight occasion. Neither is it every horse which shows signs of pain in the inside that has the colic, even though the symptoms set down in the second column of the "table," at page 90, do not appear; for, he may be afflicted with pain in the kidneys, or inflammation of the bladder, which the attendant should ascertain before giving the stimulants that may be very proper in most stages of colic, but would accelerate the diseases incident to those "urinary organs." The careful reader should therefore turn to the subjects "Kidneys," and "Bladder," before he proceeds to treat the animal simply for colic pains.

In its mildest state, flatulent colic first appears in the form of violent purging, which is in fact no positive disease, as before observed, but an effort of nature to rid itself of a collection of offensive matter, either indigestible, cold, or irritating. Of what precise kind this may be at any time is ascertainable upon the view, and requires only to be assisted in coming off, provided but little pain is evinced by the animal. If he be a crib-biter, pieces of extraneous matter are usually found among the dung, as bits of wall, of wood, litter, &c.; if an aged horse, or one that has been kept on bad hay, his food comes off undigested ; if a very fat horse, the mucous secretion comes away as described under "molten grease," just above—and all these require at most some of the milder purgatives that are least likely to irritate the bowels.

Whenever the ears become cold, after gripes have continued some hours, it is a certain indication that inflammation has taken place of some one or more organs, mostly of the intestines. This is sometimes discovered when too late, to attend a rupture of the distended bowels through the peritonæum (Vide Book I. page 46), when the protruded gut mortifies (as is found after death) in consequence of strangulation. After this, the pain seems to subside, and the animal dies quietly. The ruin that has taken place is only told on dissection. Yet do most ignorant persons pronounce horses still alive to have a "twist in the guts," and stranger still, they prescribe a remedy for it, although it is incurable. The ears act also as a good barometer, when inflammation of the kidneys may be apprehended, or inflammation of the bladder is more than suspected, on account of the difficulty evinced by the patient in passing its urine. If the water come off high-coloured, it is a sure sign of inflammation, which is further corroborated by cold ears; if of its natural colour, the ears will be warm, and the difficulty in staling is occasioned by the hard distended gut pressing upon the ureters and neck of the bladder: procuring a good stool or two, or a clyster, then restores the functions of the bladder.

The earliest symptom observable in his manner, is when the horse looks round at his flanks occasionally, whisking his tail at intervals; he looks at the attendant, if there be any breed in him, seeming to implore help. He stamps with his hind feet alternately on the ground, sometimes striking at his belly. As the pain increases, these symptoms are oftener repeated, and with more vehemence; he gathers his legs under him, as if preparing to lie down; which he at length effects, rolling about in the stall and getting up again repeatedly. It may here be remarked, that this rolling on the back is well calculated for affording temporary ease to the bowels; but should inflammation have already attacked these, or at the kidneys, this rolling on his back would but increase the pain of the animal, and his jumping up instantly upon his legs, as if the spur or whip had been applied, goes to prove the existence of inflammation at one or the other viscus.

Cure. Too much care can not be exercised in ascertaining the precise nature and amount of the disease; for, in error in this respect resides extreme danger of life, which is too often sacrificed to precipitancy, to ignorance, and presumption. As soon as a horse is pronounced "ill of the colic," the attendants, without investigation, proceed to give "something to do him good;" which is ever of the stimulating class of domestic remedies. Warm ale, with ginger, peppermint water, gin and water, whiskey and pepper, are the common popular remedies usually applied in this case; and, provided the disorder be really flatulent colic, relief from the pain must follow the exhibition of either one or the other. Frequently, however, it happens, that the doing good is carried too far, and inflammation is thus superinduced, if it do not already prevail. By such persons every internal pain is pronounced "the colic;" and they all conclude that what has removed it once will remove it again, without being certain that it is the same disorder—as they do, that whatever is good in small quantities must needs be more so in larger ones. But I have already observed, that the removal of umbilical affections, whether flatulent or inflammatory, by rough, harsh, or protracted means, scarcely ever fails to produce the other concomitant disease, and the inflammatory symptoms no sooner subside, than the jaded vessels contract spasmodic affections, as do also the continuance of flatulency, and some of the means of curing it superinduce inflammatory symptoms.

In whatever shape the horse is attacked with those disorders, the first and most obvious duty is the employment of clysters, to be repeated at short intervals, with this single variation; viz. in cases of relaxation, where the animal is already purged, the clyster is to consist of simple water-gruel only; but when the patient's bowels are overloaded with hardened dung, the addition of salts, as prescribed at page 92, will be found most effectual. In the absence of Epsom salts (for no time must be lost), a handful of common culinary salt may be employed, in the quantity of four or five ounces. Back-raking, too, should be assiduously applied, when the body is in this state, as recommended in the case of inflammation at p. 91, with the laxative drench prescribed at the same page, or the simple salad or castor oil in default thereof.

In ordinary cases, when the attack is not of the most violent kind of either description of colic, that is to say, when neither purging nor constipation prevail extremely, let the following be given.

10 *

Colic Drench.—No. 1.

Epsom Salts, 4 or 5 ounces,
Castile soap, sliced, 2 ounces.
Dissolve these in a pint of warm ale, and add
Oil of juniper, 2 drachms,
Venice turpentine, 2 ounces.

Mix well together, and give it warm ; repeating the same in four or five hours, and if the symptoms do not visibly abate, repeat once more. Tincture of opium is sometimes substituted for the turpentine to the amount of 4 drachms; but the drench is thereby rendered exceedingly nauseous, and should be given deliberately. Opium is, moreover, least proper when a tendency to costiveness is discovered to exist.

Colic Drench.—No. 2.

Tincture of opium, 2 drachms,
Oil of juniper, 2 drachms,
Spirit of nitrous ether, 1 ounce,
Tincture of benzoin, 4 drachms,
Aromatic spirit of ammonia, 3 drachms.

Mix together, and preserve the same in a bottle, and give in a pint of warm peppermint water. Repeat in three or four hours.

When the case is not very alarming, a neater manner of giving opium, in the form of a ball, is recommended :—

Sedative Ball.

Asafœtida, 4 drachms,
Opium, 4 drachms.

Make into four balls with liquorice powder and syrup, and give one every two hours. The balls may be given along with the oily laxative at page 91; immediately preceding it, or before the laxative has operated. These balls are very serviceable to travellers on their journeys, and may be given to horses that are liable to contract spasmodic colic, which is the case with heavy, fleshy draught cattle, with post horses and the like.

Colic is not often fatal, unless it terminate in inflammation; whilst it should be kept in mind, that colic always ends in inflammation if not removed in time. A day, or at most two, may pass away without danger and without relief, in ordinary attacks of spasmodic colic; and where a looseness takes place, a short time longer of neglectful carelessness might not terminate the life of the animal; but, when inflammation commences, a shaking or undulation of the tail is observable, with evident shivering of the whole frame. The danger is then great; especially when each fit of shivering is not succeeded by perspiration.

If the costiveness is not well removed when those symptoms, with cold ears and legs, come on, let the belly be fomented with warm water by means of woolen cloths steeped therein. A horse rug may be used to advantage in this way by two men, one standing on each side the horse and fomenting the belly by bringing it nearly together across the back and supplying with warm water. After half an hour's application, or more, let the coat be well rubbed with dry cloths, and the animal wrapped in body clothing

The clystering, and other remedies recommended in cases of inflammation, should then be employed with assiduity. Lastly, employ the tonic system recommended, generally, in all inflammatory cases, at page 69.

DISEASES OF THE LIVER.

1. INFLAMMATION. 2. THE YELLOWS, OR JAUNDICE.

WHEN we consider the vast active functions the liver has to perform, in cleansing the blood which takes its passage through it, and the secretion of bile, that becomes more obnoxious as this organ is more diseased, we ought to feel surprise that so large an animal as the horse has so few ailments springing out of that source, rather than lament the frequent existence of this one. For, the two names set down at the head of this article, agreeable to the general practice, have only one origin, viz. inflammation; but differing as to the amount of heat, and situation of the evil, which is scarcely distinguishable until after death. The symptoms of both are the same, and the first attack ever becomes the most lasting, if the remedies be delayed, or wholly neglected.

Having been led to enter somewhat at large into the causes and remedies for certain affections of the liver, while describing its structure and functions in the first book, p. 49, I shall find less occasion to add much more at this place. The reader will of course turn to that page.

Cause of inflamed liver.—Inflammation of the liver does not very often take place as a primary affection, but more frequently participates in the disease of some of the adjacent organs, as the stomach, bowels, &c. and according to the acuteness of the inflammation, an increased or diminished secretion of bile is the immediate consequence. The blood, in passing through the liver, acquires a portion of this extra heat, which reproduceth more at its next passage through it, more at the next, and so on, until the inflammation of the whole liver is completely effected. Increase of the bile or gall proceeds in the same ratio, until the gall duct, that communicates with the small gut, is closed by the uncommon heat of the inflammation, or by the thickening of the gall, or by both operations united, no matter which. At any rate, the bile which ought to be conveyed away by stool, is returned into the system, and occasions yellow skin—whence the vulgar name. When this occurs, I apprehend the inflammation lessens, but the communication with the bowels does not always return to a healthy state, though I believe it to be partially the case. Indeed constipation in the first instance often obstructs the passage of bile into the bowels, and thus increases the evil. Over-feeding has the same effect, and both produce slight temporary yellowness, which goes off upon the removal of the cause; generally followed by diarrhœa. The feverish symptoms also which accompany the commencement, also pass off, leaving a low, irregular pulse, until the bowels resume their wonted course, either naturally, or by the aid of medicine.

Symptoms of inflammation before yellowness comes on.—As this last and surest indication of diseased liver only appears when the evil is a confirmed one, and is extremely difficult of cure, particularly in old animals, we should assiduously set about ascertaining its commencement, so that the remedy may be promptly employed, and a further procrastinated mischief be timely prevented. And the more so, seeing that what constitutes a remedy in its earliest stages is no longer so after a time has been spent in delay.

Whenever inflammation, or extraordinary action of the kidneys, or of the diaphragm, has lasted some time, in ever so small a degree, in that degree will heat or inflammation attend the liver. It enlarges upon the accession of this heat, visibly so when this has continued a while, but may be previously ascer tained by the feel. As will be seen,* the liver extends much farther back than the last rib, and a little beyond the false one. Here a considerable protuberance appears when the liver is enlarged, and disease may be ascertained that is attended by the presence of pain only. Old horses, which have been well bred, retain chronic affections of the liver to a very great age; and this is frequently the main disease under which they suffer for many of the last years of their lives: great numbers of such animals die with a liver of so small a size, that nought but its situation could assure us it ever had any functions to perform. Horses so visited with a trifling undetected affection of the liver lose their courage, and gradually sink into lethargy the longer it lasts: we often hear such animals accused of being "used up, done for, or 'tis all up with him," and yet driven about to the last moment of a painful existence.

When the attack is rapid, and acute inflammation, arising from the causes just set down (page 99), the pulse is the sure indication of the ruin that is going on, by its irregularity, quickness, and uncertain vibration. See page 62. One lobe only suffers in this case, and then the animal turns its head round sharply to that side from time to time. Constipation always accompanies acute inflammation of the liver.

Remedy.—Acute inflammation, which comes on with dangerous strides, when the subject of attack is of vigorous habits, must be met by a bleeding proportioned to the state of its pulse, and that without delay. For, it speedily communicates to the intestines, and death ensues; or, being suffered to expend its virulence (provided the animal possesses strength sufficient) by stool, the bleeding will then be unnecessary; or being persisted in, will confirm the slighter affection just spoken of probably to the end of his days. A purgative ball should accompany the bleeding, as in all other cases is prescribed generally at page 63; but, if the animal produce a stool voluntarily, the disorder has taken a turn, and neither the operation nor the physic is required.

After bleeding, let the sides be rubbed with the blistering ointment (vide page 76), and apply a rowel to the chest. These latter, however, are doubtfully eligible, though always employed by the regular collegians. The patient will require the same treatment, as to diet and regimen, as for inflammation of the organs of respiration and general fever, before treated of at page 60, in the course of which his pulse and fæces should be watched, and a relapse provided against. Calomel is that medicament which more immediately acts upon the liver, and unless the horse scours, should be administered in the form of

Alterative Balls.—No. 1.

> Aloes, 9 drachms,
> Calomel, 1 drachm,
> Hard soap, half an ounce.

Mix with mucilage sufficient, and divide into three balls; to be given on three successive nights, unless a thin stool comes off with the second ball. But in case of scouring, give

* In the plate of skeleton, at the parallel lines H, 30, is placed the kidney of the near side; whilst the off-side kidney in the same subject would be intersected by the line 29. With this latter, the right lobe of the liver lies in contact, and when an enlargement of it takes place, it may here be seen and felt; when the access of inflammation and tension render it painful only, the doctor should press the points of his fingers (of the left hand) gently behind the last or false rth several times, whereby he will ascertain whether any and what degree of pain the patient endures. If seated high up on the liver, he will not, of course, flinch at the first slight touch

* To prevent error, I would here mention, that in the picture of a skeleton now referred to, it is the left lobe of the liver that is there represented, and this was reduced in size, in order to show a clear profile of the stomach.

No. 2.

Oil of turpentine, ⎫
Hard Soap, ⎬ of each 1 ounce.
Ginger, powdered, ⎭

Mix with flour and mucilage to form three **balls; and give one on each of three** successive nights.

THE JAUNDICE, OR YELLOWS.

Cause.—Inflammation of the liver, or any other obstruction of this organ, which, preventing the escape of the bile into the duodœnum, or smallest gut, through the gall duct, by reason of this duct being inflamed, or choked up with the thickened bile, whereby it is sent again into circulation, and thus pervades the whole system. When the inflammation is very great, the disorder quickly carries off the patient; the inference therefore is, that poor animals alone acquire the yellowness which gives name to this disorder, though it must be allowed that the same effect may be produced by over-feeding and constipation, by swallowing hard substances, or otherwise offending the said gut, or the pylorus orifice of the stomach, as described at pages 44, 45. Its situation may also be seen depicted in the plate of a skeleton at the intersection of K 26. At that place I did not choose to speak of negatives, and therefore omitted to notice the fact, that the bile or gall secreted in the liver of this animal proceeds at once, as soon as it is formed, into the gut, without being detained in a sac, or gall bladder, as is the case with all other animals, except deer ; so that, upon any revulsion or hindrance to its free entry to the bowels, the gall must at once return to the numerous cavities that pervade the whole liver, and its re-absorption by the blood is no longer problematical.

Symptoms.—A dusky yellowness of the eyes, bars of the mouth, and tongue, The dung scanty and pale, generally hard, and covered with slime; but in some few cases the horse scours; that is, when slight inflammation of the bowels also attacks an ill-conditioned horse. The pulse is that of low fever, and the same kind of drooping inactivity, with loss of appetite, noticed under that head at page 64 ; differing from it only in respect to the seat of disorder, the low fever being general, or of the whole system, jaundice of the circulation only. Sometimes, however, yellowness comes on without the other **symptoms,** after an inflammatory fever; an occurrence that can not fail to be foreknown. Genuine jaundice may **further be** discriminated by the yellow lips, yellow saliva, and dark urine. **From this** latter appearance we may draw these curious inferences—viz. that the colouring of the bile which has ceased to **impart** its property to the dung, having gone with the blood to the kidneys, **there** leaves its darkest or more earthly particles—the lighter or brighter ascending **to** the heart, and passing through the vascular system, there imparts its yellowness. By this providency of nature we see how it is that malevolent particles in the blood are cleansed at the kidneys, and pass off by urine. Thus it is that grease and other tumours are cured by judiciously stimulating the kidneys. The urine voided, as above described, which is ever done with evident pain and difficulty, leaves on the ground an appearance of blood.

Cure.—Young horses and fat ones, are easily cured: they have indulged too freely in good living, on hard meat, and require no more treatment than a good physicking. Give the purgative ball (page 63), or the alterative ball, No. 1, prescribed in page 100 Give bran mashes, green food, and succulents, according to the season. Bleeding is seldom necessary, or proper, which the state of the pulse will show.

The LIVER is also frequently affected with tumours on its fine surface, as well as with ulcers or schirrus, which are all the effects of an evil state of the blood, of over action, and probably of accidents from external injuries, communicated by the kidneys.

We can easily conceive that the thin parts of this large viscus may be diseased, and even inflamed, without causing derangement of the biliary function, further than increasing its action, and by thinning the blood over much, it obtains more bile. The animal then waxeth thin, though devouring his food as usual for a while; and we may ascertain when this evil has begun by the state of his dung, principally as to colour, which will then be of a much deeper hue. As pale dung is a symptom of suppressed bile, so is deep colour an indication of a superabundance, that is caused by over action, which is itself occasioned by the heat of the liver, from some cause or other. One of these may be "inflammation of the kidneys," or it may be occasioned by ulcer, and we set about ascertaining which, according to the instructions set down at page 100: and in the latter case give the alterative balls, the same as for inflamed liver, at page 100, 101 according to the circumstances there discriminated; but it never happens that a scouring is of a dark colour, and No. 2 would in this case seldom be required, a strong purgative never, though the bowels should be kept moderately open. When there is reason to apprehend that the adhesion of the ulcer to the intestines has taken place, as described at page 50, the animal should not be worked hard, though moderate exercise is desirable, and so is change of physic, as in all cases that require alteratives. The following balls may take place of the preceding, particularly when the coat is staring.

Alterative Balls.

Emetic tartar, 3 drachms,
Aloes, 9 drachms,
Hard soap, 1 ounce,
Ginger, 1 scruple.

Mix, and divide into three balls, one to be given on successive nights, unless two have operated.

INFLAMMATION OF THE KIDNEYS.

This being one of those diseases which bears resemblance to another, and as the mistaking and treating the one for the other generally proves fatal, reference should be had to what is said under the head "Inflammation of the neck of the bladder." Such a mistake of the disorder in the present instance is very likely to be made by the common observer, inasmuch as the kidneys, as soon as they become inflamed, secrete much more urine than in a state of health, and any one noticing this, and subsequently its defalcation, as the disease goes on, may easily imagine the bladder itself is affected at the neck. It is worthy of remark, that mares are more liable to affections of the kidneys than horses, particularly brood mares: while, on the contrary, they are less liable than the male to inflammation of the neck of the bladder, in consequence of its shortness. its straightness, too, affords easy proof of the real seat of the disorder,— that essential prelude to effecting a cure. See page 53.

Causes.—Too constant use of the diuretic powders and balls, commonly brings on inflammation of the kidneys, by the irritation and over-action of the glands which are thereby occasioned. When one kidney only is affected though in a mild degree, if suffered to continue, it soon communicates to the other, and sometimes proceeds with such rapid strides as to affect the intes tines, when mortification and death ensue; but we have no means of ascer taining when this last incurable mischief has taken place until after the animal is dead—nor would the knowledge be made available for any present purpose; though finely instructive as to future cases; then it is the kidneys present an enlarged and rotten appearance and feel, their texture yielding to the slight est impression of the finger-nails, which shows in what degree and how long they have been affected.

A hard blow across the loins will injure the kidney on the side so struck, and, as is said before, soon affect the other also. Sudden transition from an open airy situation to a stable that is close and hot; violent riding or driving, or an ill-cured affection of the bowels, whether inflammatory or spasmodic, will affect the kidneys in more or less degree. Those causes all together com bine to affect these parts more frequently than is generally supposed, the rea son for which misconceit is nevertheless most apparent to me: it is owing to the neglect of a'l the milder symptoms; some persons imagining that unless bloody urine be produced, the defective staling is caused by something less re mote than the kidneys, though in all obstructions of the liver, as we have seen above (page 101), the quantity of blood these send to the kidneys leaves some of its colouring property to the water. This class of unreflecting people gene rally fix upon the bladder as the seat of disorders that so affect the quantity of water. They almost invariably give stimulating medicines, that do but in crease the disorder and confirm the ruin it is their duty to prevent.

Symptoms.—The most evident of these has been just now alluded to, and was formerly treated as a distinct disease, under the coarse title of "BLOODY URINE:" it is, however, considered as happening more frequently to horned cattle than horses, and to the female rather than the male.

When this symptom appears, it is accompanied by a corresponding symp tom, viz. great tension and soreness of the part; which may be ascertained by passing your hand along the small of the back, over the kidneys, when the animal shrinks from the touch. No doubt can then exist that this bloody urine indicates genuine inflammation of the kidneys; and of course that we should treat it as such, and nothing else—nor by any other name. If the pain and tension can not thus be ascertained, then "bloody urine" is caused by obstruc tion in the liver. Another symptom that may be relied upon is a stiffness of the hind leg on that side which may be attacked first; afterwards, when both kidneys are affected, the animal becomes stiff of both legs. This symptom does not occur in "inflammation of the bladder," and is a good distinctive mark to go by, when we may be labouring under doubt in some other point of resemblance between the two diseases. In all stages of this disorder, the horse stands as if he wanted to stale; straddling, and making the most exer tion when he voids the least urine (then generally bloody), which shows the destructive tendency of these efforts on the gland itself. The consequences are, that the kidneys waste away, and the disease communicates to the blad der, until the final ruin—mortification, ensues. The practitioner, in this case, will not fail to look at what I have thought proper to say respecting "stone and other calculus," a few pages farther onward.

"Suppression of urine" is also a sure indication of the genuine inflamma tion of the kidneys; that is to say, the capacity of secreting it is nearly ex tinct, or it is performed with exceeding great difficulty, pain, and danger. Whereas, in affections of the bladder, the secretory function is not lost by the kidneys (or suppressed); but, when the urine has been sent into the bladder this latter has not the power to expel its contents. How this happens, see page 53, &c.

But the most prolific source of diseased affections of the kidneys, and the least perceptible of any are ill-cured pains of the intestines and of the liver. These leave behind them certain morbid effects that are not immediately felt nor easily discoverable, but nevertheless work their ruin imperceptibly; for, as previously observed, when the kidneys lose their function of secreting urine, they enlarge, and after death scarcely bear the pressure of a finger point.

Cure.—Seeing that strong diuretics are reckoned with truth, among the causes of diseased kidneys, no man in his senses would think of administering any such, after he has ascertained that this organ is disordered in any way whatever. Such, however, is too often the practice of unskilful persons, who, after noticing the defective quantity of urine produced, think of restoring the animals capacity for producing more by medicines that stimulate the parts, which already labour under a disease of too much stimulation. " As in all other cases of inflammation or fever [how often have I not repeated the same words!] when the pulse is high, let the animal be bled according to the amount of attack." See general observations at the head of this chapter, pages 59 to 63. Give warm clysters frequently as there prescribed; and with a similar view give him a loose stall, if the paroxysms are so acute as to cause him to lie down and get up again. Immediately after bleeding, give castor oil 18 ounces, provided the animal has not dunged during the last twenty-four hours, as commonly happens; less may suffice in general; but a horse that has been much addicted to diuretics, though his bowels may be in a tolerable state, will not suffer aught from a small proportion of aloes:

Mild Purgative Ball.

Aloes, 4 drachms,
Castile soap, 4 drachms,
Mix, with mucilage enough for one ball.

Should the symptoms abate nothing in consequence of this treatment, the bleeding must be repeated and the purgative too. Rub over his loins with a stimulant

Embrocation.

Spirits of wine, 2 ounces,
Soap, 2 ounces,
Camphor, 1 ounce.

Mix and apply it with the palm of the hand to the loins; cover the animal up well, and be careful how it is subsequently exposed to the air. The mustard embrocation is equally efficacious: being rubbed on soft sheep-skin, cover the loins therewith. Give the cooling decoction in large quantities, as at page 70; and if the animal is disposed to eat the sodden seeds, it may be permitted to indulge: they are little nutritious when the saccharine has been drawn out by the hot water.

The food should consist of bran mashes, green food, and the cooling regimen already recommended in all cases of inflammation at pages 61, 69, to which the reader is respectfully referred for some general directions for his rule and conduct, equally applicable in all such cases.

DISEASES OF THE BLADDER.

These are really much fewer than are commonly ascribed to it, the bladder being but the vehicle or outlet for several evils that take their rise higher up;

and among these I have already denounced the alarming appearance of "bloody urine" as a disorder of the kidneys and liver, page 103. Neither is the "suppression of urine," nor its obverse "diabetes," ascribable to the bladder, but to the kidneys; for if these secrete none or imperfectly, little or none can be sent into or escape out of the bladder; but retention of urine may be a fault of the bladder, or collapsion of its neck; and the means of procuring its escape was before recommended at page 53, &c.

INFLAMMATION OF THE BLADDER, and consequent "incontinence of urine," are the same disorder; the latter being the irritating effects of the inflammation only, and this I shall consider separately, referring those other disorders that are commonly ascribed to the bladder, to consideration under the head of "Diseases of the urinary organs, generally."

Cause. Heat and inflammation of the kidneys communicate this effect to the ureters and bladder. It may be inflamed also by the irritation of stones or gravel concreted within it; or the excessive labour imposed upon it by the great access of diabetes, after these have ceased.

Symptoms. Frequent desire to stale, the bladder contracting upon every drop of water, almost, that finds its way into it. A quick, sharp pulse, and small, accompanies, if it has not preceded inflammation of the bladder; yet bleeding would not be proper in this case, as it is the poorness of the blood which brought on the diabetes that caused the inflammation. When, however, this symptom has not preceded inflammation, the pulse will be more full, and bleeding to an amount proportioned to the state of the pulse (see page 62), would then be necessary.

Remedy.—A slightly purgative ball should of course follow the bleeding, but employ neither in the extreme. Give the cooling decoction recommended in general fever, at page 70; administer clysters of the same, two or three times in the day. Should great heat of the bladder continue, notwithstanding these remedies, give the fever powder, No 2, at page 65, and afterwards No. 2, made into a ball, daily.

DISEASES OF THE URINARY ORGANS, GENERALLY.

Besides the foregoing main diseases of the kidneys and bladder, there are several other conjoint affections of the same organs, or parts dependent thereon, which require notice, and demand attention, while we examine the distinctions that ought to be drawn between the one set and the other. Mistakes as to the actual seat of disorders are more dangerous than the unskilful administering of medicines, for these might do good by accident, the former never can be applied properly: the better the "receipt" may be, the worse for the horse. Few of these lesser diseases are original, but arise from some defect or ill-cured disorder in the other parts of the animal's system. They may be considered under the heads—1. Diabetes, or excessive discharge of urine. 2. Bloody urine. 3. Calculi, or stone. 4. Strangury. 5. Suppression of urine. I am aware that the ingenuity of some doctors has subdivided these, and added to the number of diseases incident to the kidneys, ureters, and bladder; but, omitting those which attach to the organs of generation in breeding animals, and also those seated higher up—the communication of acute pains to the more vital parts, by means of the emulgent and vena cava, to the heart itself. This last, however, is so immediately the precursor of dissolution, that no other benefit can arise from the doctor's skill in this respect, than bidding him to cease his efforts, to forbear to torture the expiring patient, and to preserve his medicines for a less forlorn purpose: the pulse, by its extreme languor, tells when hope itself must resign its place.

11

DIABETES, OR EXCESSIVE STALING.

The cause of animals discharging great quantities of urine can not in every case be traced to its right source; but one thing always happens, namely, irritability of the bladder, by reason of the absence of the mucous secretion that is to protect it against the saline effects of the urine: see page 53. A defect in the mucous secretion of the whole system succeeds the disorder termed molten grease, and the irritation just spoken of soon communicates to the kidneys, which are thus compelled to secrete urine to the utmost extent of their power, and to send it forward to the bladder. To an impoverished state of the blood, arising mostly from the use of strong medicines—for the cure of inflammatory diseases leaves more of lymph than of serum in the vital fluid, with an accelerated tendency to increase that baleful difference—may be ascribed the chief cause of this obstinate disease. Bad dry provender, with ill-usage, and the denial of green food, in season, have a similarly evil effect on the blood.

Symptoms.—Of course, the most obvious is the discharge whence the disorder derives its name, being frequent and in very large quantities. At first, the water is colourless, but occasionally comes off like puddle. Constant craving after water, a staring coat, evident weakness, and weak quickened pulse, succeed each other, and increase as the disorder is suffered to proceed unchecked.

Cure.—Change in the animal's diet, whatever they may have been. If the horse be labouring under the remains of some ill-cured disorder, attend to that first, and by removing it, the excessive staling, which in that case is but an effect thereof, will also cease. Give vetches, grass, sodden oats, water in small quantities and often. If the pulse be higher than ordinary, give the fever powders, page 65; and when the number of strokes per minute is reduced, let the oats be given dry, and resort to bracing medicines. In slight attacks, as well as for the less robust animals, the various preparations of bark will be found sufficiently tonic.

Tonic Ball.—No. 1.

Cascarilla, .   }
Gentian root,   } of each 2 drachms.
Powdered caraways, half drachm,

with treacle sufficient to form the ball for one dose. Give morning and evening.

In the more formidable cases, where greater strength or more tedious symptoms require to be combated, give the

Tonic Ball.—No. 2.

Venice turpentine, 1 scruple,
Sulphate of copper, }
Ginger,   } of each 1 drachm.

Mix, with liquorice powder sufficient for one dose, and give twice a day for two or three days. After this, a return to the use of No. 1, would be desirable, until the disorder is subdued. Should costiveness ensue, give a clyster, which will also relieve the irritation of the parts; castor oil, one pint. must also be administered, if the costiveness appear obstinate.

Above all things, the horse-owner should avoid the use of such excessively ignorant prescriptions as are recommended, in this disorder most particularly, by every village quack: they are mostly the horses of hard-working people that are attacked with this disorder, and those people more than any other lie open to this kind of advice.

Incontinence of urine is of the same nature as the last-mentioned, only differing in the discharge being involuntary, and the amount, or quantity produced. The disposition to stale frequently, or the urine coming away with scarcely an effort, proves that great irritability of the bladder is the proximate cause, and we may infer that the quantity would be greater if the animal had more in his system. For this feature of the diabetes attacks only old worn up horses, in whom the quantity of blood is small, and its course slow. Diabetes of the younger animals sometimes terminates in this mode of producing water by driblets and in small quantities, but to which the moderns have given a distinct term, though both are the same disease ; a small degree of inflammation prevails when the animal is greatly affected with incontinence. See page 106.

The treatment should be the same, nearly, as directed at page 106. Give occasionally the tonic ball, No. 2, page 70, for two or three days. A run at grass for a week, and generous feeding afterwards, generally complete the cure, no other obstacle intervening.

BLOODY URINE,

I have already said, is but one feature among many other symptoms of inflamed kidneys ; and the only reason why I deem it worthy of separate notice is, that real "inflammation of the kidneys" is not always present when bloody urine appears, especially when no other symptom thereof accompanies this single demonstration of disorder. Its causes may be traced to excessive labour, as drawing in a cart or wagon, whilst a slight cold of the kidneys may obstruct their proper action : the office of separating the blood from the water is in this event performed with much difficulty, and of course imperfectly ; and small portions of the former, instead of ascending towards the heart, descend to the bladder with the urine, while the animal is straining every nerve and vein.

Rest and a cooling diet are the best remedies for this apparent affliction. Should tenderness of the kidneys be evinced upon the touch, or other symptoms of augmented pain appear, give the tonic ball, No. 2, page 70, occasionally employing No. 1 instead : the alteration will be found beneficial. If these symptoms increase (which I should not apprehend), then of course the attack must be met with strong appropriate remedies. But I have never known one case of bloody urine out of several score, where the appearance thereof ceases with the day of rest, and comes on again with hard labour, that did ever terminate in genuine inflammation of the kidneys : it will return at intervals (upon hard work) during the animal's whole life probably, without any further ailment attending it.

CALCULUS; OR STONE IN THE CŒCUM, KIDNEYS, URETER, AND BLADDER.

When we consider for a moment the vast circulation that passes the (liver and) kidneys, there to undergo separation, as before fully described in the first book ; and recollect, that hard extraneous substances pass through these organs, and find their way even into the blood, our astonishment ought to cease at discovering earthy particles, often hardened into stone, in some one or other of those parts.

Cause.—The first particle that is deposited or left behind is no doubt very trivial, as the bisection of many such stones most amply proves. Want of

vigour at the time of its access, and the consequent inability to expel the intru
sion, appear to be the immediate cause of this otherwise inscrutable disorder
Subsequently, other congenial materials reach the original evil, mostly in the
liquid form, and thus add to its size, increase the number of striata, and height-
en the danger.   The water that is drank by quadrupeds is abundantly impregnated with fit materials for generating calculi : soft river water, and that of
turbid pools, convey the softer or earthy particles into the animal's system,
whilst that drawn from springs contains the elements for forming stone, as
perfect as any geologists find in the strata of our earth.   The softer kind of
these concretions are found in the blind gut, or cœcum ; the harder, or stony
kind, in the other viscera above named.

Heat is the power that separates these elements, and hardens each additional lamina that has accrued, or grown over the preceding, from time to time,
as the animal may have been exposed to drink so impregnated.   This is visible on the section of those stones which have been found in horses and other
animals, and preserved by the curious, and cut in two by the lapidary.   Every
such concretion so found, of whatever nature it may be, exhibits in the centre
the nucleus or commencement of the evil, which proves itself to have been
either originally stone, or some soft substance, as a bit of chaff hardened by
the heat ; but much oftener it presents a perfect pebble, that must have been
borne along by force of the current, and in the cleansing function of the kidneys got detained and deposited there.   If not entangled, as it were, in the
cellular membrane of this gland, such a pebble will detach itself occasionally
and descend through one of the ureters into the bladder.   For full information as to the structure and functions of these several viscera, the reader is
again referred to the second chapter of book the first, which treats alone of
such matters ; as regards the cœcum, at page 46 ; the kidneys at page 51 ; the
bladder at page 53.

One original cause of such concretions has been ascertained beyond contradiction, and as the information may prevent its recurrence among a numerous
class of horse proprietors, I quote my authority much at large, by way of pre
ventive advice, seeing that a cure is at present beyond the reach of art ; remedies worse than useless.   Let us hope, notwithstanding, that the mite which
is here contributed may not be thrown away, but incite some future close observer of nature and her ways to add hereto the result of his own inquiries,
and so increase the sphere of his utility in one respect, since imperious circumstances have contracted it in another and more obvious line of his profession—the desire of gain.

Millers' horses are most liable to contract this disorder, and for obvious reasons ; being large heavy animals for the most part, their owners opulent if not
rich, and grain and pulse ever at hand, dry food is invariably given to them
with a liberal hand.   To render these substances more agreeable, to hasten
digestion, and thus produce a fine coat with a well-filled carcass, their corn is
passed through the mill, the beans also are usually broken ; and, thus pampered, they eagerly devour the ready feed, and with it whatever extraneous substances it may have acquired in the process of grinding.   These are not few
in quantity, it seems ; for such articles are invariably ground between stones
of a soft nature, that easily part with their rough surface, and these stony
particles all find their way into the stomach and intestines ; some, here and
there, pass on through the circulation, by means that are neither uncertain
nor inscrutable in the minds of those who have studied such subjects, and will
refresh their memory by turning to what I have said thereon in the second
chapter of the first book.

Dr. Withers, of Newbury, Berks, having many years before given to Dr.
Hunter a large intestinal stone, which proved fatal to the horse whence it had

been taken, communicates to the Medical Society of Crane Court, London, a similar circumstance which had come under his observation—both being cases of millers' horses. He then describes "the case of a very valuable horse belonging to Mr. Andrews, another miller, which lay ill of the colic," as the owner supposed. "I told him (says Dr. Withers) that if he would examine the intestines after death, he would most probably find a large stone, which was the cause of the horse's illness." This, the miller, of course, neglected to do; but his dogs made the discovery for him: it was a large round stone, broken, from which circumstance I infer that it had been at first a soft or earthy concretion, and proceeded from the cæcum. Four such instances all together were remembered at the same mill, besides many others elsewhere; but, with characteristic negligence, the millers in no case thought proper to furnish the doctor with the when and the where found, nor does the doctor say why.

The symptoms of calculous deposite throughout apparently resemble colic to the view of common observers, as in the case of Mr. Andrews' horse, just quoted; the animal looking at his flanks, straddling when a kidney is affected, as if he would stale, which he does with great difficulty, and sometimes a little bloody. This last appearance also occurs when the bladder has been affected for any length of time, so that the anguish of acute pain had communicated to the kidneys by means of the ureters, in which manner alone blood could possibly have been produced in the celebrated case cited by two contemporary writers from M. La Fosse, the elder. When stone resides in a kidney, it may be ascertained by pressure of the hand thereon: I will not exactly say you can feel the stone, for it lodgeth underneath, but the greater tension and enlargement of one kidney beyond the other, leaves that notion on the mind; besides which, the animal will shrink, or rather start, a little quicker that in case of "inflammation of the kidneys"—the symptoms whereof, as set down in a preceding page (103), the reader should consult in order to shape his practice accordingly.

Calculous, or earthy deposits of substances in the cæcum may be ascertained and distinguished from simple colic or gripes, by passing the hand along the lower part of the belly, as described in the first book, at page 46. While such an obstruction remains deposited near the blind part of that gut, no immediate danger or inconvenience is to be apprehended; but when the lump, by any means whatever, moves to the orifice, and obstructs its only passage, the most distressing consequences ensue. One of the causes hereof is the exhibition of hot, strong, or drastic medicines, which are usually given in cases of genuine spasmodic colic; and as the symptoms that attend both are alike almost throughout, with the exception just made, no mistake is more general, probably, than people treating this disorder as they would colic, which course endangers life.

The ureters, it will be seen, are but of small capacity, and in its descent from the kidney, whence it has been detached, the stone sometimes meets with an insurmountable obstacle; the irritation it thus occasions communicates to the adjacent parts; entire suppression of the urinary secretion is the immediate consequence, and mortification of the intestines and death ensues, without the possibility of relief. Indeed the remedies that seem most proper do but accelerate the catastrophe.*

Much perspiration attends the first hours of the suppression, and it affords evident relief; but painful efforts to void urine, which comes off in very small quantity, and ultimately ceases altogether; and then cold ears, cold legs, tremor and an alarming irregularity of pulse, precede but a short time the

---

* I say seem, for none can say precisely what is taking place. He whose judgment brings him nearest the real cause of pain being most likely to apply the proper remedy.

dissolution of the functions of animal life. This is the most dangerous species of disorder, arising from calculous deposit, that I know of.

In the kidney, however, little danger to life is to be apprehended from the stone, unless the animal is put to severe work, so as to produce the symptom of bloody urine before described. They are mostly fat horses that die with stone in the kidney; in fact, all that I have ever seen or heard of, and these have been numerous; for I have long made a point of inquiring after such cases of calculus, where they seldom escape notice, viz. the horse-slaughterers' yards, of which it is proverbially and truly said, that not a hair enters but is turned to profit. The probability is, that when the stone detaches itself and descends into the ureter, the fat which partly enveloped it and the residue of the kidney had been withdrawn, through disease or poor living, and the membrane which supported both had divided. I once thought I had made some observations on this part of my subject which would be worthy of public perusal; but these are not sufficiently mature to find place in this little volume, devoted as its pages are to matter of fact, and fair deductions therefrom, and wholly exclusive of theoretic speculation. Nevertheless, in aid of what others may think fit to say in any other place (out of a spirit of controversy), I would just add, that only one kidney is affected at a time, or one ureter; that the calculi found in either of these are invariably of the hardest kind, whilst those of the bladder are softer, and those of the intestines softer still, or little more than concrete earth. Lastly, that none of those horses which I have found troubled with either kind of calculous disorders suffered under a second at one and the same time.

CHAPTER II.

EXTERNAL DISORDERS.

Abscess and Tumours.

SWELLING, with inflammation of the solids, the glands, or simply pustules on the skin, are all tumours, have been divided into eight classes, and according to their situation, are termed superficial, or deep seated abscess. Superficial are those which appear on the skin, as farcy, &c.—Deep seated are those which more generally are hidden amongst the muscles, ligaments, &c. as poll-evil, fistula, &c.—A few general observations on the remote causes thereof seem necessary to a right understanding of each particular complaint.

All those disorders in common, together with several others, I have no hesitation in attributing their remote cause to constitutional defectiveness at least, or incapacity in the function of circulation, better known by the homely expression, "a bad state of the humours," as before insisted upon, principally at pages 53—61. Both series are referable to the same predisposing cause. That species of inflammation of the whole system which we have agreed to term fever, frequently terminates by concentrating its latent humours, and depositing the same critically in some fleshy part of the carcass or limbs, producing matter (or pus) which, with heat, constitutes the disease. Whether abscess or tumour supervene, both have immediate connexion with blood-vessels of no small consideration, though the disorder may have commenced with the finer vessels (capillaries), as insisted upon at the pages above referred to; and hath been repeatedly proved. First, as regards tumours, these being probed, the patients have bled to death, with arterial blood. And secondly, in every case of abscess, in proportion as they increase in size, so does the patient's strength invariably diminish. When nature makes an opening to the surface, after long-protracted illness, the patient is usually so exhausted, and the parts ad jacent rendered so unfit to re-unite, that the strength of the constitution appears to run off at the orifice: life is seldom preserved, health never completely restored.

Tumours sometimes appear of tolerably large size, that become indolent, without feeling, and are moveable under the skin. These are caused by the same evil state of the blood, or its vessels, and the inflammation or irritation having ceased at some time or other, the enlargement remains, though the connexion with the system of animal life has long ceased. Although very unsightly, the animal feels little inconvenience from these protuberances: they receive the name of wen, and might be taken off by dividing the skin, and pressing out the wen: it is then to be drawn forth with the forceps, and the healing of the wound is effected by strapping down the skin with adhesive plaster; the cure is thus said to be effected by the first intention. The usual precautions of taking away the hair, and afterwards keeping the patient's head up for a few days, would of course be adopted.

The genuine tumour is soft and tender, and is contained in a membranous case, or cœstus, that has been likened to the finger of a glove, or to many of them, when it acquires the distinctive name of fistula. The case, or cœstus, having been formed by the disorder, and matured by heat, acquires strength the longer it is suffered to continue unopposed, seeking its way inwards, until the knife alone can afford relief. At the shoulder the fibrous and membranous construction is exceedingly strong. Look at page 11. Generally speaking, all swellings of a circumscribed nature are tumours.

Some objections which have been raised against the view I have taken of the origin of this whole series of diseases must not go quite unnoticed here, though I dislike controversy as much as any writer who has gone before me on either side the question. At the very commencement of this book (page 59), and without adverting to either set, or indeed thinking at all of the con-troversy, I assigned a reason why the apparently triumphant proof of Mr. White, at page 29, is no proof at all, but the contrary, as to the thickness or viscidity of the blood increasing with the continuance of inflammatory fever. Every writer on this subject allows that the swelling and discharge of matter that frequently occurs after a fever, or inflammation of the whole system, de-notes the crisis or termination of that disorder; and insists that it must be considered as but an effort of nature to throw off something that is offensive to the well-being of the animal. The same happens often after "inflamma-tion of the liver" has been reduced; but this kind of occurrence, though it adds nothing material by way of argument, leads us directly to the point at issue. General inflammation (fever), it is allowed on all hands, begets something of-fensive, and so does partial or local inflammation of any organ through which the blood passes, particularly of the liver and kidneys, through which the whole mass gets filtered, as it were: and nature's efforts to get rid of this of-fence against her rules are evinced in swelling of the external parts, in the in-flammation thereof, and subsequent escape of the offensive something, where by a cure is effected.

All this is agreed upon by those who deny the necessary pre-existence of a general ill state of health, as well as by those who already know, or have yet to learn, that the liver, that acknowledged cleanser, permits much grosser ma-terials to pass through it than those offensive matters, or gross humours, which we contend reside in the blood, and constitute disorder of one kind or other or the surface, or at least predispose the animal to acquire such, according as circumstances may determine one way or the other. Seeing that such gross substances as bits of straw, chaff, &c., have issued from a vein on blood-letting, it is too much to concede the ultimate point that the feculent humours, which constitute tumours, farcy, &c. may not in like manner escape into the circulation, and be detained at that particular part which is rendered by some accident less capable of continuing the harmful matter in a fluid state? A blow, a gall, a ligature, or bruise, are known to occasion this disability and bring on disease in one of its varied shapes. So does "a cold" produce fever in some animals sooner than in others; according as the circulation may be more languid, or more predisposed to inflammation, or otherwise unfitted for its purposes; whilst some again acquire inflammation without any such accidents or cold, the fever being lighted up occasionally by warm stabling alone, though the air they breathe may be perfectly innoxious.

How it is that those external diseases, enumerated at the head of this chapter, are generated, I shall not here repeat: the reader may consult the principles upon which my opinions are founded in the twenty-ninth section of book the first, page 30: to which I will here merely add, that the tumours we perceive on the body that are not of a nature to break and discharge their contents—as farcy, grease, &c.—are usually, if not always, accompanied by corresponding tumours on some vital organ, as the lungs, liver, &c. But single tumours, containing matter, as the whole tribe of fistula, &c. are designed to counteract and carry off obstructions and all baleful affections incident to the organs just mentioned, and of all others: an owner ought therefore to deem himself fortunate, when some inscrutable long illness of the inside terminates in this manner. The appearance of these latter on the surface may be taken as a good assurance that none then exist internally; nor, indeed, any other disorder whatever, the natural strength of the animal system enabling it thus to cleanse itself. Again, we may remark in general, that as it is the better bred animals that are most liable to affections of these organs, so is it the "country-bred cattle," without any breeding in them, that mostly suffer those external attacks. To the reflection of every man of experience I refer this material point of dissonance between the two varieties of horse, which serves to prove that those having great lumps of muscle at the parts liable to such attacks are most disposed to contract local inflammation, and that puffing up of gland or lymphatic which we call tumour of various kinds. Local inflammation alone, however, could not effect the evil, without some corresponding cause; else, how comes it to pass that none but aged horses, that are heavy in the hand and low in blood, contract fistula or abscess; young and lively horses, and those with some breeding in them, never? Once more,—if the disorder reside not in the blood, how does it come to pass, that a horse having contracted one species of tumour, he is never known to undergo an attack of any other species—and there are a dozen at the least? For example, give a horse the poll-evil, and see how little he will be disposed to contract the glanders.

Fleshy horses, those of the cart breed and of indolent habit of body, are most liable to contract poll-evil, fistula, &c.; indeed I might say, the ready disposition thereto is confined to that breed, though either could be inflicted upon higher bred cattle, which might not be so predisposed by a bad habit of body or by the gross humours before noticed. When the animals are young, and feed ravenously, the strangles carry off those humours; when youth leaves them and more doltish habits comes on, these humours appear in some other varied shapes: besides those diseases just named, the farcy, grease, &c. all come on from the same indolent habit of body. They are always ravenous eaters, gross feeders, and consequently lethargic in their movements, that acquire poll-evil; for they demand harsh treatment to keep them at their work which frequently devolves into ill-usage, unless the drivers possess the patience of Job.

Hence the duty of attending to the health of such horses, as much as may be consistent with the avocations of the owner; of avoiding the infliction that is often the immediate cause of either species of ailment; and, these being discovered of applying the necessary remedies for their instant dispersion—if the symptoms are mild, and thus promise success, a low regimen follows of course. But delay too often confirms the disease; it approaches towards maturity, and will not be repressed : then does the duty of "bringing it forward" to suppuration present itself as the only means of obtaining a radical cure; and I may add, that this is always the safest, the best, and the most certain means, when the disease yields not to the first efforts at dispersion. In ordinary cases of saddle gall, the swelling and heat will bend before an assiduous and early application of the repellent lotion; not so easily, however in case of "fistula in the withers," which lies deeper and is more obstinate. Least of all will confirmed poll-evil give way before the strongest repellents; or, if the resolution be apparently effected, the least external injury, or none whatever, will subsequently reproduce the disorder with more than its original virulence. Perhaps, in no part of the farrier's art has he the opportunity of evincing his judgment more, than in choosing the precise period when he will quit all attempts at suppressing the abscess or tumour, and set about bringing it forward to suppuration and a radical cure; when he will also quit the low regimen which was proper in the first attempt, and adopt a more generous diet, that is better adapted to the painful discharge his patient will now be compelled to undergo, either by dint of medicinal applications or the knife.

Abscess in the more fleshy parts of the body, or under the belly, are far less dangerous or troublesome situations than on the parts just named; they also prove to be symptomatic of the actual state of the blood, of which they then form the crisis or point of cure, and therefore the repression of such (as recommended in other cases) should not be attempted, neither should the animal system be lowered, but the contrary. If, however, the tumour appears near a joint or just above it, as the hock, so as to impede its action, in which case it would soon assume an ulcerous appearance, by reason of the movement of the muscles of the limb in going, repression should then be resorted to with assiduity and skill. Artificial inflammation, excited upon the skin and cellular membrane, near the part, by means of blistering, or rowelling higher up, has the good effect of drawing off the heat and tension from the more important joint, nor does the animal by this application undergo so much pain as he would were the tendon affected, whereby the limb would become irremediably stiff and useless.

CRITICAL ABSCESS

Is that swelling or tumour which is occasionally thrown out on the body or limbs from no apparent accident, but what may be traced to that derangement of the system we call fever, and is sometimes attendant upon protracted inflammation of the liver, when the disease appears on the fascia of the muscles of the belly, on the jowl, or other glandular parts.

The cause and the effect thus become manifest together; and when great tenderness is evinced upon touching the parts in ordinary cases, nothing more is required than to make an opening in the lowest edge of the swelling, and expressing the contents; the cure is effected by means of the common "digestive ointment," which is prescribed under the article "Poll-evil," farther down. But the proper time at which the opening is to be thus made requires close observation. In general, this may be ascertained by a change in the animal's manner: he will eat more heartily as the matter increases: which proves that the disease of his habit has accumulated at this precise spot. He should not be allowed long to remain in this state, lest the offensive matter should penetrate inwardly or laterally. If the disease is thus distinctly known to proceed from the remains of ill-cured fever or inflammation, poultices should be applied to bring it forward to the surface, and the animal receive increased feeds of dry oats, of beans, or sodden oats, according to his former habits, in order to encourage the access of matter; for nature, exhausted by the violence or the continuance of the disorder, is incapable of expelling this last remains of the enemy, and stands in need of support. Should the horse have been lately laid up with fever, or for some time past shown languor in his gait, and heaviness about the eyes, or it may be concluded from his recent hard labour and hard mode of living, that he has been long ailing inwardly: in this case the abscess being evidently a critical symptom of the general evil state of his blood, nature must be assisted in getting rid of the offensive matter; and for that purpose bring the tumour to a head by means of a poultice. The head is most commonly the seat of swelled glands.

Drawing Poultice.

White bread, the crum of a 4lb. loaf.
Onions chopped, 2 lb.

Boil the onions in water, and pour the whole on the bread: mix to a tolerable consistency, and whilst blood-warm apply copiously to the parts in a cloth. Support the application by means of a bandage of stout linen cloth, with ligatures tying over the forehead and across the poll thrice, as described in the annexed sketch. Some persons have recommended the use of a solution of gum to render the cloth impervious to liquids. See page 79.

Should circumstances require a more extended application, or that the patient restlessness might rub off the bandage, let a more extended bandage be employed. For such a one, and as to further particulars, the reader may consult page 79, where a bandage for sore throat is depicted.

By those means the swelling will come to a head, and give signs of being about to burst, but which I have reason to believe seldom happens spontaneously by reason of the thickness of the skin. Apply the knife, or bistoury, as directed much at large in the case of poll-evil; give a mild laxative the same day, and lower his diet. When it so happens that the opening has been made too soon, before it has accumulated sufficiently, the orifice may be kept open by means of a seton passing through it to the lowest or most depending side, and the running continued for several days, until it assume a healthy appearance and the swelling subsides. This plan must be always adopted with the slow or sordid tumour, which will not come forward, though heated with the onion poultice, and even with a blister: then let the seton be applied, changing it daily and soaking the tape in the irritating mixture, as in case of poll-evil, page 119. That other critical abscess, called strangles, comes under a distinct head, farther down.

Deep-seated abscess, under the fascia of the muscles of the belly, is scarcely ever curable, being seldom discovered to the eye until too late to render assistance in bringing it to the surface by means of strong drawing poultices, as in case of obstinate poll-evil. On passing the hand over the part, the animal may be observed to flinch from the touch; but this symptom is seldom attended to, and it makes its way inwards, bursts in the cavity of the abdomen, and kills the patient.

POLL-EVIL.

Causes.—Next to a diseased habit of body, as just above noticed, which predisposes a certain description of horses to contract tumours in various parts of the body, the poll-evil is frequently occasioned by a blow, or gall, of a very trivial nature, if it do not come on without this kind of excitement. The action of the head is very great with some horses, arising probably from an itching in the upper part of the cervical ligament, where it is attached to the vertebræ of the neck; and this causing irritation, we need not hesitate long in accounting for the inflammation that affects the muscle which interposes between it and the poll-bone, in a cavity that is greater with some breeds of horses than others. This variance in conformation is exemplified in the whole length portrait of a skeleton which is prefixed to chapter i. wherein the cavity that should form the seat of this disease is scarcely perceptible; whilst the small figure, inserted at section 16 of that chapter, to illustrate the uses of the cervical ligament, has this cavity of the usual extent. Of course, this latter would be still more predisposed to contract poll-evil than the former, which was a peculiarly formed horse in another respect also; and it is more than probable, that, if the two were to fall into an equally bad habit of body, whilst the latter might acquire poll-evil thereby, the constitution of the former might throw off any offensive matters that might accrue by some other means.* The reader will do well to turn back to the section referred to (p. 20), as well as to the skeleton [at A 5].

The wheelers, in a set of horses, will frequently throw back the head in

* These might appear in shape of grease and farcy; but it has been generally observed that a disposition to farcy abates, if it do not subside entirely, upon the appearance of poll-evil. Again, horses that are most liable to contract the grease, are precisely of the same disposition as those which are afflicted with tumours, &c. viz. of indolent habit, heavy in the hand, and slow of blood, fleshy and dull:

warm weather, or after brisk work, at feeling the reins that run through their head-harness to the leaders: this action is performed, as the reader will have learnt, by the action of the cervicular ligament, the upper end whereof terminates where the ear-band rests, and perhaps pinches the part. Horses that are given to shy are likely to contract poll-evil when hanging back, and throwing up the head with a jerk.

But the most prolific cause of poll-evil I am inclined to attribute to the low stable door-way, whereby the animal gets many a trivial hit at going in and coming out; next in point of frequency is that brutal mode of attacking restive horses about the head with the butt end of the whip. Education of the lower classes has effected the abatement of this as well as many other unfeeling practices. Ofttimes, the edges of the ear-band, being sharp, create a painful itching, then soreness and irritation about the part, as does also the showy tip, or "cutting at a fly," practised by our flashy four-in-hand men, who may have discovered that touching up the animal in such a vulnerable part is "sure to make him go along." Stage-coach horses, however, do not now acquire poll-evil, so far as I can learn, like what they did formerly; for the great expedition these vehicles are constrained to, compels the proprietors to use better bred cattle than their predecessors—those that are less indolent, not so heavy in the hand, nor sluggish, consequently not so liable to contract diseases incident to a bad habit of body, or vitiated state of the blood, like poll-evil and its nauseous train of co-existent evils, that we shall proceed to take into consideration one after another.

Symptoms.—At first the animal appears restless, throwing his head back and returning it to the former position, as if the efforts had occasioned pain. Soon after, it droops the head, holding it now on one side, now on the other; appears dull about the eyes, and becomes sluggish in its movements. In this state it continues a longer or shorter time (even weeks) as the violence may have been greater or less that brought about the evil; the time depending also in some measure on the height of the pulse: a languid system making of course the slowest advances towards bringing the abscess to maturity. This uneasiness of manner is accompanied by heat, swelling, and shortly by tension of the part, and increase in the pulsation. As it goes on, a disposition to flinch from the touch is evinced whenever the part is approached with the hand; if the evil be deep-seated, the swelling is wide, but not so high; but when nearer the surface, it presents a point, is circumscribed within a well-marked circle, and ultimately tells how necessary it is that the contents should escape, by a throbbing which may be felt at this point. Again, to ascertain that the matter is near the surface, apply two fingers alternately on the sides of the tumour, and the matter will recede from side to side. Let it out.

Cure.—At first, this may be attempted, in the earliest stages of the disorder, by repression or dispersion, provided the disorder be not deep-seated near the bone; which will be the case if it has been brought on by violent means, or it be a second attack, when endeavours to repress it would be vain indeed. On the contrary, if we can trace the cause to a hurt of no long standing, or of trivial import, and we know the horse was in good health before the swelling took place, then our duty is to carry off the evil through the animal system, by means of active physic. Foment the part well with bran and water, warm; rub it dry with cloths, and apply the

Embrocation.

Spirits of wine, half a pint,
Camphor, 2 drachms,
Goulard's extract of lead, 1 drachm

Mix, and apply the same two or three times a day, gently rubbing the part as much as the animal can bear. Give also at the same time the

Alterative Ball.

Aloes, 4 drachms,
Castile soap, 2 drachms,
Calomel, half a drachm.

Mix with mucilage, and give one every third day, provided the embrocation is applied so long.

During these applications, a cooling regimen should be observed, the feeds being reduced to half the usual quantity of oats, and ultimately discontinued altogether. There will be no propriety in clothing up the patient, nor need he be exposed to the cold air, if it prevails. When the disorder has been brought on by simple compression of the ear-band, and is recent, I have never known the foregoing treatment to fail; and in cases of vigorous constitutions, the swelling, heat, and tension have been reduced so quickly (i. e. in four or five days) as to leave certain careless observers in doubt whether the animal had really laboured under a genuine attack of poll-evil.

Remove the halter, and if the animal be put to work, contrive to keep back the ear-band. A good and valuable embrocation will be found in simple vinegar three or four times a day, or the sediment of very stale beer. Old verjuice answers the same end; and all this kind of embrocation must be laid on warm, by means of cloths soaked and applied repeatedly.

. The same treatment and observations will apply to all the other species of abscess in its milder state, fistula, warbles, quittor; but of these I shall speak more particularly under their respective heads of information.

Second method of cure.—Very few cases present themselves to recollection of even recent poll-evil, that would admit of being completely dispersed, and a radical cure effected, by any means whatever; and it is due to candour to acknowledge, that some of the most stubborn attacks were found to have relapsed after a while, which proved that the cure so effected to all appearance was not radically good, but had left a violent predisposition to renew its ravages afresh. Probably, the time of inflicting the injury had not been accurately marked, nor its degree ingenuously reported to the owners in those cases of relapse.

However this be, when the disorder is found to baffle the endeavours employed to disperse it, the whole course of proceedings must be changed, as before hinted in the concluding sentence of my general observations on this topic. Instead of putting back the swelling by those means, let us pursue a direct contrary course, in order to bring it forward: the mode of feeding must be changed along with the medicines that now become proper to procure suppuration, or a discharge of the offensive matter; a full habit being mainly conducive thereto, and proving how closely connected is this disease with a gross habit of body, which in all fleshy animals superinduces a diseased habit, vulgarly but accurately termed "full of humours." After having found useless your efforts to disperse the tumour, or, mayhap, finding at the first view of it, or by the first touch, certain symptoms that prove it ought never to be dispersed, the practitioner will of course seriously set about permitting, or forcing, the offensive matter to escape. Every hour's delay in putting this resolve into practice serves but to render the ultimate cure still more difficult and hazardous; for the evil is all this while extending its baleful effects inwards and sidewise, and forming around it, in every direction, the fistulous case or cæstus before spoken of, which is a film, or skin-like substance formed

12

of the cellular membrane, thickened by the disorder. (See Book I. Sect. 27. page 28.) In this event, the tumour has become decidedly fistulous, and is to be treated as such, when the great length of time it may have been suffered to make head, and its now extended surface, warrant that conclusion. The knife is almost the only remedy, notwithstanding the superficial tumour will in some cases break and discharge matter of itself; this, however, never happens with the deep-seated abscess, which lies close to the bone, and destroys not only it, but the muscular substance of the poll, and the end of the cervica. ligament also. In these series of abscess or fistulous tumour, nothing but the knife can ever reach the disorder, and it must be employed fearlessly, but with a commensurate share of skill, after the skin has been prepared with fomentations, &c. Let the parts be softened and drawn with poultice of oatmeal, put on lukewarm, twice a day; and if the effect be not visible to the eye and touch, as before described, increase the powers of the poultice by the addition of onion chopped and mixed with the poultice whilst warm. Or, a mere change may be adopted, and a bread poultice applied instead; for, notwithstanding oatmeal is stronger, yet I have occasionally found the milder have more effect when the former had not succeeded entirely according to my wish. The poultice should be provided in sufficient quantity to cover the whole swelling two inches thick at least, having a small quantity of sweet oil, hog's lard, or oil of turpentine mixed therewith. Fix it on by means of a contrivance that is sufficiently explained by the annexed cut, in which it will be seen that the girth is to have a web breasting, to which the lateral corners of the cloth are to be attached by broad tapes, as was explained in another similar case at pages 79 and 114.

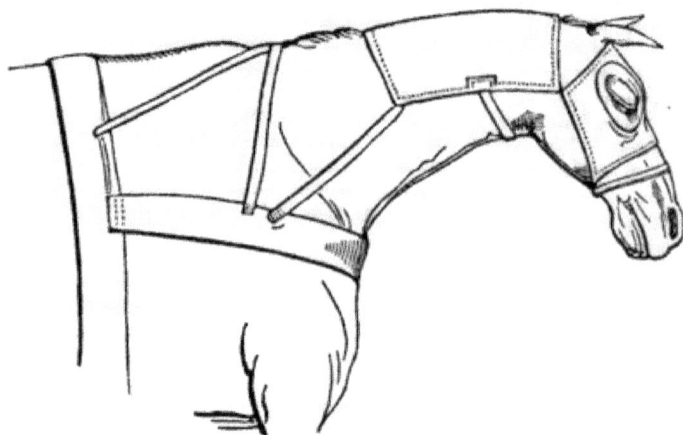

I have here represented the bandage rather longer than requisite, under the presumption that it may occasionally be applied to other affections farther back; a prolongation of the bandage may be affixed at either end, either plain or plaited, according to the amount of the swelling.

When the symptoms above stated inform our senses that the matter ought to be so "let out," an opening is to be made the whole length of the abscess, a little below its centre; taking especial care that the knife do not pass crosswise, lest the attachment of the cervicular ligament to the first (vertebræ) bone should be severed; in which case the animal would droop its head ever after as may be learnt by consulting its construction at p. 20, of Book I. On the escape of the matter, after ascertaining by a probe whether it runs in pipes, or sinuses, this way and that, or with small bits of diseased fibre or membrane stretching across the cavity, so as nearly to divide it into unequal parts—let each be just touched with the knife or scalpel. There is no propriety in the old practice of squeezing out all the offensive matter from this kind of abscess, although it be very proper in that deep-seated sort where no pipes, nor the small cavities just spoken of, are to be felt or seen, for the following reasons : the first mentioned kind have the case or cœstus before described, which con tains the matter, and if laid open before the evil be sufficiently ripe, it doe not come away freely. This, however, the operation effects in two or three days, if kept running by means of a seton, or other contrivance placed at the orifice ; but the application of tow, or any other substance, that obstructs the escape of this matter, is ever to be avoided. On the contrary, when the abscess is very deep, reaching to the bone, which may be felt, and presenting but one large cavity, then the matter should be expelled by pressing gently on two sides of it at once. Let the lips of the opening be dressed the first time, and as long as it may be found necessary to keep open the wound, with any ointment hereafter mentioned, on which has been strewed sulphate of copper, powdered. Should the lips adhere together, or appear much diseased, wash with muriate of ammonia, taking care it does not run upon the sound parts, nor into the cavity. In either case, wash off the dead parts with warm water, before each new dressing is laid, sponge it well and dry, after inserting the probe on every side into the fistulous sinuses, and continue this mode of treat ment until the parts assume a healthy appearance.

The seton should never be neglected in bad cases of either description, but be introduced at the lowest or most depending side of the abscess, after being wetted with the following

Irritating Mixture.

Spirits of wine, 2 ounces,
Corrosive sublimate, 1 scruple ;

Mix and saturate the tape therewith daily. This will keep open the orifice until the offensive matter has run off, and is succeeded by the more healthy issue of a thicker consistency, and nearly white. On this appearance the seton is to be withdrawn, and the parts dressed with the digestive ointment, the animal physicked once or twice with a moderate purging ball of six or seven drachms of aloes, and the cure will complete itself with the usual dress ings, viz.

Digestive Ointment.—No. 1.

Yellow wax,
Rosin,   } of each 1 pound.
Burgundy Pitch,
Turpentine (common) 4 ounces.
Linseed oil, 20 ounces.

Dissolve over a slow fire, and spread upon leather or stout linen cloth, suffi ciently large to come over the undiseased region of the evil, after the wound has been well cleansed. Fresh dressings hereof should go on daily, but in no case until the matter assume a healthy appearance, which it never can be brought to, unless the whole recess has been reached with the knife or by the operation of the "scalding mixture" of the old school of farriery. This remedy, so applied, though at variance with our modern notions of pathology, has been adopted by the collegians of St. Pancras, and with good reason, for it never fails to effect a cure, by effectually cleansing away the diseased parts. Three several mixtures are adopted in different parts of the country—the Hertfordshire and midland county farriers employing No. I.; No. 2 is that recommended by Gibson; and No. 3 is Ryding's.

Scalding Mixture.—No. I.

Tar,
Mutton suet,   } of each 2 ounces.
Rosin,
Bees wax, 1 ounce.—Melt slowly, and mix in
Spirits of turpetine, 2 ounces.
Verdigris 6 drachms.

Mix and pour into the orifice hot, and close it with stitches. The next two have the recommendation of being more scientific, and are withal better adapted for penetrating into the sinuses.

Scalding Mixture.—No. 2.

Corrosive sublimate,
Verdigris,   } of each 2 drachms.
Blue Vitriol,
Green copperas, half an ounce.
Honey, or Egyptiacum, 2 ounces.
Oil of turpentine,   } of each 8 ounces.
Train oil,
Rectified spirit of wine, four ounces.

Mix, and apply as before directed. The difficulty of retaining this last in its proper place, is its only defect; but Gibson appears to have prescribed a quantity sufficient to allow for spilling a good portion. Since writing the above, however, I have inserted the sketch of bandaging for poll-evil remedies at page 118, to which the reader will refer, when requisite, and introduce such modifications as the nature of the applications may demand to prevent the loss of any part.

Scalding Mixture.—No. 3.

Oil of turpentine, 2 ounces.
Verdigris, 1 ounce,
Ointment of yellow resin, 6 ounces.

Mix and apply as above. In using any of those hot mixtures, a piece of tow should be so placed as to surround the orifice and prevent its running over the sound parts—which would be injured thereby, as would the operators fingers, &c. if he neglect the proper precautions. These he should not fail to take as regards the acrimonious discharge from the abscess, as absorption thereof might take place at the root of his nails; so, if the discharge be allowed to rest upon the sound parts of the horse, it will be found to corrode and produce ulcers.

    Frequently it happens—and I believe the old farriers always "repeated the dose," that a second application of the "scalding mixture becomes necessary, for their cases were always very bad ones. In this event, opportunity is afforded of employing both prescriptions in succession; but whichsoever is first adopted, let it remain undisturbed from sixty to seventy hours, if the stitching do not sooner burst. Sponge out the parts with warm water; cleanse away the adjacent filth, and either repeat the same or proceed at once to the cure— a determination the doctor will come to, according as the rottenness may have sloughed off, and the inside of the abscess may present a healthy appearance, or otherwise. If it be quite clean, the adhesion of the parts will follow with very little further care than applying the digestive ointment according to the receipt in page 119,—or the following

Digestive Ointment.—No. 2.

Common turpentine, 4 ounces,
The yolks of two eggs.—Mix these well, and
    add
Myrrh, in powder, 4 drachms,
Mastich, 2 drachms,
Tincture of myrrh sufficient to bring the whole to a proper
    consistence.

Should the cure of the wound proceed too fast, the over luxuriant granulations of new or proud flesh must be touched with caustic.

But notwithstanding all that has been said above, it sometimes happens that a totally different course becomes necessary, when abscess in the poll is connected with another disease arising from the same vitiated state of the animal's system, and the remedy for one of these will cure the other. Farcy is the correspondent disease to which I allude, or rather I should say a tendency to farcy, visible in certain scanty lumps or tumours on the body and legs: these will run off sometimes by means of a copious discharge at the poll. More frequently, however, the farcy is of too inveterate a description, and proves that the whole mass of the animal's system requires correction, and that it must be treated with medicines proper for the farcy, as well as the local affection of the poll.

People in general like to be borne out in their most novel opinions by those of longer standing in society, and I confess myself one of those sort of people as regards the doctrine of a vitiated or a corrupted state of the animal's system, which it is absolutely necessary to correct by medicine before the cure of some disorders can be effected. I strongly touched upon this topic in the first book, and at page 59, to which probably the inquiring reader will turn, and become convinced with me that poll-evil may be no other than the critical abscess of farcy; which farcy is a disease of the system, and is correspondent with glanders, as poll-evil is with quittor, &c. The writer I shall quote as agreeing with me, mainly, in this view of the subject, is Richard Lawrence. He says, "the poll-evil is sometimes connected with a disposition in the habit of body to farcy; this may be known by the animal appearing universally [i. e. generally] unhealthy in his coat, the tightness of his skin, and also by small lumps or swellings in different parts of his body, and particularly on the insides of his legs. When it is ascertained, therefore, that the poll-evil arises chiefly 'rom a disposition to farcy, the mere operation of opening the abscess, and using the dressings usually recommended, will not prove sufficient, without the aid of medicine given internally; because the abscess, not being then a

local affection arising simply from partial injury, it will be necessary to correct
the general habit of body, before a cure can be effected. The medicines best
adapted for this purpose will of course be found under the head of "Farcy,"
a few pages farther down.

FISTULA IN THE WITHERS.

Cause.—Although closely resembling poll-evil in so many respects as to
seem the self-same disorder arising from precisely the same causes, but differing in situation only, I must here premise that some other distinctions are
proper to be taken, which it will be necessary to keep in mind. We have
seen, a few pages higher up, that poll-evil may be produced without external
violence; this never happens with fistula in the withers, which is always
brought on by external injury—namely, the galling of the saddle: in the first .
case the tumour frequently turns out a simple abscess, in the present case
never, but becomes fistulous at its very earliest stages. This arises from the
quantity of the membrane which is found in the shoulder and whole forehand
of the horse, in the cellular structure whereof the offensive matter finds an easy
receptacle, and spreads its ravages from side to side; and accumulating in
quantity, by its own specific gravity, finds its way, eventually, amongst the
muscles, and forms sinuses. How this operation of nature is performed, the
reader is instructed in the first book, at sections 26, 27. I have also descant-
ed somewhat at large, in the general introductory observations on this whole
series of disorders, as to the distinctions proper to be kept in mind between one
kind of tumour and another, how they are formed, and what description of
horses are mostly liable to this or that species of the disorder. At page 112
will this information be found, and which the reader would do well to consult
once more before he sets about treating his horse for fistula in the withers
The symptoms are most obvious to the touch, as in all inflammatory tumours,
the animal shrinking when the hand is passed over the shoulder from the
mane downwards. But the ill-formed saddle, or one that fits the particular
animal like nothing, or one that is so badly girthed on, that the poor beast may
be perceived going in great pain, shall be set down as the symptom of all symptoms, that the animal is destined to contract this particular disorder of the
parts so injured. Sometimes he tumbles down, or seems to trip frequently,
which should admonish its inconsiderate rider or driver, that his carelessness
is very likely to cost him a broken neck.

Cure.—As soon as the journey can be brought to a close, which has been
thus improperly pursued, remove the cause and bathe the part well with the
cold saturnine lotion, and when the saddle has undergone the proper alterations, the journey may be pursued, if necessity demand such an exertion.

Cold Lotion.

> Subacetate of lead (goulard), 2 ounces,
> White vinegar, 4 ounces,
> Water, 3 quarts.—Mix, and apply with a sponge.

Should not this prevail, and the horse evince pain at the touch, with increased heat and tension, and swelling of the part commence, the disorder is
confirmed; and if not repelled in its very earliest stages, suppuration must ensue. Let it be taken in time however—that is to say, in the course of a day
or two, or a week, with healthy active horses, is not too long—and the heat
and inflammation will be reduced by employing the embrocation, recommend
ed in incipient attack of poll-evil, at page 116, and giving at the same time the

alterative ball there set down. Success more generally attends this first me
thod in the present kind of tumour than in that to which I have just referred,
viz. poll-evil; but this method of curing both is so exactly similar, that it would
be a waste of words to go over the same grounds again, or make the same observations which I thought proper to set down under that head of information.
At page 116, the reader will perceive, that when he is attempting to repell the
tumour and allay the inflammation in its earliest stages, he is to employ a
cooling regimen; that when the disorder has been brought on by a trivial
cause, this method of cure seldom fails, if taken in time; and also that fistula
is easier prevented hereby than is poll-evil. "However this be, when the disorder is found to baffle the endeavours employed to disperse it, (as I before observed), the whole course of proceedings must be altered;" the regimen, or
feeding must be higher, the parts encouraged to collect matter and come to the
surface, instead of making inroads upon the adjacent muscle and bone, which
it will effect more hideously as the animal may be afflicted with a gross habit
of body.

After having found all efforts useless, the practitioner will change his plan;
and force the matter to escape as soon as may be; for the disorder is every
hour extending its baleful influence. For this purpose the knife, or common
bistoury, is to be employed when the tumour is sufficiently ripe, which is a
state it may be brought to, by means of the application of a poultice. Of these,
I prescribed two or three kinds, with the method of fastening them on, but in
this latter respect, a material difference arises in consequence of the different
shape of the parts. The bandage in this case must be allowed to come farther
back, and be there detained by tying the tapes short behind and lengthening
the front ones. See figure at page 118.

Fomentations of warm water, in which cloths have been steeped, slightly
wrung out and applied to the parts, will be found highly serviceable, and may
precede the application of poultice. When by these means the tumour appears ripe and ready, open the most prominent part with lancet or bistoury,
and insert a whalebone probe to ascertain the direction that the fistulous sinuses or pipes extend, in order that these may be laid open, and the whole matter suffered to escape. In some cases a stiffer and larger probe may be employed, and when a sinus lies favourable, introduce the probe, and cut down
upon it. But as to the lowermost sinus, when it tends towards the shoulder,
so as to interfere with the action thereof, the knife is not to pass through it,
but a seton is to be inserted in its lowest or most depending part, so that the
matter may escape through.

As directed in the previous case of poll-evil at page 119, the knife should be
fearlessly applied in severing any small bits of muscle that may appear to grow
across the cavity; a touch of the knife will be sufficient for any purpose, as
by keeping open the lips of the wound, all that belongs to this diseased part
will slough off, and should be wiped away as before directed, every time new
dressings are applied. Let the seton be soaked in the mixture of corrosive
sublimate and alcohol as directed at the page just referred to; and in the worst
cases apply either of the scalding mixtures in the manner mentioned at page
120, and repeat the same if the first does not accomplish all that is desired

The operator in this case will not fail to use the proper precautions as regards the application of those scalding hot remedies, nor neglect to remove the
matter that is discharged from the wound, in the manner set forth at page
121. Most frequently the lips or edges of the sore are thickened, and assume
a very inflamed and ulcerous appearance; this should be reduced by the knife
or caustic, or it becomes so luxuriant at times as to close the orifice, and to
cause a renewal of the fistula, in which case you have all your trouble to go
over again. At Alfort, they have a very neat method of cleaning out fistulous

ulcers, by rolling up pledgets of linen cloth, the edges whereof have been scraped out thin, so that when introduced dry to the bottom of each sinus or pipe, and being twisted round, it brings forth the offensive matter and any residue of blood which may have got into them during the operation. They also employ gentian root to keep down the swelling or thickening of the lips of the sore. The healing is not to be suffered to go on too fast, nor until all the offensive matter has been expulsed, and a more healthy discharge, whilst it manifests the change that has taken place, and warrants your closing the sore. Blue stone spread on any plaster of digestive ointment will effect this; or take

> Ointment of nitrated quicksilver, 3 ounces,
> Oil of turpentine, half an ounce.

Mix, and apply as long as may be found proper to keep the orifice from closing, to which it will be ever too much disposed.

Something was formerly said about scraping the bone when the long continuance of the disorder, its virulence, or the bad state of the horse's general health, hath been such as to affect its surface; but this part of the operation is rather showy than useful, as the rottenness so occasioned will come away as the discharge is kept up, there being a constant disposition throughout the whole system to throw off all such offensive matters.

SADDLE GALLS; viz. WARBLES, SITFASTS.

The first of these partake of the nature of the disorder just above treated of, viz. fistula, and are caused by the same means, bruise of the saddle; but being situated farther back, less scope is allowed for the spreading of the original tumour. Consequently, the smallness of the affliction renders it much less formidable, though, if suffered to suppurate, they become most troublesome sores. The means to be adopted for the cure of warbles are similar to those recommended for other tumours, viz. at first try to prevent the accumulation of matter by repellants, such as the embrocation prescribed at page 116, the domestic remedies in the next page, as verjuice, made hot and applied by means of cloths soaked therein, and repeatedly changed. Or apply, in the same manner, the following

Cold Lotion.

> White vinegar, } of each 3 ounces.
> Spirits of wine, }
> Super-acetate of lead, 2 ounces,
> Water, 6 ounces. Mix.

Should not these succeed, change your treatment, adopt the direct contrary mode, and bring the tumour forward to suppuration by means of poultices, &c. as before recommended; and finally, when ripe, open the tumour with a lancet, promote the escape of the offensive master, and then proceed to healing the sore, as in the former case of poll-evil and fistula in the withers.

Sitfast is an indurated tumour, one that has neither matter nor motion in it, and may arise from either of two causes. The first is simply a gall or bruise, which has produced no inflammation, and consequently no matter has been engendered; the second comes of an ill-cured warble, that has closed, leaving a hard insensible swelling behind. Blistering is the favourite remedy with most farriers, though fomentations and poultices will frequently achieve as much good in very little more time. When suppuration has taken place, the cure is to be completed by dressings of detergent ointment, taking care that the sore does not heal too fast. Should this be the case, put blue stone powdered, upon the plaister once or twice, or merely touch it with lunar caustic as often. Sometimes the callosity does not come off of itself, though the edges rise up; it is then to be taken away by force, separating it from the living parts with the knife. The small portion of blood that comes away does no harm, but the contrary. If, however, any one objects to the use of the knife, or doubts his skill in this operation, mercurial ointment will effect the same end, as follows:

Ointment for Sitfast.

Oil of turpentine, 10 ounces,
Blue ointment, 8 ounces,
Gum ammoniacum, 4 ounces;
Mix and apply to these and all hard tumours.

QUITTOR

Is a disease of the foot, at the coronet, but is so decidedly fistulous, that I choose to treat of it in this place, rather than in the chapter devoted to the foot in general, that the student may more readily remember the general observations I thought necessary to prefix to this whole class of diseases, at page 110; &c.

Cause.—A tread which the horse inflicts on itself, for the most part, seeing that it generally occurs on the inside of the foot. This tread or bruise may either be inflicted upon the coronet, or lower down, by over-reaching, or even at the sole; by taking up a stone or other hard substance; also by a prick or blow in shoeing. A quittor is also sometimes occasioned by gravel working up into an aperture left by an old nail, acting upon the sensible laminated substance, separating it from the insensible, leaving a cavity from the aperture quite up to the coronet, where it lodges, inflames, and produces abscess, which is frequently very difficult and troublesome to cure; if not early attended to, sinuses form, sometimes reaching to the coffin bone. The blood vessels at the coronet cease to perform their proper function of secreting new horn and the consequence is frequently the loss of the inner quarter of the hoof.

Cure.—The sore is always very small, but admits of a probe being introduced, by which the extent of the evil may be ascertained, and this is generally very extensive and ruinous, according to the time it may have been allowed to make head. The probe will pass readily forward and backward to the whole course of the disease, and sometimes it will be found to have penetrated to the coffin bone, every where forming sinuses or pipes, as in fistula of the withers before described. But in this case situation makes considerable difference: unlike the former, poultices are rendered inapplicable, whilst the employment of the knife or lancet is dangerous in the extreme. Besides which, the diseased part is already open, and seems to invite the only species of remedy yet known, in the shape of escharotics, that by irritating the case or cæstus, which forms the sinuses, shall cause it to slough off. In slight cases, those which are found not to have penetrated deep, the simple application of a wash will prove sufficient, and may be employed in this manner. Dissolve blue vitriol in water, and charge a syringe therewith; this is to be discharged into the orifice, and suffered to remain, as much as can be retained. A poultice of bread or oatmeal is to cover the part, and the cure will be completed after two or three days. But unfortunately for the owner and the animal, the disease is seldom taken in hand thus early, but is suffered to proceed until much stronger means become necessary. For this purpose take a long narrow slip of thin paper, and moisten it with muriate of antimony; over this strew powdered corrosive sublimate, and roll up the paper, so that it may not be too big for the pipe which it is intended for. Generally it happens that the opening requires to be enlarged before this pledget so charged with the escharotic can be fairly introduced. Take especial care that the pledget reach the bottom of the pipe, cut it off close, and pass a similar one into as many sinuses as may have been ascertained forms the disease. As considerable irritation of the part will quickly ensue, a poultice sufficient to cover the foot should be previously got ready, and applied immediately.

In three or four days, the bandage being removed, the diseased parts will slough off, a considerable opening presents itself, leaving a healthy looking sore. Let this be sponged off with warm water, and when dry apply tincture of Benjamin, which will effect a cure. A solution of white vitriol is used with advantage, especially when a disposition to secrete unhealthy matter is at any time perceptible. Physic the patient after the operation, according to the actual state of his bowels, the motion whereof will alleviate the pain necessarily attending the escharotic quality of the pledgets applied to the foot. If the horse's bowels be found in the ordinary state, give two balls on successive days, thus:

First Alterative Ball.

Aloes, } of each 2 to 3 drachms,
Hard soap, }
Oil of cloves, 6 drops,
Calomel, 1 drachm ;

Mix, with mucilage sufficient to form the ball for the dose.

Second Alterative Ball.

Aloes, 4 to 5 drachms,
Soap, 6 drachms,
Oil of anise-seed, 10 drops;

Mix, and give one dose the day following the first ball.

VIVES.*

This is the term given to swellings of the glands just under the ear, towards the angle of the jaw, that mostly attack young animals. The tumour is easily repressed or driven back into the system, and by more simple means than those employed in more inveterate complaints of a similar nature, showing themselves in other parts of the body. In some respects this disorder bears near affinity to the strangles.

The cause of Vives may be distinctly pronounced "a cold," that prolific source of so many other disorders incident to man and horse. The vives usually comes on after hard work and sweating, by being then exposed to a current of air, or cold rain.

The season of shedding the teeth, when the contiguous parts are unusually tender, is that in which swellings similar to vives pervade animals of any species. Nevertheless it sometimes attacks horses at an advanced age, notwithstanding they may have previously got over the most healthful form of stran-

* From the French "avives," and the verb avirer, to be brisk and lively as if it were conferred ironically upon the animal in its dullest state.

gles, when we might reasonably suppose nature had ridded itself of a disposition to secrete any more such pestilent matter.   Want of the usual head clothing is then the immediate cause of vives.   The violence deemed necessary in breaking colts also causes the vives, when the pressure on the parotid glands, at reining up the animal, irritates the parts.

Symptoms.—Swellings under both ears, generally, that occasion manifest pain when touched : the animal coughs more than one which has the strangles, and a difficulty of swallowing soon becomes evident.   Stiffness or aridity of the neck follows, and the patient makes frequent efforts to swallow the saliva, which it is the proper function of these glands to secrete, but which they are soon disabled from performing, by reason of the cold checking or chilling those functions.   Of glands generally, their construction and uses, the reader will find many instructive particulars in the first book, at page 29; these of which we now speak being called "the parotid glands," from their situation; and as they now refuse to perform the office of secretion, the watery humours flow from out the animal's eyes, which it partly closes, as if he were about to sleep.   For want of the same supply of saliva, inflammation of the mouth and gums takes place, producing what is vulgarly called "the lampers," or swelling of the roof of the mouth near the front teeth, which I shall speak of separately a little further down.   Sometimes the swelling of these glands, if not assiduously subdued, continues a fortnight or longer, becoming more troublesome every day, and evidently occasioning very much pain; all this while the horse loses condition, is feverish, and at length so weak as to totter when he moves even in his stall.   Spreading downwards under the throat, they at length terminate in strangles, and are then to be treated as such.

The cure of the vives that arises from simple cold is very easy, but not so that which is connected with a general bad habit of body; for then the swelling and subsequent suppuration of the abscess must be considered as an effort of nature to relieve itself from something that is offensive to it, and must be treated as a disease of the whole system, nature having adopted this or that particular spot for demonstrating its offence.   But I have already explained my opinion on this interesting point of veterinary pathology, much at large, when treating of other tumours and abscesses.   Vide page 110, &c.   Oftentimes it happens that the vives depend upon glanders or farcy, of which they are then a correspondent symptom, and will only subside when the virulence of these are reduced.   However, no harm can come of fomenting the part with warm water at least; and after it has been well dried, clothe the head so as to keep off the air, upon the principle of "remove the cause, and the effect ceases of course."   The application of the bandage described at page 114, will sufficiently clothe the part.

Much of the pain and tension of the tumour will be alleviated even by this treatment, and a slight attack will be removed by following it up with fomentations of marshmallows; or, anoint the parts with ointment of marshmallows, and cover the head as before.   A bread poultice affords relief, and bleeding in stubborn cases of simple vives is often necessary, with purgatives.   Indeed, the body should be opened, whether we bleed or no : always leave open the main road for such humours to escape by.   This alone will carry off a recent attack, provided the head clothing be kept on at the same time, nature performing the remainder by absorption.   To assist nature however, employ the following

Lotion.

Sal ammoniac, half an ounce,
White vinegar, 6 ounces,
Goulard's extract 1 ounce;

Mix and rub the part well twice a day.

Low diet, a plentiful supply of water gruel, and bran mashes, to which an ounce of nitre may be added daily, will reduce that thickened state of the blood which ever attends this species of tumour. But, as in the preceding cases of tumour (poll-evil and fistula), it is sometimes found impossible to remove the vives by those means or any other; matter is formed, the tension and inflammation continue upon the increase, and plainly indicate that suppuration must ensue, and all our labour is rendered vain, if it ought never to have been so employed. In this event, apply a meal poultice, restore the animal to his ordinary diet, and promote suppuration, which effects the cure in the same manner as all other abscess mentioned before. See page 114, &c.

False vives, or imperfect ones, that are hard and insensible, sometimes cause a good deal of needless trouble. They neither come forward nor recede, do not seem to cause any particular pain, but still continue an *eye-sore ;* and give reason to apprehend disagreeable consequences, and always prevent an advantageous sale of the animal. Stimulating embrocations are well calculated for reducing these hard tumours, and the blistering liniment, made of cantharides and oil, never fails.

LAMPERS, OR LAMPAS.

Cause.—As just said, lampas is occasioned by inflammation in the mouth. This is brought on by inability in the parotid glands to secrete the saliva necessary for lubricating the throat and gums. These glands, though liable to the disorder we term vives, yet the derangement of their secretory function does not always show itself by the vives : it may continue to flow, though not in sufficient quantity to meet the increased heat of the animal. Idle or illworked young horses are most liable to lampas.

Symptoms.—A swelling of the bars of the mouth follows the rising vigour and heat of the animal; they then project below the surface of the teeth, and interfere between them while feeding. The pain is necessarily very great on feeding, and the animal ceases to chew of a sudden ; it afterwards commences anew, with greater caution ; but as the disorder becomes worse, it refuses food entirely, and starvation would be the consequence if something did not intervene which is always sure to happen.

The cure would be effected of itself, if the horse lived in a state of nature, or more probably in that state he never would have contracted the disease. Over-gorging and consequent fulness of habit having occasioned the blood to flow luxuriantly towards the region of the head and throat, so that the disorder is thereby produced, the reduction of that full habit follows this compulsory abstemiousness which the afflicted animal practises much against his will, and might teach man himself a monitory lesson he is usually slow in attending to, until too late. Reducing the system is the neatest method of removing lampas, and purgatives should be employed ; bran mashes, in which an ounce of nitre daily has been introduced, may also be given until the pulse becomes more natural. If the lampas be not lessened, by these means, the projecting part is to be cut with a lancet, but some people commence operations with the searing iron, as the readiest way, and give physic afterwards. This application never fails.

CHAPTER III.

EXTERNAL DISORDERS.

Purulent Tumours: Diseases of the Glands.

STRANGLES, GLANDERS, FARCY, GREASE, as they owe their origin to the same predisposing cause so evidently, that the appearance of either is good assurance that no other disorder is then to be apprehended—neither of the above nor those treated of in the preceding chapter, a few preliminary observations should occupy attention, before we treat of any one in detail. Both series of diseases are in like manner constitutional, or residing in the blood; and the whole class agree together so nearly in cause, symptoms, and effect, that the situation of each on the various parts of the body constitutes the main distinction between them; as this does also affect the appearance and consistency of the matter produced.

What I most strenuously maintain is, that the latent cause of all tumours, inflamed glands, and spontaneous discharge of matter by skin or membrane, is entirely attributable to the actual state of blood of the individual animal. Whence I infer, that some horses are more liable to incur contagious diseases than others, and this in a degree proportioned to the state of the blood at the time of communication; so that some might escape with impunity, whilst others meet with certain death from the self-same cause. This accounts for the greater virulence with which some horses incur glanders, for example, compared to what others suffer, which catch the disorder at the same moment of time; as was proved on a largish sort of a scale, and that pretty well known among practitioners, during the late war on the continent. The case was briefly this:—A transport with cavalry horses on board, on its way to the Low Countries, met with bad weather, so that the hatches were battened down, and in this manner were part of the horses suffocated. Of those which survived, amounting to some twenty-two or more, scarcely one escaped the glanders, but, notwithstanding we may conclude that they infected and reinfected each other at the same moment, and under precisely the same circumstances as to heat, respiration, and privations, yet the symptoms varied greatly, and some few recovered so readily as to leave great doubt whether they really had received the glanders or not, whilst others exhibited real glanders in the highest degree of virulence. Between these extremes, we are informed, the remainder were variously affected: all which circumstances prove incontestibly how much depended upon the previous health of each individual, the vitiation of its blood and its co-fitness or adaption to receive the infection. I imagine this to be conclusive of the doctrine I have all along laid down. But I will adduce another authority—a veterinary writer of France, who carries the principle even farther than I have adventured to push it.

With that specious ingenuity which attends all affairs of research in that country, an author named Dupuy, who also quotes the *rapport* of another called Gilbert, deduces the disposition to contract such disorders from the progenitors of the afflicted, or, as I should have said, from the blood or breed, and he recommends a corrective kind of regimen for brood mares and stallions; that is to say, in other words, an airy situation for the breeding stud, with

13

pastures rather elevated, where they will have sufficient norriture during the period of gestation, and can find occasional shelter from the weather. "By these means (says M. Dupuy) the disorder may be prevented in great measure." The disorder he here speaks of he calls "scrophulous tubercle ;" to which "all cattle whatever, bred in marshy situations with scanty allowance to the parents, are very liable." This disorder of the blood or breed, according to M. Dupuy, "predisposes the horse to contract those diseases that are known to us under the terms strangles, bastard strangles, farcy, and defluxions from the eyes ;" which latter, it will be seen, at page 127, is a corresponding symptom and never failing attendant upon the vives, as it is of all other glandular swellings about the jaws. The Frenchman thus converts a single symptom into a disorder!

In England, moreover, we do not talk or write of scrophula in horses, or a disposition thereto, this being a symptom of a vitiated system in carnivorous animals. For, the mange in dogs, scurvical or scrophulous eruptions in mankind, and the farcy or grease in the horse, although appearing very similar to the eye of a common observer, and all originating in a depraved state of the system ; yet the immediate cause of each of these differs greatly, by reason of the manifest difference in the structure of the capillary vessels or tubes that deposit the offensive matter of either kind, demand a very different treatment at our hands, and we reject the anomaly of M. Dupuy as inapplicable to horse-medicine. But when this gentleman represents the general predisposing cause as a "tuberculous or fistulous affection, that is capable of being alleviated, prevented, and in some cases cured," he brings his arguments quite within the range of our conceptions ; and I, for my part, take all that he subsequently adduces, as being in perfect consonance with my own doctrine respecting the predisposing cause of diseases. As to ancestry, and breeding from a good stock, in favourable situations, of which this writer appears to entertain correct notions, I had already anticipated him, as the reader may perceive at pages 18, 19, which is a part of my book that appeared in the Annals of Sporting for 1822.

THE STRANGLES.

The Strangles, as the name imports, is first indicated by a coughing, and difficulty of swallowing, as if the animal would die of strangulation. It is a disorder of youth (like our hooping cough), is inherent to the nature of the animal (as is our small pox) once only, and its virulence may be abated by inoculation, whereby we choose a favourable period for meeting the inevitable attack, after duly preparing the patient.

Cause.—Repletion of the system of life, and the deposite of blood in the glands under the jaw ; which failing to be taken up and reconveyed back again into the system (called absorption—see book the first, p. 21), the glands become inflamed, swell, and burst, the discharge of the offensive matter being the cure. I have always considered it a critical disease, and treated it as such, encouraging the formation of matter, and assisting nature in throwing off a something that is evidently obnoxious to the constitution. Indeed, I have never heard of any other practice ; the impertinent attempts at repression, so frequently adopted at the request of proprietors in other cases of tumour, never having extended itself to this. Strangles, strictly speaking, are incident to the young animal only—that is, from two years old, until five or near six ; when the circulation (as the blood is called) has attained its fulness, and, perhaps, slight cold has first detained any portion thereof in the glands, whereby the inflammation is engendered that constitutes the disease. When these glands swell and discharge at a more mature age, the strangles must then be considered as the effect of constitutional depravation, and would as properly come under the general description of critical abscess, treated of at a former page, 113.

Symptoms.—A swelling commences between the upper part of the two jaw-bones, or a little lower down towards the chin, and directly underneath the tongue. A cough, and the discharge of a white thick matter from the nostrils, follow; with great heat, pain, and tension of the tumours, and of all the adjacent membranes, to such a degree that the animal can scarcely swallow. The eyes send forth a watery humour, and the animal nearly closes the lid : this is mostly the case when it happens that the two larger glands under the ear are affected also, which frequently happens ; but when these latter are disordered without the animal having the strangles, we then say he has the vives. The swelling increases and usually bursts of itself, sometimes without any medical aid whatever, and even without being perceived by any one. This last happens to colts and fillies at grass, when their wants are little attended to, and they seldom fail of doing well. Consequently, it follows, that those attacks which take place in the open air are of a milder nature than those more obstinate cases we so frequently meet with among in-door cattle, which serves to prove, once more, my doctrine as to the cause of all tumours or "tuberculous affections," as M. Dupuy has it. The horses that are kept in-doors accumulate gross humours, by this mode of living on dry food and lying on soft beds, the exercise they take not being sufficient to carry off the effects of either. Enervation generally accompanies this mode of treatment ; the glands and membrane suffer relaxation, the pampered animal is not exposed to the air sufficiently to occasion that check, or slight cold, which is generally the immediate cause of strangles, and the accumulation of these humours proceeds, until they overcome the capacity of these organs, and the strangles then become a formidable disease.

When this is the case, the feverish symptoms run high, loss of appetite follows with constipation of the bowels, the horse can neither drink nor eat, and the pulse increases. The tumours in these bad cases will be found to have risen nearer the jaw-bone than they do in a mild attack, and are longer in coming to maturity than those which begin more towards the middle. The disorder is seldom fatal ; but when this does happen, the animal dies of suffocation, in which case it stands with the nose thrust out, the nostrils distended . the breathing is then exceedingly laborious and difficult, and accompanied by rattling in the throat.

Cure.—For this last mentioned extreme case, no other remedy is found than making an opening in the windpipe, through which the animal may breathe. For this spirited operation, I must refer to Mr. Field, the veterinary surgeon, who has performed it frequently, and says his practice was to cut an aperture the size of a guinea, which nature afterwards supplies in due time. With this exception, perhaps, I might be justified in saying that we have little or no business to meddle with the strangles ; unless, indeed, unfavourable symptoms arise, and the previous habits of the horse, his present fleshy or gross habit of body, with the unfavourable situation of the tumours near the bone, give good reason for believing that the disease will turn out a tedious or dangerous case. And yet I should be very loth to recommend purging or bleeding for strangles, as I have seen done with no good effect ; for, although the symptoms are thereby lowered, yet the continuance of the disorder is protracted to an unmeasurable length, and I have heard of the strangles devolving into glanders by this course of proceeding—the subject of this case being a five year old mare.

On the contrary, the disorder being constitutional, that is to say, an effort of nature to relieve itself of some noxious matters, the strength of the animal system should be sustained in some degree proportioned to what it may obviously require. Therefore, horses that may be in good condition at the time of the attack, and withal highly feverish and full of corn, will only require opening medicine, whilst a brisk purgative might do harm by lessening the access of matter to the tumour, and the system would still retain a portion of the offensive cause of disease, which would break forth at a future period in some one or other of the correspondent diseases dependent on tubercular affections. In this case give the following

Laxative Ball.

Aloes, }
Castile soap, } of each 3 drachms,
Ginger, 1 scruple. Mix for one dose.

If difficulty of swallowing is already perceivable, a drench would be found the more desirable form of arriving at the same end. Then give the

Laxative Drench.

Castor oil, 6 ounces,
Water gruel, 1 quart,
Salts, 6 ounces. Mix.

Meantime, at the first appearance of the disorder, let the hair be clipped off close at the part affected, and a little way round, to allow of greater effect from any application that may be deemed necessary. The head being clothed, will restore as much warmth as hath hereby been abridged. Mild cases will require no more than this, probably, and the assistance of a poultice and fomentation of marsh-mallows daily to the throat, to bring the swelling to a proper state for opening. This will be shown by its pointing, or becoming soft and peaked in the middle. But a premature employment of the lancet is to be avoided, for the reason before assigned, viz. to give time for the whole matter to collect; when this period arrives, the whole swelling will be soft and yielding to pressure, unless the animal be a very thick-skinned one, with a great chuckle-head. In these cases, the part should be rubbed with a stimulating liniment, and if the tumour is working its way inwards so as to threaten suffocation, blistering ointment may be applied. After each and either of these applications, the poultice is to be again put on with care; and as much of its efficacy depends upon its remaining in contact with the throat, the adjusting of it properly requires great pains and some adroitness for the thing.

Much difficulty being experienced in making serviceable bandages, I have annexed a descriptive sketch of such a one as would be proper to keep on the applications. It needs no further explanation than is given in cases of sore throat at page 78, and of abscess, at page 114.

Stimulating Liniment.

Mustard, powdered, 1 ounce,
Liquid Ammonia, 3 drachms

Mix and apply assiduously to the part.

The suppuration may be further promoted by steaming the head over warm water, or fumigating it as follows: give him bran mashes frequently, placing the vessel that contains this, well secured, in another vessel larger than the first, into which much hotter water can be introduced, so that the vapour may rise up all round the mash, and constantly envelop the head. When the tumour is ascertained to be ripe, and not before, an opening is made at its most depending part, and the matter expressed gently; wash it off clean with warm water, and if the sore appear healthy, it will heal spontaneously, or with the application of adhesive plaster.

Inoculation for the strangles has been recommended above and was partially practised. About 1802, M. La Fosse, the younger, mentioned the affair in his Manuel d' Hippiatrique, which book I translated into English the following year, and we hear that two or three country practitioners in England afterwards adopted the suggestion. The method was merely to scratch the inside of the nostril, and then smearing the sore with matter from the abscess of a diseased horse—it never failed. In careful hands the practice was feasible enough; but great danger would accompany this imitation of variolous inoculation, inasmuch as the matter might likewise convey a disposition to farcy or glanders.

STRANGLES OF THE GULLET. Sometimes we find those symptoms of strangles reduced to one only, viz. an obstinate running at the nose, which usually lasts a long while, and occasionally ends fatally, by the animal wasting away in pulmonary consumption, as I am informed from good authority, but never witnessed such a termination of this species of strangles, which La Fosse calls "strangles of the gullet." Many people mistake this disorder for glanders, but it may be distinguished from that contagion by a rattling in the gullet, whence its French name; also by the quality of the running, which

13 *

is neither so white nor of so much consistency as the true sort; but water
and curdled.   The animal scarcely ever is troubled with a cough, and then
is very feeble, but to make up for this exhibits frequent contractions of th
larynx.

After the tumour is opened, give gentle physic, for which purpose the laxa
tive ball recommended in page 132 will answer every desirable purpose; o.
you may add thereto one drachm of emetic tartar, and give another ball after
an interval of one day, unless the pulse is low.

BASTARD STRANGLES is a favourite term with some persons, who would
soften down the real fact of their horses having the glanders, which it really
is, and not strangles.   But bastard or not bastard, it is always infectious, and
the animal either dies of strangulation, or the disorder becomes the glanders,
producing a sanious discharge from the nose when the cough ceases. Apply a

Fumigation.

Take the leaves and root of marsh-mallows, an arm-full.
Water, 6 quarts.

Boil them, and put the whole into a nose bag, and hang it round the head of
the animal to make him inhale the steam.   The bag may be made of stout
cloth, but hung with the upper part quite open, to avoid suffocation.   Leave
the bag at the animal's nose until no more steam will arise.   This will be
found a very proper remedy in all cases of strangles, the first stage of glanders, and obstinate colds.

GLANDERS.

For about twenty years I apprehended that we had arrived within a short
space of finding the true cause of glanders, and that we should then soon ascertain the means of preventing our horses from engendering the malady, if
we could not avoid their catching it, nor discover a specific remedy.   But, lo!
we were not yet agreed even as to the symptoms of true glanders; as to that
which was communicable and dangerous, compared with another affection of
the glands and pituitary membrane, which was but a temporary disease, not
easily communicated, and was asserted to come within the reach of the curative art.   Most small proprietors, unwilling to destroy their afflicted horses,
maintained that they belonged to the latter description, and in this they were
frequently supported by the cupidity of practising farriers, who administered
medicines and performed operations with a confidence which never could belong to any department of science—and least of all to that of medicine,* which
is, alas! ever uncertain.

During this state of the question, we turned to the French veterinarians,
who up to a certain period enjoyed the reputation of being superior to all Europe besides in this and a few other pursuits [war and chemistry, *videlicet*],

* Much inhumanity was shown by the country practitioners in their mode of treatment·
they scraped the bone after slitting the nostril; and also seared the swelled gland with a hot
iron.   A late writer applauds the practice of searing ulcers and abscesses, generally, "whereby (adds he triumphantly) they are reduced to common scalds:" he was then speaking of the
doctors in Morocco! mere Turks.   The gentleman, probably, did not distinguish between abscess and indolent tumour.

It is related by La Fosse, that in 1801, several regiments in Alsace and Loraine employed the
actual cautery as a cure for glandered horses.   Some "applied fire to the jugular gland in three
lines; others cauterized the bones of the forehead and nose; but the most ridiculous affair of
all was, to see forty horses together which had fire applied round their eyelids to cure the running," that is common to all glandular affections about the head'

and found one of their most respected names had arranged the boundaries and distinguishing qualities of the two into three divisions. Nor was this all. M. La Fosse, the younger, who enjoyed the post of "principal farrier to the French army," and was withal a member of the Institute, insisted with much energy that "glanders of the first species, the real glanders, glanders properly so called, absolutely consists of nothing more than the loss of the sense of smelling," and is "a curable disorder, if treated early, but incurable when confirmed." His treatment was very simple, and worthy of calm consideration, as are also the means he proposes for ascertaining by the symptoms when it is a horse is afflicted with communicable glanders, and ought to be destroyed. It is in this latter respect that I reprint here the substance of La Fosse's researches on this highly interesting subject; for, since none of us can offer a remedy that ought to be relied upon—unless the animal be submitted to our measures earlier than is usual*—I think a useful particle may be added to the new study of medical jurisprudence, by showing the line of demarcation that divides health and contagion—the point at which destruction ought to commence, by authority, or, in common humanity, to prevent the thoughtless from immolating the property of others, who are usually little proprietors.

I am the more determined in this course of proceeding, because all my inquiries on the subject are already in print, and these coincide so nearly with the speculations and reasonings of two or three respectable writers now before the public, that I could add but a small portion of novelty to what has been so elaborately discussed by others. On all those points on which I differ from them, I shall offer a few practical observations, in the hope of being serviceable, whilst I shall sedulously avoid the "debateable land," which some contend for a little unamiably.

"So great has been the destruction of horses which have either really been, or reputed to be glandered, through the prejudice of ignorant persons, that it has been said, whoever can point out the distinction between the communicable disease and those which bear some affinity to it, will confer a benefit on society, and serve the cause of humanity. Much labour had been bestowed on this subject by the elder La Fosse, who threw a great deal of light on a disease, which, of all others that attack the horse, is least known, and therefore most misrepresented."

What M. La Fosse proposes to show is, 1. "That it is easy to confound this disorder with others that resemble it, in some particulars. 2. That among the different sorts of glanders (so reputed), some are infectious, whilst there are others that are not so; and 3. That some of these are curable, and others incurable.

"There are few veterinary practitioners who do not know that strangles, bastard strangles, pursiveness, or asthma, and other pulmonary complaints, exhibit the same external appearance as the glanders, properly so called; consequently it is easy to be deceived on the subject, and the farrier will fail in his endeavours at a cure if he has not previously ascertained the distinguishing symptoms of the disorder. What then ought he to do in order not to be led into error, and to ascertain with precision that species of the disorder with which the animal is affected?

"After much experience on the subject of the diseases of horses, we are convinced that it is necessary to distinguish three sorts of glanders, viz. The first sort, which is the glanders, the real glanders, the glanders properly so called; the second is nothing more than some disorder circulating in the mass of blood; and the third may be denominated the farcy glanders. Glanders

* From the number of experiments which have lately been made at the Veterinary College in London, and the beneficial results arising therefrom, we may at length hope that this dreadful malady is brought within the sphere of curable diseases.

of the first kind is not infectious, except it be complicated with other disorders ;
but this is seldom the case, though we may daily witness horses thus attacked abandoned as incurable. or with little more humanity put to death. On
the contrary, glanders of the second species is communicable, because the
horse, besides running at the nose, and becoming glanderous, has likewise ulcers, and these ulcers appear to be the only proximate cause of contagion.

"The third species of glanders is in like manner contagious, because it not
only occasions a running of the nose, but the tumefied glands and the cartilage of the nose are ulcerated, and likewise certain parts of the body are covered with lumps and ulcers, which latter characterise the farcy glanders, the
most dangerous disorder of the three, but not the most common. These two
latter species of glanders are infectious, because the disease resides principally
in the blood ; but the glanders of the first species, the real glanders, the glanders properly so called, is not in anywise contagious, although it most frequently occurs.

"The second and third species are incurable, but the last only is mortal.
But as to glanders of the first sort, it is neither incurable nor mortal. In the
first place, we repeat, this disease is not mortal in any case, and a horse attacked by it is in the same situation as a man who has lost the sense of smelling ; it is the loss of a sense, and the loss of a sense prevents neither the man
nor the horse from fulfilling all the animal functions; for, as we daily observe
men affected with ulcerated noses preserve an otherwise sound constitution,
and even look jolly, so we may observe a glandered horse preserve his strength
and health.

"Secondly ; it is incurable only when inveterately confirmed; but when
taken in an early stage, its progress may be stopped with very little trouble.

"Thus we find that glanders of the first species, the real glanders, glanders
properly so called, absolutely consists of nothing more than the loss of the sense
of smelling. Its cure may be readily effected by frequent bleedings and fumi-
gations. Hence may be estimated the little necessity there is for killing
horses attacked by this disorder ; and what important services may be rendered to society or to a regiment, for instance, by an intelligent farrier making a
proper distinction between this species of glanders and all other affections and
diseases resembling it."

So far M. La Fosse : his table, prefixed to the translation, "Veterinarian's Pocket Manual," is sold separately by the booksellers, and may be consulted with profit by those who would push further their inquiries respecting
"true glanders."

Cause.—The glanders is a contagious disease only when it has lasted for
some time. Original glanders may be acquired by horses being shut up close
together, in hot, damp stables, in swampy situations—as in the case of the
twenty-two cavalry horses adduced higher up (page 129), which were confined damp, under hatches, but were variously affected, according to the predisposing cause in the constitution of each individual. Those animals were
improperly condemned, because the disorder had not continued long enough
to render it contagious, and they might have recovered if treated as for a
simple cold.

A sudden transition from cold air to a hot stable, as well as from heat to
cold, will occasion a running at the nose; or a blow there, as well as a drench
clumsily administered : either of those causes being foreknown, should render
us chary of pronouncing the running contagious, and thus subject the property to destruction, as proposed. Almost any running, from whatever cause proceeding, or however healthy the previous state of the animal's system, causes
the glands to enlarge and inflame ; after a while, remaining uncured, they
usually adhere to the bone, when alone we should pronounce the glanders con-

firmed and incurable.   This is "the second species" of La Fosse, which may
be communicated by contact, or by respiring the same air, in the stable;
though it does not appear until eight or ten days after the infection, in the enlargement of the gland, accompanied by running.   The third species is caused
by farcy being in the system, or by inoculation, in which way the glanders is
often communicated by experimentalists : the running at the nose and swelling of the glands are then symptomatic of farcy, and must be treated as such.
What inference is to be drawn from all those premises, but that we should
endeavour to ascertain the length of time the patient has been afflicted; whether
he has received any external injury to cause it, or, has he been brought in
contact with infected horses, and when ? and out of the answers hereto we
form the resolution of condemning the animal to solitary keeping, at the least ;
and setting about the remedies that are likely to restore him to health. Crowded towns, posting stables and barracks, are most subject to contain glandered
horses, on account of their closeness, and the frequent succession of inmates
to which they are liable ; for some horses will bear it for a good number of
years, the discharge almost subsiding (though the swelling of the glands remains) upon changing to country quarters, or to a succession of regular living
and regular work.

Symptoms.—No cough accompanies real glanders in any of its stages; and
this though a negative piece of information, shall be taken as a good and positive criterion that must not be neglected : a running may make its appear
ance, as it does at the left nostril usually, in the glanders, and the glands under
the jaw may adhere to the bone, as they do in real glanders, but no cough accompanies these symptoms of glanders.   When cough supervenes, the disease may be a catarrh, or a consumption, the asthma, or strangles, but these
are not contagious, unless they last a long time, and adhesion of the glands
takes place : in these last mentioned disorders the discharge commonly proceeds from both nostrils alike ; whereas, the running in incipient glanders is
chiefly confined to the left,* and the gland of one side only is then affected.

As the disorder proceeds, it affects both sides alike ; ulcers appear all over
the pituitary membrane, occasioned by the corrosive nature of the discharge.
This assumes a different appearance as the constitution of the individual
may have been more or less gross or vitiated ; the appearance or quality of the
discharge differs also, according to the manner in which the disease may have
been acquired ; i. e. whether it has been engendered or caught by infection.
If it come of the first mentioned, through a depraved system, the glands are
harder, often smaller, and always adhere closer, than in those cases which are
derived from infection, at a time when the animal is otherwise in comparatively
good health.   Again, with the infected horse, the matter comes off copiously;
it is curdled, and may be rubbed to powder between the fingers when dried.   It
subsequently hardens, and becomes chalky when submitted to acids ; whereas
the animal that engenders the disease without receiving infection sends forth
matter that is party-coloured, less in quantity, blackish, watery; and mixed
with bloody and white mucus.   Finally, if the animal that receives the disorder by infection be previously in a bad state of health, those symptoms are complicated and more intense, the ulcers are more numerous, the cartilages of the
nose become rotten, and the bones likewise in a short time : the creature seems
to have combined together the evils of its own system with that of the sufferer
from whom he had received it.   In both cases the swelled glands are simply
hard tumours without any matter in them.

In addition to the preceding tokens for discovering at an early period the true

---

* Of eight hundred cases of glanders that come under the notice of M. Dupuy only one horse
was affected in the right nostril.

glanders from another disorder, having some of the same symptoms, let the
nostrils of the animal be examined, and the left or running nostril will be
found of a deeper colour than ordinary, whilst the other or dry nostril is of a
paler colour than ordinary, or almost white.   At this period the discharge is a
white glary fluid, and the maxillary gland of that side is but just perceptible
to the touch; but these being symptoms that belong equally to a catarrh, it is
best to be guided by the varied colour of the two nostrils, remembering that in
catarrh, or cold, both nostrils run.

It has been remarked by some, that when horses in a tolerable state of health
first receive infection, they show mettle, and are full of freaks theretofore not
experienced; as the disorder proceeds in its ravages, this mettlesomeness goes
off; other acquired diseases have the same effects on all animals—the venereal,
for example, on man.

The remedies that have been applied to the afflicted animal in this forlorn
disease are found of no avail, unless taken at an early period.   As hath been
said higher up, infection is not to be apprehended at first, and therefore the
precaution some use to prevent its spreading, by bleeding and purging all the
rest of the horses in that stable, is unnecessary, unless the animals require
that process in other respects; but some practitioners must be doing something, and some owners will not rest satisfied unless preventive measures be
undertaken.

As soon as a horse is suspected of glanders, it should be kept separate from
all others, and the fumigation of marsh-mallows applied, as prescribed at page
134, repeatedly; a purgative or an alterative ball may be given, according to
the state of his body, and the usual remedies as for a catarrh, continued for
a week or ten days.   If the disorder does not lessen in this time, but the
symptoms increase in virulence, the horse should be destroyed; but unfortunately for healthy animals, this measure is not compulsory, no statute existing
upon the subject.*   Besides which, disputes might arise as to the precise nature of the symptoms, and the executioner subject himself to heavy damages
for his temerity.   Something of this sort happened near Woburn, in Bedford-
shire, early in the present century, to a lately deceased statesman.   A neigh-
bouring farmer having a horse in a state of confirmed glanders (in my opinion), persisted in keeping it in an old shed on the roadside: his obstinacy was
highly provoking, and Mr. W. the gentleman alluded to, went with his servant and shot the animal; at which the venal part of the periodical press set
up a great clamour, from which none defended him, for the transaction was at
variance with his public professions.

Stables that have been occupied by glandered horses retain a long time the
taint, and the means of communicating the disorder, which nothing will remove but washing with soap and sand, and scraping with sharp instruments,
every part of the rack, manger, and all other parts that may have come in contact with the diseased horses.   After examining the cases reported by various
writers, particularly St. Bel, I have come to the conclusion that there is no
analogy between glanders and the venereal disease, but the inoculation of
sound animals; and that the exhibition of mercury in any form is utterly falla
cious.   One case, in which this mineral was employed with asserted success,
at the Pancras College, is proved unworthy of credit, by the failure of the
same medicine in every succeeding attempt.

* The common law, however, a fully sufficient to prevent improper exposure of animals afflicted with a contagious disease in horse-markets, fairs, and other assemblages of cattle. A
case of this sort was adjudged at Guildhall, London, the facts whereof were detailed in the Annals of Sporting for March, 1826.

FARCY.

Causes.—General ill state of the blood, vulgarly, but most appropriately, termed "corruption of all the humours of the body;" and, by prevalence of the farcy buds in the course that the veins run, all over the surface of the body, no doubt can exist that it resides in the blood. In fine, the original cause has been already defined at the head of this class of diseases, to which the reader who is fond of research would do well to turn back for a few minutes, at pages 129, &c. However, infection is frequently the immediate cause and (as observed of the glanders) the animal will be afflicted more or less severely, as his constitutional health may be sound or otherwise at the time of receiving the infection. When this disease is engendered or created—which is easily supposed to have happened at first, and capable of being so produced at the present day, the blood being overcharged with offensive matters unfit for its proper purpose, it becomes stagnated at the lymphatics which follow the course of the veins (see Book I. page 30), and these corrode the parts, inflame, and appear on the surface in the form of "buds." And I should feel surprise if they do not also pervade the large glands of the viscera, though I have had no opportunity of examining: such tumours on the skin of mankind have been found similarly seated on the inner surfaces, on dissection. "I feel all over as if pins were running into me, observed a patient; and so, poor fellow, he might, for on dissection I found the same sort of tumours even on the heart," said Mr. Abernethy in one of his lectures. As regards the horse, I take this upon credit, and by analogy for a while, purposing to satisfy myself more closely upon the first favourable opportunity that offers.

A predisposition to farcy must exist in the system, for it is cured by means of correctives of the blood; and its connexion with glanders has been proved, for the one will produce the other by inoculation: and without it running at the nose is one of the symptoms of farcy. A certain inability to perform its office, termed "debility," that leaves the finer vessels filled with the vital fluid, which exercise might have carried off—is one main cause of farcy; and a sudden check by cold after exercise stops at once the perspiration, and the blood that would otherwise be taken up, or absorbed into the circulation, remains in those fine vessels, as aforesaid. Hot and crowded stables relax the vessels, and indeed the whole system, when sudden exposure to the air inflicts the same evil I have just now contemplated. The reader may recollect, that I referred the cause of fever to the same want of ventilation, to the same exposure to cold air, and made the remark, that the state of the animal's bodily health at the period of the attack would determine whether it should acquire this or that particular disease; the quantity and kind of cold, or chill, would also determine whether the horse should be afflicted with inflammation of a certain part of his inside, or of his whole system, which we term fever.

Symptoms.—Though too well known to be mistaken, we yet may describe them, as, in the first place—skin tight and dry, for want of perspiration, as just said, when some swelling is perceivable about the hind legs, and on the insides particularly. This symptom increases to an extremely large size in the course of a night, when the genial heat of the animal's system, and of the stable, appear to have matured the disease. The lymphatic vessels, and the more perfect glands, that run in the same direction as the veins, rise above the surface; and it is easy to be seen that they are sore when touched, the glands in particular, which feel hot, light, and hard at first, similarly to the glands of the throat, as described in the glanders. A few hours confirm the exact nature of the mischief: the inflammation of those glands proceeds, they become softer, and each throws out an ichorous, unhealthy discharge. They are then termed farcy buds. The edges have a chancrous appearance which it is

found impossible to heal with ointments. As the disease advances, a glanderous running at the nose takes place, with swelling about the nose, lips, and
all over the body nearly.

According to the constitution of the horse at the time of receiving the infection, so will the progress of the disease be rapid and disastrous, or unaccountably slow and uncertain as to the result. In this latter case it retains its appetite, and bears up its strength for a considerable time.

Cure.—Three stages of the disorder present as many methods of cure: first,
when the effects are slight or partial; second, when it resolves itself into large
tumours, and a more copious discharge from a smaller number of ulcers carries
off the disease: third, when it is confirmed, general, and diffused over the
whole system.

In the first instance, when the glands only are affected, it may be treated as
a local disorder, which has not yet found its way into the animal's system;
and if appearing on one limb only, the natural inference is, that the system is
indisposed to carry on the threatened evil. This happens mostly to animals
in good condition, that are strong and vigorous, and of good habit of body;
with such the farcy is not brought on spontaneously, or by being engendered
in the animal, but has been acquired by infection. The limb affected is generally so to a good extent, and the corded veins scarcely visible; this happens
mostly to a fore-leg (not always), and has been considered of a dropsical nature, in a slight degree, or rather, lymphatic, the glands still continuing their
functions. In this case give a purgative ball, and repeat it in three days
after; but should the animal be a very fleshy one, and full of condition, with
full pulse, this may be preceded by bleeding to the amount of three quarts,
or four.

Purgative Ball.

Aloes, 8 drachms,
Castile soap, 1 drachm,
Liquorice powder sufficient to form the ball for one
dose.

Prepare the animal with bran-mashes; let his drink be chilled, and he may be
moved about, under shelter, with body clothing on that covers the affected
limb. The limb should be fomented with warm water, or the chamomile decoction, taking care to rub the part dry, and wrap it up warm. This course
usually prevails against an ordinary attack of farcy, and the cure is aided
when it affects the fore-leg only, by a rowel inserted under the chest. But
the absorption or taking up of the disorder into the system, and carrying it
off by stool, is by far the neatest manner of managing the cure; for this purpose give the following ball, after the animal has been reduced by the foregoing treatment.

Alterative Ball.

Camphor,
Emetic tartar, } of each one drachm,
Asafœtida,
Ginger,

With mucilage sufficient to form the ball for one dose. Give one of these for
three successive nights; then stop one night between each dose, until the disease is removed.

If these efforts to absorb the disease prove insufficient to conquer it, recourse must be had to the use of mercury, as recommended for the third or most virulent stage of the disease, which is also the most common of the three. Meantime, we come to consider of that particular kind of farcy which is the least common of all three; and this is wherein the tumours are larger than usually happens, and smaller in number. The disease then partakes very much of the nature of critical abscess (page 113), and of the strangles (page 130); both of which, the reader will perceive, are but the efforts of nature to relieve itself of an accumulation of offensive matter; and, this escaping, the cure is effected. In this second kind, or stage of farcy (as I call it), nothing more is requisite than to promote suppuration, as directed in the diseases just referred to, and follow it with the physic prescribed for the strangles.

The third, most common and virulent kind of farcy, that which comes on quickest, lasts the longest, and requires the most powerful means for its removal, is that which is spread minutely all over the body and limbs, and has penetrated the whole system. In whichever manner the animal may have acquired the disorder, we may safely presume that the mass of humours is hideously depraved, and mercury, in one or other of its varied forms, is the only antidote to be relied upon for its extinction. Previously, however, the farcy buds and ulcers must be reduced to the state of common sores, by means of the actual cautery freely applied to each. When these slough off, and the sores assume a healthy appearance, less of the mercurial preparation will be required; but if these retain a livid and therefore unhealthy hue, accompanied by a poisonous discharge that ulcerates the adjacent parts, a thorough course of mercury is the only certain remedy, and this must be managed with caution.

Mercurial Ball.—No. 1.

Æthiop's mineral, 2 drachms,
Opium, 10 grains,
Liquorice powder and mucilage to form the ball for one
dose.

Give twice a day, until the patient's breath smells very offensive, and then discontinue the medicine a day or two, as you should also when the animal is found to stale inordinately, or the bowels be very much disordered. But, when the bowels are only slightly affected, increase the quantity of opium to twenty or thirty grains.

Mercurial Ball.—No. 2.

Corrosive sublimate, 10 grains,
Emetic tartar, half a drachm,
Opium, half a drachm.

Mix, with liquorice powder and mucilage sufficient to form the ball for one dose. Give as before, at night and morning.

Feed the patient generously during the operation of this strong medicine, watch its progress closely, and lessen the quantity, or discontinue it altogether a day or two when he is agitated greatly within, particularly if a kind of sickness or gurgling be discernible, and the horse is off his appetite. Let him be clothed completely. Malt mashes, sodden corn, and coarse sugar mixed with his corn, dry, are good assistants to the proper operation of mercury. That is a mistaken notion, which induced some farriers to give the edible roots, as

14

turnips, carrots, &c. to the horse under a course of mercury. They war with its operation, and cause that very commotion in the bowels we should most sedulously avoid.

In recommending the free application of the actual cautery to the farcy buds, in the last page, I do but follow the common practice, being altogether the safest means in ordinary hands, who apply fire in many other cases, with much less reason than is done in that of farcy. Butter of antimony, or sulphate of copper, effects the same end, and has the recommendation of being used exclusively by the French veterinarians. What La Fosse says on this point is emphatic, and shows his opinion of the predisposing cause of farcy: "Do not apply fire in any manner to lumps produced by farcy, under an idea of stopping the disorder. The disease being in the blood, treat it accordingly, and as for the lumps, cut them off: apply blue stone, dissolved in water." When he forefends the "idea of stopping the disorder," doubtless in saying this he only allows that to be the true farcy, which I have considered as the third stage, or confirmed kind. But the earlier or milder stages, which would ultimately end in the third or most virulent kind, if not stopped, being occasioned by the cessation of the lymphatic function—when the attendant glands refuse to communicate with the system (the blood,) can not have yet carried the consequence of that stoppage into the blood. In making this remark, I have not overlooked what was said of the practice in Morocco at a preceding page, 135, *note.*

ANTICOR

Is more prevalent in France than in this country, and is so named from its position, *anti* against, and *cor* the heart. The French words *ante-cœur* have the same meaning, and are derived from the same origin. It consists in an inflamed swelling of the breast near the heart, and the name is extended to any other swelling from this part back under the belly, even unto the sheath, which also swells: in this event anticor is decidedly dropsical.

Cause.—Full feeding without sufficient exercise, similarly to this whole train of disorders which I have been just above considering. Hard riding or driving, and subsequent exposure to the elements, or giving cold water to animals that are very fleshy in the forehand, as is the case with the greater part of French horses; these, combined with a vitiated state of the blood, which is then sizy, produce those extended swellings that partake somewhat of the nature of swelled limb in grease, and yet terminate in abscess when the case is a bad one.

Symptoms.—An enlargement of the breast, which sometimes extends upwards to the throat, and threatens suffocation. The animal appears stiff about the neck, looks dull and drooping, refuses his food, and trembles or shivers with the inflammation, which may be felt. Pulse dull and uneven. By pressing two or more fingers alternately, the existence of matter, or a disposition to suppurate, may be ascertained (as in poll-evil) by its receding from side to side as the pressure is withdrawn. On the other hand, if the disease owes its origin to dropsy, each pressure of the finger will remain pitted a few seconds after the finger is withdrawn. Consult "Poll-evil" in its two stages.

Cure.—As in other cases of tumour, that do not partake of critical abscess after fever, &c., this disorder admits of being repressed, readily, by the means before prescribed, or of being otherwise cured, as it may be allied to some disorder of the constitution. To repress the swelling, bleed the patient copiously; give purgatives and clyster him; give bran mashes, and let the chill be taken off his water. Foment the throat and breast with bran mash or marshmallows, every four or five hours; and when these have reduced the symptoms, give an

Alterative Ball.

Emetic tartar, 2 drachms,
Venice turpentine, half an ounce.

Mix with liquorice powder enough to make the ball for one dose. Give one every eight-and-forty hours. On the contrary, if the swelling depend upon dropsy, as aforesaid, let a fleam or horse lancet be struck into the skin at four or five places distant from each other, and in the lowest or most depending part of the swelling. From these punctures a watery discharge will take place, that relieves the patient of his affliction hourly, and the issue of matter is to be promoted by keeping open the sores as directed in the case of fistula, &c. at page 119: again, when the swelling indicates the collection of morbid matter, let it be fomented, poulticed, and opened as directed in cases of critical abscess, in poll-evil, fistula, &c.: the whole series of these diseases are of the same nature, but differing principally as to situation, which sometimes affects the disease mainly. In this case, for example, the swelling sometimes ascends along the throat, and goes nigh to choke the patient: recourse must be had immediately to poultices, and let these be changed twice a day. The modes of bandaging may be learned by consulting those I have given sketches of, in other cases, at page 79, &c.

GREASE.

Causes.—This is another of the diseases that take their rise in a tardy circulation of the blood, and consequent indisposition to take up and carry back again to the heart that which has been sent into the extremities for their nourishment and renovation. In Book I. at bottom of section 44, this process of taking up, or absorption, is spoken of, whilst the few pages that are there bestowed on the manner in which the circulation is carried on, show the importance of this function, and point out the principles that should guide us in promoting it, when aught has occurred to retard its action. When great age and consequent lethargic habits cause the blood to circulate slowly, our art can but ill supply the remedy, though the evil may certainly be alleviated by stimulants. A small portion of beans given to aged horses admirably assists the circulation of the blood, especially towards the heels, whilst this very species of food given to young horses will promote humours of the hind legs in particular, where grease is mostly situate. That is to say, at the part of the animal that is remotest from the heart is the effect of a slow circulation most frequently recurring, and to heavy fleshy cart horses oftener than to those that are lighter and freer from flesh about the heels.

Trimming the heels of the hair, which was intended to keep them warm in winter, is a very prolific source of grease. Thorough-bred horses never incur this disorder, so far as I can learn; and the chances in favour of those which are produced by crosses from blood stock, is in proportion to the amount of their breeding.

A cold in the heels is caught by walking the horses through water whilst they are hot: or being put into the stable with wet feet at nights; or lying in a stable that imperfectly keeps out the wind; all conduce to that stagnation of the blood, or tardy performance of its function, that causes the animal to generate this disease. They term it debility, but I think we had better say "want of ability," or of vigour to drive on the circulation of the blood; so that if the blood that is left behind in the fine capillary vessels be ever so good and proper for its purpose at first, yet the very circumstance of its remaining idle causes an inflammatory heat, that attracts towards itself all such congenia' particles of the blood which may have been sent through the arteries to the part for the propagation of new horn, or the supply of marrow—of the nature whereof the matter partakes. Indeed, I have very little doubt that the marrow is concerned in the production of grease; for I have successively examined twenty legs which were affected with grease at the time life was extinguished, and the marrow was invariably confined to the lower part only, as if it were fallen down there for want of vigour, whilst the upper part of the bone was hollow, in every instance : healthy leg bones are always full to the top of each, and I have reason for thinking that this is the case with all debilitated horses. Again, the glutinous substance that pervades the surface of the coffin-bone, and to which I have attributed the formation of new horny matter of the hoof, is always found scanty in greasy-healed subjects. See my observations on the foot in the next chapter.

One of those legs parted from the knee, having the skin removed, but otherwise untouched, was hung up in the yard whilst the sun was at 70 degrees (July, 1825). In three or four days the grease might be seen to give a colour to the lower part at the fetlock joint, and every day the greasy nature of the colour was evident to touch and smell, whilst the articulation of the large pastern and sesamoid bones remained unaffected in either way. Upon breaking the bones nine months afterwards, the marrow had all escaped without a puncture, i. e. through the bone.

The following ingenious suggestion I find among much voluminous Veterinary Memoranda, but whether it be my own, or I owe it to some friend, I have no means at hand for ascertaining, nor does my recollection serve me sufficiently to say who. "Horses with one or two white feet are more liable to the grease in the feet that are white than in the others; and if the proposition be true that white feet are weak ones, we come to the same conclusion, that the want of colour having occurred through want of vigour in the part: then weakness and grease have the same cause."

Symptoms.—First perceptible by a swelling at the heels, mostly of the hind legs. This is occasioned by local inflammation, and is soon followed by a slight issue of greasy matter, whence the name; but it is sometimes more watery, ichorous, and offensive, which will depend principally on the constitutional health of the patient. The swelling sometimes extends much higher than the fetlock joint, even towards the hough, and occasions stiffness of the limb and indisposition to move. He can not lie down, by reason of the unbending nature of his joints, and therefore stands to sleep, which renders the disorder more virulent by the accession of fresh matter to the part; the skin cracks at various places, and ulceration ensues. The hair sticks out like furze, the discharge is darker than originally, is thin, acrid, corroding, and stinking.

Remedy.—The grease is one of those disorders about which we should employ our ingenuity in prevention rather than the cure; and this indeed is the case with nearly all the diseases that depend upon constitutional defectiveness, or rather inability of some of the organs of life to perform aright the functions of nature. How these ought to act I have spoken at large in the second chapter of book the first; and pointed out the free circulation of the blood as the principal cause of health, as would also the want of a good circulation prove the harbinger of disease. Now this affair of grease being produced entirely by such inactivity, it seems clear that exercise would be the best preventive of it; and the horse-keeper should also keep the heels dry after work is over, and hand-rub him a little with as much industry as he can afford. He should also let the hair remain on the heels of his heavy horses, and give to the large ones sufficient depth of stall and bed, so as to prevent such from throwing their long legs half way out in the stable (as too often happens) upon the cold floor, of winter nights.

In slight attacks, a wash made of a solution of alum, as under, will correct the disposition to grease, and a dose of physic set all to rights in a short time; both, however, regulated according to circumstances. When considering these, we should inquire into the preceding habits of the patient, as to his usual evacuations, and whether these have been stopped; for it frequently happens that grease is caused by the suspension of the urine balls, to which many proprietors are so very much addicted, that they give them without reason, or suspend the giving through the same whimsicality. In this latter case give the diuretic powder, and the horse will require very little more physic. Again, if the animal require opening physic, give him the purging ball as under, and in-door exercise; but should his debility be then very great, the commotion this would occasion might reduce him too much, and therefore, the alterative ball will do better, with the same attention to in-door exercises if he can bear it. Sometimes, however, the heels are so cracked and chapped, that every step the animal takes only makes the matter worse; we should then assiduously apply ourselves to keeping the heels clean, with water of which the chill has been taken off, and with a brush get rid of as much of the running as possible; and after drying it well with cloths, use the alum wash of the stronger preparation; provided always the inflammation be not too high at the time, but which the warm water without the alum wash has a tendency to alleviate.

I will now set down the several articles just recommended above, premising this much as an apology for the numerous recipes here prescribed, that the grease requires we should be always doing a something for the animal, either of topical application, or in devising the means of carrying off the cause of the disorder by stool, by urine, or by perspiration. For, by keeping one or other of these evacuations a-going, we enable the animal system, to take up, or absorb (as before described) the watery particles of the lymphatics, which remaining indolent constitute the disease.

Alum Wash.—No. 1.*

Alum, 2 ounces,
Blue stone, 2 drachms,
Water, 1 pint.

Mix and wash the part two or three times a day.

Strong Alum Wash.—No. 2.

Alum,
Sugar of lead, } of each 2 ounces,
Vinegar,
Water, 1 pint. Mix and use as before.

Strongest, or Mercurial Wash.—No. 3.

Corrosive sublimate, 2 drachms,
Muriatic acid, 4 drachms,
Water, 1 pint.

Mix and apply in inveterate cases.

* Instead of this, the following is preferred by some persons, and those good judges too—

Goulard's extract, } of each 1 drachm,
White vitriol,
Water, 1 quart. Mix.

14 *

Diuretic Alterative Powder.

Nitre,
Powdered resin, } of each 2 ounces.

Mix, and give in four doses, of mornings. To be continued until its effects are visible.

Purging Balls.

Aloes, 9 drachms,
Hard soap, 3 drachms,
Ginger, 1 drachm.
Mix with mucilage sufficient to form the ball for one dose.

The Alterative Ball.

Aloes, 6 drachms,
Hard soap, 8 drachms,
Ginger, 3 drachms.

Mix with mucilage sufficient to form the mass, and divide it into four balls. Give one every morning until the bowels are opened sufficiently.

The perspiration must be promoted by the following

Diaphoretic Ball.

Emetic tartar, 2 drachms,
Venice turpentine, 4 drachms.

Mix well, with liquorice powder sufficient to form the ball into one dose; and give every other night for a week or ten days, taking care to clothe the patient, or put a rug on his body at least, regulating his sweats according to the weather. Some persons do not think it too much trouble to divide the foregoing ball into two parts, and give one every night for the periods just mentioned, which would bring the whole quantity of emetic tartar to the same amount in the end. Be careful to buy it genuine; and if the horse be taken care of while in his sweats, it will mainly contribute to his getting well. The patient is not to have this sweating ball whilst he has other physic in him; but it may be given alternately with the foregoing powder of nitre and resin; and is better administered thus, when it happens that the individual requires to be set a staling, and we think best to sweat him at the same time. If the medicine makes his bowels grumble, add to the ball

Opium, half a drachm,

which some do put into the prescription, whether or no. But then the opium having a tendency to bind the body, it counteracts our labour in this respect, and is not desirable in case the animal requires opening physic.

Regimen.— A good generous feeding should be allowed, with a few beans for the elder patients only; and in all cases where the disorder has lasted a long while and the cure is effected with difficulty, a run at grass is greatly conducive to complete recovery; especially if the convalescent can be allowed the onion of a field, or covered shed, lying high and dry, or the advantages of the homestead, with an allowance of corn and hay. This change of regimen is greatly assistant of absorption when the physicking has ceased, espe cially where the disorder has terminated with a tedious ulceration which causes lameness, on which event he should not be exercised; but let the parts be poulticed with a turnip poultice, or it may be made of oatmeal and the grounds of stale beer, or both may be employed alternately; and then the parts, if luxuriant or thick, may be washed with a solution of blue vitriol in water, or the wash, No. 3. Dress the cracks with the following

Ointment.

Oil of turpentine, 3 drachms,
Hog's lard, 6 ounces,
Litharge water, half an ounce. Mix.

This may be varied by substituting Venice turpentine, half the quantity of the oil. In these inveterate cases we have now under consideration, a change of medicine is desirable, if but for the change which it occasions in the animal's digestive powers; for this purpose the blue pill has been given as an alterative, as well as that other preparation of mercury, the well-known calomel. Both act upon the kidneys, and set them in motion for the production of urine; and calomel chiefly effects this, by previously stimulating the liver, which again is very desirable by way of change.

Alterative Balls.

Calomel, 1 1-2 drachms,
Aloes, 3 drachms,
Castile soap, 6 drachms,
Oil of juniper, 40 drops.

Mix; make into three balls and give one daily for a week; but should it gripe the animal, discontinue it, or add opium from half a drachm to a drachm.

SURFEIT—MANGE.

Both of these diseases of the animal's system, and the first-mentioned proves its connexion with the second by sometimes ending in the mange. Overfeeding, or too much of it, or gross feeding, as it is the cause of these twin diseases, so are the two appellations it receives in the different stages of the attack descriptive of the cause: both are of French origin, as I apprehend; *surfait* or overdone, being tantamount to *mange*, in its imperfect tenses, the effect of eating too much, which has brought on the disease. A surfeit, or sur-fait, is not an uncommon disease with reasonable man, and is alike caused by eating improperly, if not too much, and sometimes from the preparation of viands that are over-luxurious for the stomach that is to receive it, and is in fact incapable of digesting it. This is a state of the stomach that is by no means uncommon, and up to a certain extent happens every day to the full feeders of every genus of created beings. If, during this full and over-replen ished state of the alimentary canal, and its then active state of lactification (o making of new blood), a sudden check be put upon the said process, by drink ing cold water for example, what happens but the rapid propulsion of some part of the blood through the arteries, whilst the mouths of the offended lac teals close up for a period, and the blood, already filling the capillary vessels of the surface, becomes extremely irritable? Perspiration ceases; the lympha tics refuse to perform their office of absorption, and the blood so deposited in a due course of nature, forms innumerable small tumours under the skin, or become scabby, and throw off a dry scurf. The first has received the appro priate name of surfeit, the second is the more loathed mange, both having but one common origin. They are of the class of tubercular diseases, spoken of by M. Dupuy quoted higher up (as partaking of glanders, &c.), are akin to grease, and to other accessions of matter on the surface, differing only as to situation, and like the grease, require that we should promote absorption and the application of repellants.

The cause of surfeit is thus distinctly met by the means of cure. The symptoms, however, frequently announce the disorder that has taken place within but a few minutes ere they subside again, to the utter surprise of all beholders. On such occasions mischief is supposed to lie in wait, and it is generally understood that the pustules, or tumours, only retreat from the skin to infest some more vital internal organ; but I always considered that such an attack had subsided through its own weakness, for nothing ever came of it after thus retreating spontaneously. Like surfeit in man, these tumours are attended with a pricking pain, the animal appearing restless, flinching from the touch, and looking round sharp at his legs and sides as if he were spurred trivially. Whenever he can bring the parts to bear against the stall, the bail, or the wall, the animal will rub violently, until the hair comes off, and the skin is raw. Instead of tumours that emit a sharp, acrid, and stinking humour, like grease, a dry scurf appears, resembling scabs, and this is mange in some animals: whilst other subjects exhibit no eruption whatever, though every hair is affected in a small degree, the skin becomes dry, and he is then hidebound.

Cure.—Surfeit is easily removed by a cooling purgative; but if the pulse be high, he should be bled also. Promote perspiration by means of the diaphoretic ball recommended at page 146, with the same precautions as are there set down. If the animal be fat, he must be reduced; give bran mashes, sodden oats, and good exercise; and should moisture be found to discharge from the skin, wash it with the

Surfeit Wash.

Blue vitriol, 1 ounce,
Camphor, half an ounce,
Spirits of wine, 2 ounces.

Mix in a quart bottle, and fill it with water. Wash with soapy water warm (as in grease), rub dry, and apply the above wash once a day, and at the same time give one of the diaphoretic balls, as above. Let the diet be cool and opening, as scalded bran, sodden oats, or barley: and if the horse is low in flesh, mix an ounce of fenugreek seeds with his corn daily for a fortnight at least.

THE MANGE

Sometimes succeeds an ill-cured surfeit; and is moreover an original disease, arising from filthiness, hard living, ill-usage, and the consequent depraved state of the system. It partakes of the nature of itch in man, is communicable by means of the touch, by using the same harness, clothing, &c. and probably by standing in the same stall as a diseased horse may have left.

The symptoms are stated in the preceding pages, and from its cause we may rest assured never attacks horses in condition. As in surfeit, the horse is constantly rubbing and biting himself: great patches of the coat are thus rubbed away, and ulceration frequently supplies the places. Scabs appear at the roots of the hair of mane and tail; large portions whereof fall away. When eruptions appear, they form a scurf, which peels off, and it is succeeded by fresh eruptions.

The cure is to be effected by topical applications of sulphur, and giving the same internally as an alterative; but mercurials are mostly preferred by our moderns; and there is not such a variety of opinions and prescriptions at this moment in practice for the most momentous diseases, as for this loathsome malady: neglect and ignorance having brought on the evil, ignorance and stupidity engage to effect the cure. I shall subjoin a few forms of those which are in most repute, and have been found effectual: even alteration is frequently found beneficial, though it may not at first seem to have been for the best.

Mange Ointment.

Prepared hog's lard, 2 pounds,
Sulphur vivum, 1 pound,
White hellebore, in powder, 6 ounces.

Mix with oil of turpentine sufficient to make a soft ointment, rub the animal wherever the eruption and scurf appear, with hair cloths, or a new besom, so as to get rid of the loose filth before applying the ointment. Rub it in well every other day, and give the following

Alterative for the Mange.—No. 1.

Tartarized antimony, 1 ounce,
Muriate of quicksilver, 2 drachms,
Ginger and } powdered, of each 3 ounces.
Anise seeds, }

Mix, with mucilage sufficient to form the mass; divide it into six balls, and give one every morning until the eruption disappears.

Alterative for Mange.—No. 2.

Antimony in fine powder, 8 ounces,
Grains of Paradise, 3 ounces.

Mix, and add Venice turpentine to form the mass which divide into twelve balls. Give one daily whilst the rubbing is continued.

HIDE-BOUND.

The cause of hide-bound is commonly the same as that which produced the last-mentioned disease, viz. poverty, only that the particular animals may not both be in the same state of general health, and the more depraved would incur mange, whilst another would become simply hide-bound. This is less of an original disease than the effect of some other, and of bad digestion and consequent defective perspiration beyond all others, as may be inferred from what I have said concerning the intimacy that exists between those two operations of the animal system in my second chapter of book 1, at pages 23—25. The justness of this view of the cause of hide-bound was further proved by a series of dissections of this particular malady undertaken by me in May 1820. I invariably found tumours had formed upon the larger lacteal vessels of the peritonæum, on the gut, or the like kind of attack on the pleura that covers the lungs. The formation of those tumours was no doubt the mediate cause of hide-bound, and had been brought on (I have every reason for believing) by the inordinate use of diaphoretics, the stimulating nature whereof, as is usual in all such cases, had thus defeated itself.

Horses that are so affected with tumours, are they which become distressed easily, though in good apparent health, upon being pushed on a journey, or at a heavy drag, particularly when the belly is distended. Thus the cause of hide-bound exists long before we can perceive it, and is the reason why I recommend the arsenical preparation hereafter prescribed, as a tonic, previous to administering the sweating remedies. The appearance of hide-bound is frequently ascribed to the worms, botts, &c.; but in all those subjects examined by me, amounting to half a score or more (for I kept no notes), no worms were to be found of any consequence, for it would be ridiculous to talk of two or three such stray insects occasioning such an extensive disorder, so remote from the seat of their supposed ravages. What is very well worth remarking (though such a thing is not very singular) is, that the writer who has most ustily cried out "Worms, worms!" upon every occasion of disordered skin and staring coat, has recently suggested that after all, worms are necessary to the horse's digestion; and the gentleman seems to think that these insects act upon the horse's stomach much in the same manner as pepper on that of mankind! Thus he blows hot and cold with the same breath, or rather worms and pepper with the same pen; for, whilst the worms are so lauded in one volume of his works, the malediction remains uncorrected in the other.

Symptoms.—As the word implies, the hide or skin seems bound or glued to the bones; the animal is always very low in flesh, or we might aver that the skin adhered to the flesh. The pulse is low, and great weakness is manifest in every step the patient takes. As the tightness is first observable at the sides of the animal's body, before it reaches the limbs, and every hide-bound subject examined by me proves the fact, I have no hesitation in ascribing hide bound to disordered digestion, which includes the negation of wherewithal to digest, or starvation and hard work. Again, one of two extremes attends the bowels: they are either relaxed greatly, or much constipated—usually the former; which may be the effect of a long fit of illness from inflammation or fever, and the use of strong medicine, or much of it.

Cure the animal by the direct contrary conduct to that which brought on the illness. If its stomach be empty, as commonly happens, fill it nearly with food that is easy of digestion; if it be too full, empty it; give alterative laxatives and tonic alteratives afterwards; restore the perspiration by the diaphoretic ball recommended at page 146, and let the curry-comb and brush be assiduously applied to his coat. He may then be exercised, but not before, as it is nearly impossible without inflicting great pain. Besides which, forced exercise, or sweating, as hath been strongly recommended, would in this case only aggravate the disease; for if the animal did sweat, it would be caused by internal pain; probably the tubercles which had formed upon the membrane would suppurate and burst, and thus confirm the disorder internally by the inflammation of the particular viscus where the disorder began.

Alterative Laxative.

Aloes, 8 drachms,
Hard soap, 7 drachms,
Anise seeds, powdered, 1 oz.

Mix with mucilage sufficient to form the mass into four balls. Give day after day until they effect the purpose of bringing away a good stool. Then give the arsenical tonic alterative, thus proportioned for a large horse, with care.

Tonic.—No. 1.

Prepared arsenic, 10 grains,
Ginger powdered, 1 drachm,
Anise seeds, powdered, 4 drachms,
Compound powder of tragacanth, 2 dr.

Mix with mucilage sufficient for one dose. Give daily for a week, preceded and followed by mashes, and then give the bark, thus:

Tonic.—No. 2.

Cascarilla, powdered, 4 ounces.
Ginger, 8 drachms,
Salt of tartar, 10 grains.

Mix with mucilage sufficient to form the mass into four balls; give them daily. If the preparation of arsenic in No. 1 is disliked, substitute the alterative ball at page 147, and follow it up with the bark as above (No. 2).

The following ball is calculated to improve the coat, and will be found beneficial when the animal is recovering, if given in these proportions for ten days or a fortnight.

Alterative Balls.

Tartarized antimony, 3 ounces,
Powdered ginger, 2 ounces,
Opium, 5 drachms.

Mix with mucilage sufficient to form the mass, to be divided into ten balls.

WORMS.

As remarked in a preceding page, 150, so many other disorders, external as well as internal, have been charged to the existence of worms in the intestinal canal by veterinary writers, that we find much difficulty in persuading ourselves that this is not the precise ailment which afflicts the animal when his coat becomes staring, and his skin sticks to his ribs. Most frequently, however, that ugly appearance which denotes hide-bound, and other similar symptoms that depend upon suspended perspiration, arise from tubercular diseases of the mesenteric canal (see page 46), and not within the gut or stomach; for the excess or the suspension of perspirable matter must alike depend upon somewhat of a more general affection than worms, that fasten on this or that part of the stomach or intestine (as we are told), and can only influence the part they immediately occupy. Unfortunately, we know of no specific cure for worms, the remedies that are usually prescribed being of a hot, burning and destructive nature, that are as likely to injure the intestine as the worm, it becomes our primary duty, therefore, to ascertain when the disorder be really the worms, so as to prescribe the proper remedy when we have ascertained that the fact is so. It is very easy to say a horse "has the worms," and to give him worm medicine; but much more difficult to ascertain the real fact, than to remove it when well authenticated. Our inquiries, then, should be directed towards this point as much as to any other unsettled question—the existence and quality of true glanders, for example; and yet more fine learn ing has been bestowed upon the uncertain knowledge of botts and other worms than has attracted the attention of our veterinary writers to any other portion of their labours.

Causes.—Indigestion and consequent stoppage of the aliment in the stomach and cœcum; which again may be occasioned by bad corn, musty hay, or hay made from rank grasses,—if all hay whatever does not contain the means of generating insects, when used without sufficient water; also, when either substance be swallowed, as often happens, without being properly masticated, through wearing away of the teeth (see page 17), the lampers, &c. Much pampering of the appetite, by dealers and others, to produce fine coats by means of stimulants, as eggs, wine, ale, bread, diapente, linseed, &c.; when the effects thereof are worn away, these leave the lacteals (see page 47), impaired or offended at being deprived of a short-lived energy. The articles just enumerated form indigestible crudities that become the appropriate *nidus* or generating worms in the canal so deprived of its natural functions by artificial means. Consult again what is said at the conclusion of the first book, at page 54, &c. Irregular feeding also tends to the lodgment of crudities in the cœcum, or second stomach.

Symptoms.—A staring coat, with emaciation and weakness, were formerly deemed sufficient indications of the existence of worms to warrant the doctor in pouring into the animal his monstrous mixtures; for a worm case was esteemed by the professor like a little annuity, pro tem. Those symptoms, however, are at first rather the presage than the concomitants of worms; since they are also symptomatic of several other internal diseases, some of them producing worms in the sequel, whilst other some are found still more rapidly destructive of life than worms are, and therefore demand more immediate consideration. Slight affection of the lungs, as well as of the liver, being of long continuance, occasion partial roughness of the hair, and slight hide-bound of the integuments nearest the seat of disorder, that spreads progressively all over. The cough which accompanies severe attacks of the worms differs from cold in the organs of respiration; the first being more deep and cavernous, leaving a shake or vibrating heave of the flanks, whilst the former comes off with a wheeze, as if not fetched from so deep a recess.

As the disorder proceeds, and the worms may be supposed to extend their ravages, the patient's appetite is subject to extreme variation; he being some times ravenous after food, at others not caring to eat at all; which shows that the stomach is affected, and is frequently succeeded by vertigo, or staggers. A horse with worms that give him uneasiness in the bowels will leave off eating sometimes for two or three minutes, when a cavernous rattle may be heard coming from his inside, and he resumes his feeding. If he endeavours to kick his belly, it has been construed by the worm advocates into the pain occasioned by worms gnawing his bowels; but neither symptom is an invariable indication of worms, for he does the same when attacked by any other pain of the belly—whether colic, tight girth, injury of the sheath, &c. When the worms appear coming away spontaneously, with successive stools, no matter of which kind, it affords proof that the animal has taken grass or hay that contains grasses of an anthelmintic property, and points out the propriety of continuing him on the same food.

A yellowish ordure appearing about the fundament something like flour of sulphur, shows the death of a good number of small worms (ascarides) has been occasioned by some such natural means as the preceding. Some worms come away as soon as generated in the aliment, but if no other sign of their existence is manifest, the solitary fact should excite no uneasiness. When botts, having been detached by similar natural means, leave the stomach—where they do not always cause inconvenience, we find them adhering to the large intestines and rectum, to which they adhere and cause the animal to rub his breech against the wall or upright of the stall. Should those symptoms continue, and the generating of worms remain unchecked, the horse falls into profuse sweats on the least exertion, and when these cease, he exhibits a weak and languishing condition, scarcely notices a brisk application of the whip, his skin adheres to his ribs and flanks—hide-bound has commenced. Cough more or less hectic according to his remaining strength, accompanies him to his end; for, as to a cure being practicable when hide-bound arising from such a cause has fairly laid hold of him, 'tis clean out of the question.

Regimen.—As the commencement of this disorder is mainly attributable to the coarseness of the animal's food and consequent incapacity of its guts to expel the hardened materials, so will an entire change in the mode of feeding him do more towards effecting a cure than all the medicine we can prescribe, and all that the most liberal hand would bestow. I think it would be too much to expect that generous treatment alone should effect a cure of itself, but I certainly have known worms voided after a few days' casual good keep; and in these cases I apprehend we may attribute the coming away to the change or alteration that was so effected in the state of the patient's bowels. Hence the propriety of any change of his usual diet, as well as the advantages of alterative medicines. In the first place, try a run at grass, or give green food in-doors, or succulent and agreeable vegetables. If poor living has not been the original cause, some defect in conformation has; and the above change, with plenty of water-gruel, bran mashes, boiled potatoes, bruised corn, and the like, by 'ubricating the parts, may detach the worm, or at least assist the medicine, which ought to have the same tendency.

Cure.—Since the worms are not always to be killed even by strong poisons, nor brought away by brisk purgatives, for a certainty, but are frequently discharged in a few days by an alterative regimen, reason dictates and nature beckons us to follow her course, in affording to the horse which can not be spared from work, or a run at grass be obtained, to adopt the means nearest thereto that lie within our reach. Laxative alterative medicines then obtrude themselves upon our notice, and in all cases are found to do good, more or less as they may be addressed to the actual seat of the disorder: in pills, if the worms lie in the intestines; in powders or liquid, if they occupy the stomach —in all forms alternately when we are uncertain. The various preparations of mercury and of antimony, with Barbadoes aloes, as being more drastic in operation; also common salt, box, sulphur, savin (a vegetable poison), and sal Indicus, offer a sufficient variety for the bases of as many varied prescriptions; and variation here is most desirable, inasmuch as some kind of worms which resist the effects of one substance may be detached and hurried off by another.

Water-gruel, as it relaxes the parts, and prepares them and the worm for receiving the antidote, should precede every other remedy, particularly the mercurials; a course of which should be followed by a purgative, but not be given together, as is commonly practised. For ascarides, which usually infest the large guts, I have found great service in calomel to the amount of a drachm or more, given over night twice, followed by a purgative next morning after the second.

No. 1.—*Mercurial Bolus.*

Calomel, 1 1-2 drachms,
Anise seeds, 5 drachms.

Mix with treacle for two doses.

No. 2.—*Purgative Ball.*

Barbadoes aloes, 4 drachms,
Gamboge, 1 1-2 drachms,
Prepared kali, 2 drachms,
Ginger, 1 drachm,
Oil of amber, a tea-spoon full,
Syrup of buckthorn sufficient to form the ball for ：nr
    dose.

Particular care should be taken of the horse, but he should not take any gruel for the two days that the mercury is in him, as directed by White, but give him bruised corn or other dry food with little water, the calomel not having entered the system. Neither does he require any of the exercises usually forced upon patients "in physic." Let a week elapse ere the same bolus and purge are repeated as before, when they seldom fail to bring away whatever worms he may have in him. Instead of the foregoing, some persist in the following old method, by way of laxative mercurial, which, however, I must premise, seems much too strong, notwithstanding the high character some bestow on it.

No. 3.—*Laxative Alterative Balls.*

Quicksilver, 1 ounce, and
Venice turpentine, 2 ounces.

These being well rubbed together in a mortar, add

Aloes in powder, 2 ounces,
Ginger, 1 ounce.

Mix with syrup of buckthorn, and form the compost into four balls, one to be given with intervals of five or six days. Water-gruel or a bran mash to precede each ball, as before, and give the same when the physic may be working off.

Some horses, however, can not bear the bolus No. 1, calomel having a tendency to gripe; in that case the quantity should be divided into three balls and given on three successive nights, followed by No. 2, on the fourth morning. In like manner, if the horse be not a very strong one, the above quantity of No. 3, may be divided into six or eight balls, and given at intervals of two days each until purging is produced. Indeed, neither of these medicines should be given, least of all continued, when the animal dungs loosely. From those precautions, it is manifest that my opinion, so often expressed regarding the misuse of strong medicines, remains unaltered; and if I have been successful in impressing the reader with the same wholesome and humane truths, he will at once perceive the absolute necessity of attending to the symptoms, to assure himself that the patient really has the worms, and not some other affection of the liver, kidneys, cœcum, &c. as remarked by me at the head of this article. Mistakes in these respects often prove fatal, or at least affect the animal's future health.

If worms do actually exist, they can not fail to come away with the foregoing course of medicine; and the patient, though a little weak at first, will come out of hand with a good appetite, brisk in his manner, and bright as a ruby. These considerations, however, should not influence us to neglect a trial of the milder medicines, before enumerated, as containing anthelmintic properties, less powerful indeed than the foregoing, but not therefore less likely ？

prove serviceable in ordinary cases. Of these, the Indian salt (sal Indicus) deserves the first consideration, though denounced as differing very little from common salt, with a small portion of sulphur, both of which are known to be goodly anthelmintic. Be its virtues what it may, the following substitute will be found to contain all the properties of the genuine salt, and may be employed when this can not be readily procured.

Laxative Powder.—No. 1.

Sublimated sulphur, 4 ounces,
Emetic tartar, 4 drachms,
Liver of sulphur, 1 ounce,
Bay Salt, 4 ounces.

Mix for six doses, one to be given daily in the corn, which should be previously moistened with water-gruel. As soon as the bowels are tolerably opened, desist for a week at least, but should it fail to produce this effect, give the following

Laxative Balls.

Barbadoes aloes, 4 drachms,
Gamboge, 1 drachm,
Hard soap, 3 drachms,
Anise seeds powdered, 4 drachms,
Oil of cloves, 6 drops.

Mix with syrup of buckthorn enough to form the mass, and divide into two balls. Give them on two successive mornings, unless the first prove effectual. I have found these balls, without any other aid, produce worms, a few, by repeating as often as five or six times. Another preparation of antimony may be substituted for the first mentioned powder, viz.

Laxative Powder.—No. 2.

Liver of antimony, 3 ounces,
Cream of tartar, 4 ounces.

Mix for six doses, one to be given daily until the body is opened. But should not this happen, the laxative ball just advised should be given.

Savin (the leaves pounded, and a spoonful given twice a day in the horse's oats for ten days, and then laxative balls above, bring away slimy matter with the dung, and worms alive.

Arsenic has been tried, to the amount of ten grains a day, for a week, but its powers are tonic only : it is a dangerous remedy in unskilful hands. All bitters are anthelmintic and tonic; thus wormwood, rue, and chamomile flowers, have been attributed the faculty of killing the worms, but the fact is not exactly so; those effects are produced by bracing the stomach, and restoring its tone, and thus disposing the parts to throw off the intruders.

STAGGERS, APOPLEXY, MEGRIMS, VERTIGO, FITS.

STAGGERS is the common or vulgar name given to all those disorders of the head, which consist in vertigo, or "swimming of the head." Drowsiness attending this symptom confers the distinction of sleepy staggers upon this kind of attack, whilst mad staggers is that affection of the brain which causes the animal to kick, to tumble, and plunge about: both are occasioned by diseased stomach, brought on by inflammation of that organ, or simply by the retention of a great mass of indigestible food there and in the intestines: constipation attends every species of staggers, and in some cases the hardened dung may be felt or observed by applying the senses to the proper parts. The breath is offensive, the respiration impeded, and the pulse high and sharp in mad staggers, whilst in the sleepy it is slow, heavy, and full, without vibration [see page 62]. When these latter symptoms continue a long time, the blood determines towards the head, and the pulse increases, if the animal be one in good condition: and unless bleeding and purging be employed effectually, sooner or later ends in apoplexy, or one paroxysm only, which terminates fatally. High-bred cattle, stallions, and brood mares, which are pampered in their food with stimulants, frequently fall victims to this kind of attack, as do their progeny whilst under training, sometimes. In some cases the animal makes one effort, in others it drops instantaneously; so the reader may perceive that he does not stagger at all: and I infer that a manifest difference exists between the two, although both arising from the same cause; for, the one we may afford some assistance to, and usually succeed in performing a cure; in case of apoplexy, the only symptom is remediless—death. To prescribe for such an event would be utterly useless.

Under the head of "costiveness" I have already considered the origin of staggers, and prescribed the remedy at page 68; because that is the disease, whilst staggers, &c. are but the accompanying symptoms.

THE MEGRIMS is an occasional attack on the sensorium or brain, in which the animal drops down as if shot, lies motionless awhile, recovers slowly, and is next day fit to go and do the same thing again, if pushed in his work. This disorder originated in a foul stomach, in one case that came under my care, and was at first a fit of the sleepy kind, which afterwards degenerated into megrims; the morbid state of the head, I apprehend, continued in a trivial degree, which any great exertion brought into activity. Sometimes these megrims are preceded by a short warning, when the animal rears up before it falls, or rambles like a drunkard; it then tumbles and plunges about with considerable danger to those who may collect around it. The muscles of the eye are usually affected, much in the way of horses in locked jaw, or the human subject in a "falling fit;" but all those symptoms disappear upon employing the proper remedies, some of them so quickly and by such means as to appear the effect of a simple mechanical operation.

The cause of staggers, and the symptoms that distinguish the one kind from the others, being thus settled, without distracting the inquirer with needless distinctions of agriculturists or the fanciful reveries of the doctors, let us proceed to the

Remedies.—Farm horses that live much in the straw-yard, and work hard on bad hay, &c. will sometimes stand still at once, as if struck motionless in the midst of their work, which is a sure sign that some great leading function has been suspended for the moment by reason of the great exertion. The driver has nothing more to do in this case than let the tired creature rest for the space of a minute or two, and then proceed in his work a little more leisurely. Prevention is better than cure.

In all ordinary cases of staggers, simply opening the bowels will effect a cure nine times out of ten; and when the animal shows symptoms of a disordered stomach, the coming disorder may be warded off by a dose of physic. In violent attacks, let a clyster be first employed, of warm water, in which common salt has been dissolved, and the hardened dung brought away by manual assistance—as more fully detailed elsewhere—see the mode of doing this effectually, at page 69. I have known violent cases of staggers cease by this remedy alone, and the cure was completed with a purgative ball, as prescribed at page 63.

The fits that constitute megrim, or the more genuine staggers, will require the lancet, and let the quantity of blood taken be commensurate with the violence of the animal, his bulk and fleshiness. From four to six quarts will thus reduce his powers, and aided by the back-raking and purgative just recommended, a cure is soon effected.

LOCKED JAW

Is rather the effect of other diseases, of the acute kind, than an original attack, and is symptomatic of approaching death. A prick in the foot and docking the tail, are fruitful causes of locked jaw. Hot weather is most conducive to this manner of dissolution, which is brought about by great excitation of the nerves, and accompanied by imperfect digestion. The remedy would of course be found in restoring the tone of the former, and opening the main outlet of nature. I have seen a case of locked jaw proceeding from inflammation of the intestines, of a very aggravated nature.

Symptoms.—The case to which I allude was that of an old horse, from twelve to fourteen years of age, just off from hard work, which seemed to have lived badly and suffered severely the ills of a protracted life. Date, May 14, 1820, when the weather was prematurely hot. As usual, it began by the animal thrusting out its nose and eating with some difficulty, which increased as the stiffness of the neck became worse. The ears stuck up, and the sufferer could scarcely move a foot, and this with the greatest pain. Thus, every hour the malady is found to extend itself towards the more vital parts, until reaching the heart, life is then extinguished. The brain appears to be affected at the very earliest period of the attack, when the animal evinces unusual apprehension, and will neigh and prick up its ears at the approach of any one, as the last effort of nature to obtain the notice of man. The pulse is then increased to about 70; but in the future stages of the disorder it falls again below 40, and lower still until its final extinction.

In a few hours, the balls of the eyes of the animal just alluded to were turned back, showing the nerve which retained the ball in position in a very disgusting manner; he appeared to suffer much pain, respiration had ceased, the abdomen was drawn together, and immediate dissolution was expected momentarily. When the subject was opened, I was struck with the inflamed state of the mesentery, and all the lacteals assumed a bloody appearance. Previously to this catastrophe, I hit the animal hard on the forehead with my fist, once: the blow shook his whole frame, which before was as stiff as if made of wood; its eyes immediately returned full one-half way back again towards the proper situation, and I was not mistaken when I imagined that its jaws, which had been knit together, seemed to relax somewhat, and the rigidity of the neck gave way.

Remedies have been prescribed, and Mr. Wilkinson of New-castle reports several cases of successful practice upon young horses which had acquired locked jaw by being nicked, or docked, or pricked in shoeing. The chief obstacle to the administering of any medicine being the closeness of the teeth, which defies the introduction of a horn, it may not be amiss to observe, that profiting by the foregoing experiment, I have in several cases caused a little relaxation in this respect, by placing a piece of wood upon the forehead and striking a smart blow upon it with another piece or a small mallet. Some substance might then be placed between the teeth to prevent their return to the original closeness, whereby the remedies recommended by Mr. Wilkinson

15 *

may be employed with much prospect of success, for he only failed in fou cases in which the jaws were immoveable by any means which he then knew of; and as he has treated this particular subject more happily than any vete rinarian of our time, I think I can not do better than follow the example o. copying his account of a well-marked case successfully treated.

When called in, he observes, " I found the symptoms were a spasmodic af fection of the muscles of the jaws, head, neck, back, hinder extremities, and abdomen, which occasioned them to become rigidly contracted, and the abdo men was much drawn in; the pulse was about fifty, with some irregularity, the breathing a little quickened, the jaws were considerably shut, but not so close but medicine might be administered as a drench with a small horn : the appetite not diminished, but she could not masticate hay; the head somewhat raised, and on elevating it a little more, the haws covered great part of the ball of the eye, the nose was thrown out from the chest, the nostrils expanded, the ears erect or perched up, a great stiffness of the neck and back, the tail a little elevated, and, upon a little fatigue, a shaking of it, a straddling of the hinder extremities : the animal was very costive, and the urine was somewhat diminished. The mare had been shoed about three weeks before, and the farrier had driven a nail into the sensible part of the foot while shoeing her. The lameness thus produced was soon removed; and the disease came on after performing a journey ; that is, about three weeks after the injury in the foot had been inflicted. Two quarts of blood were taken off; a purgative drench and an emollient clyster were given; considerable friction was used over the muscles of the jaws, head, neck, and back, particularly where they were found most rigid ; a stimulating liniment of turpentine, hartshorn, mus tard and oil, was well rubbed over those parts, which were afterwards covered with sheep skins, as recently taken off the sheep as they could be procured, which soon brought on sensible perspiration. The diet was principally thin bran-mashes and oatmeal-gruel, of which she frequently took a little. The next day, pulse the same, breathing a little quicker, jaws not more locked; a constant perspiration had been kept up by the sheep-skins ; the purgative drench not operating, another clyster was administered, which promoted its action ; the liniment was repeated. Next day (the ninth), symptoms nearly the same, perspiration copious : the purging having subsided, the anti-spasmo dic medicine, composed of opium, camphor, and asafœtida, was given with a small horn morning and evening, and a similar mixture, with the addition of three pints of a decoction of rue, was administered as a clyster, morning and evening. The drench and clysters were repeated morning and evening till the 14th day; and during this, the quantity of opium, viz. 1 drachm, was increased or diminished according to the violence of the spasms, which at times were very severe. It was always administered in such a manner as to have its effects constantly in the system, without producing much restlessness ; during this time, there was also a most copious perspiration going on under the sheep-skins. The bowels becoming costive again, another purgative drench and an emollient clyster were administered. On the 15th, the drench not operating, a clyster was given which produced the desired effect. Pulse and breathing a little hur ried and irregular, jaws not more locked, still perspires under the sheep-skins, appetite good; but can not masticate hay. 16th, Pulse more regular, breathing more calm, perspires freely under the skins: the purging having subsided, the opium, &c. were administered as before, and continued until the 21st, when another purgative drench and emollient clyster were given. The jaws were now more open, and the mare could masticate hay; the muscles of the head, neck, back, and hinder extremities became considerably relaxed, and on rais ing the head, the haws did not cover much of the eye. On the 23d day, the purging having subsided, the anti-spasmodic medicine was again employed until the 10th of April, when another purge was administered. On the 12th, the purging subsided; the anti-spasmodic medicine was again used a few days longer, when she was completely cured of the complaint. After this, tonics were given, which, with a nourishing diet and suitable exercise, soon **restored** the tone of the muscles, and the animal became as useful as ever."

Of the twenty-four **cases** described, nine came on after docking or cutting off the tail, from ten days **to a month** after the operation. In such cases, the tail **was** fomented with warm **water, and** the sore dressed with detersive ointment. It should be remarked, that in all the successful cases the jaws were not so completely closed but medicine could be given with a small **horn, or** introduced as a bolus by means of the cane. In some instances, there appears to have been considerable difficulty in giving medicine at first, but by persevering carefully, both medicine and food were introduced in sufficient quantity. With respect to cold application, Mr. Wilkinson says, he has only tried it once, when the whole of a mare's body affected with locked jaw, except the nostrils, was immersed in snow for some time, without producing any relaxation of the muscles: on the contrary, the symptoms afterwards gradually increased, and she died on the third day. In four cases that terminated fatally, the jaws were so completely closed, that neither food nor medicine could be given by the mouth. On examining these horses after death, there was some degree of inflammation in the lungs, stomach, and bowels. It was generally found on opening the spinal canal, that the membrane covering the marrow exhibited a very inflamed appearance, and the marrow itself was tinged of a still deeper colour, whilst the membranes of the brain exhibited some marks of inflammation.

HYDROPHOBIA.

No notice whatever would have been taken of this dreadful **malady, but for** some additions to the **stock of** information already before the public **as to the** means of discriminating **the true** from the false rabies; which I am enabled to furnish from authentic sources. A disease confessedly incurable requires no more to be said of it; **but** this having been at **one** time or other, the case with several other subjects treated of in this volume, I must not, consistently with the duty I have imposed upon **myself, pass it** by in silence. Even the names of authors who have written on canine madness would be serviceable to such of my readers as **may** be desirous of extending farther their inquiries **concerning** this melancholy and appalling disease. Preceding authors have all confined their information to the dog itself, **with** mere casual **notices** of his attacks upon other animals, and on man. Their **researches extended not** to the horse, **or** but trivially so. But, inasmuch **as the symptoms of madness** discoverable **in** dogs so affected are good to be known **to** those **who would keep** their horses out of danger, I am thus further induced to bestow a **page or two on** the distinguishing character of the true symptoms, and add a **hint or two as to** prevention, since cure is nearly hopeless at **present.**

Causes.—The bite of a rabid animal, universally **of the dog, and in every** case that I hear of, on the lip. The bull-dog, the lurcher, the mongrel, the Danish dog, and the shepherd dog, are the kinds most disposed to run at horses, especially when so affected (the first-mentioned, on other occasions, usually fighting at the throat), jumping repeatedly at the horse until they get hold, and the two first pertinaciously holding fast a long time, even until killed off, as we hear and believe.* This will happen mostly with horses tight

* On the morning of September 9, 1826, as Mr. Hawkerford, of Bilston, Staffordshire, was d. i ng two ladies from Willow-hall, a bull-dog, which was with his master in the road seized

reined, or which we bear up in harness, whilst those having the head loose rear and paw off the offender, or being at large, evade or trample upon him; but however slight the bite, the mischief is already committed, so that avoidance by flight is the only preventive of an irremediable evil, unless we are prepared to shoot the caitiff, or to run him through. We hear the free use of horseflesh for keeping dogs in England, charged as one main cause for engendering rabies, or at least quarrelsomeness; add to this, the denial of water to which some of them are subjected at a season when dilution is most required—"what time the dog-star reigns," and we think the suggestion is not very far removed from the fact. At least, we are informed that this appalling disorder is comparatively small in other parts of the world, where horseflesh is less plentiful, or water, the antidote, is found in abundance, and Lisbon is adduced in proof, where dogs perform the office of scavengers, and further are supplied with water by individual housekeepers.* Our own towns, too, in which water is easily obtained, are much seldomer subject to epidemic visitations of rabies than others more arid, yet lying open to an access of carrion in abundance. Dogs invariably take water with much eagerness in every stage of the disorder, so far as I have seen, or heard of, orally; some printed accounts differ. Man dreads it; but when he can get it down, which has been done within a day or two of his dissolution, he finds the raging heat of his stomach alleviated by the effort.

Symptoms of hydrophobia. In the dog, its approach may be known by a marked deviation from the general habits of his kind, amounting to dislike of former friends, a symptom which ought to be particularly regarded. They have been seen to eat their own excrement, and lap their own urine, besides other marks of depraved appetite; though at this early stage of the complaint they are less likely to attack a horse than to resent an affront, or be guilty of treachery towards friends. But as the disorder increases, he shows an inordinate desire to gnaw any substance whatever, and evinces augmented antipathy to cats. Even the dog called Danish, though mostly kept with, and very fond of horses, would, as soon as affected, be the most likely to snap at his old companions' noses. As the malady increases, his eyes become inflamed, and are affected with a blearing from the lids. He howls horribly when the throat is inflamed at the larynx, or part where the voice (barking) proceeds from; the sound of which whoever has once heard, he can never afterwards forget or mistake, unless he himself be bit, or become deaf. The confirmed mad dog now usually sits upon his rump to howl his obstructed bark, through very pain from apparent intestinal inflammation. If suffered to range about as the last stages approach, he seems bewildered and devoid of sight, and should be either avoided or attacked with clubs and other weapons to extirpation; feeble opposition is obviously dangerous.

The symptoms of hydrophobia coming on upon the horse are direct and positive; blood on the lip, and other marks of violence, convey the first intelligence that the mischief has been inflicted; for neither horses, sheep, nor neat cattle incur rabies without inoculation. We are further told, by M. Huzart, that they do not possess the power of communicating the disease by bite to other animals, even though labouring under the highest degree of hydrophobia at the time; a fact I do not further vouch for, but which, when proven by well-marked cases, would go far towards inspiring confidence and certainty in applying any of the alleged remedies. What man is bold enough to administer a ball, for example, whose own life is at stake, ingloriously, by the

the horse by the nose, and retained its hold, though the horse ran away, overturned the gig, and threw the party into a hedge. Still the ferocious brute retained its hold, until its throat was cut on the spot. Vide Annals of Sporting, No. 56, page 233.
*"In "Annals of Sporting," No. 46, page 217, signed J. B.

feat? Increased pulsation, inflamed throat, and evident thickening of the membrane that lines it; soon after, the stomach being also inflamed, rejects food, or the patient is at least indifferent to it, which may occur about the eighteenth day after the inoculation; four or five earlier if the animal be in good condition, so still sooner if high fed and full of blood. Shortly after, i. e. from five to eight days, the bitten parts enlarge, and difficulty of swallowing evidently proves that the disorder is making progress; the patient rubs the part against the manger, stall, or wall, increasing in vehemence from the twentieth or twenty-third day. He does not drink water freely, as usual, though this is by no means a certain criterion, for his power of swallowing is already imperfect: he does not flinch from water when sprinkled over his face, but will even drink to the amount of a pailful, when occasionally he can find free passage for it, and the whim may be said to seize him. Some rabid horses will take to water, and one in a very high state of excitement was known to have run into a river. Suppression of urine next proves that the inflammation has reached the region of the kidneys, which is effected by way of the stomach; perspiration and excessive exacerbation ensue, with inflammation of the parts of generation, accompanied by contraction in the male—yet a gelding was found to have protruded its sheath, and staled with much pain to the amount of half a pint, about the twenty-fourth day.

Weakness of the back and loins sometimes is observable at any period of the disease; some quadrupeds being thus attacked, and falling down mad without previous indication of rabies.

The eyes glassy, fiery or red—loss of vision; tongue sometimes shoved out, and then gnashing of the teeth. The raging symptoms increase from the twenty-second or twenty-fourth day to the twenty-eighth or thirtieth day after being bitten, when the animal will beat itself to death, unless the owner more humanely puts it out of pain with a musket; for 'tis dangerous to approach within reach: the interposition of a strong gate across the stable, and the application of a strong rope well fastened, are good preventives of accident during this final operation, or a cart that will bear some kicking might be employed.

Regimen.—None will afford any permanent relief, though it has been usual to place before it water as a test of its madness—though now known to be a fallacious one in any state of the disorder with any animal whatever. All horses continue to feed up to a certain period—until the stomach is attacked—and some eat voraciously in the intervals of the fits, and drink too, but no good can be expected from either, unless made the vehicles for the introduction of some nostrum. If a cure be attempted, certainly nutritious food, easy of digestion, and cooling, must assist it. The stomach being very much inflamed in this disorder, points out the propriety of bran mashes, marshmallows, and of water gruel, given cold, which will afford the means of alleviating the anguish of that organ, to the coats whereof the last food taken by the expiring patient has been found to adhere after death; that is to say, the fibrous coat of the stomach of the subject alluded to identified itself with the food so intimately, that it stripped off, whilst the insensible coat still adhered.

Remedy.—Every possible remedy, some of them of opposite tendency, has been tried on the dog, and on man. Sea-bathing, the Ormskirk medicine, copious bleeding, excision of the part, the actual cautery, and cupping the parts, have been each employed—successfully, we are told; but no reliance can be placed on either, since they oftener fail, though there is no reason why the horse should not undergo bleeding and cutting off the laceration as soon after the accident as possible. When we consider that the part bitten is ever observed to enlarge previous to the horse showing other signs of confirmed hydrophobia, it seems clear that the cutting off the immediate cause of incipient rabies presses itself upon our notice as the most efficacious measure for warding off the disease. Six months is no unusual time for dogs to conceal rabid infection, a quarrelsome disposition being for a long time the only indication perceptible; but the horse seldom goes beyond the twentieth day in developing all the symptoms before enumerated; which shows that the peculiarly rapid circulation of the blood, noticed elsewhere (page 59) as the harbinger of inflammatory complaints of every kind in the horse, naturally demands early and copious bleeding as a good accessary remedy for this particular one. In this case alone we should not be solely guided as to the quantity of blood proper to be taken by the quickness of the pulse, or actual inflammatory indication, but its fulness, and habit of the patient's body: empty his body subsequently, as directed in cases of fever, with a brisk purgative, as follows:

Purgative Ball.

Aloes, 7 drachms,
Calomel, half a drachm,
Hard Soap, 3 drachms,
Oil of caraways, 12 drops.

Mix with mucilage sufficient for one dose. If the animal seem not otherwise to require purging physic, omit the calomel, and omit it also if the bleeding has been trivial on account of the previous low state of the animal's system.

The application of the plant *Scutellaria laterifolia* is lately reported from North America to have succeeded in several cases; but the symptoms do not accompany the report made to us, and we rest in doubt as to its efficacy. However, let it be tried. So we say of "any mineral acid," which a certain medical gentleman recommends may be applied—a few drops on tow to the wound whilst fresh. This may be tried in the form of oxygenated muriatic acid, which has the property of being destructively detersive; it decomposes the virus, and acts as a styptic. Salt water bathing has been employed upon a large scale, and has been loudly commended; then, let common salt be also tried to the amount of two or three ounces a day mixed with the patient's corn. It is but fair to add, that sea-bathing failed of effecting any good, when tried upon the canine under the best auspices. The king's stag-hounds, in 1823, being more than suspected of rabies, were taken to Brighton, and the ablution well performed under the directions of Mr. Sharpe, the huntsman, but to no good effect; they were all destroyed. Dr. Fayerman, of Norwich, published a case in the spring of 1825, of the cure of hydrophobia in a man of forty-two years, by giving him superacetate of lead (Goulard's extract) in doses of from fifty to ten drops on lumps of sugar. He also bled the patient, who was at one time raging mad. Strong soap boilers' lye, or solution of potash, in either of its varieties, has been used frequently as efficaciously detersive of the virus left by the bite or bites inflicted by a rabid enemy; besides which, the seat of all the wounds may thus be discovered, as they usually lie concealed by the hair; and thus, if excision be deemed necessary, every injured part may be similarly treated: let the eyes be guarded against the lye, and the wounds should be quickly pressed and assiduously washed. An eschar forms and completes the cure. The once celebrated " Ormskirk Medicine " is unworthy of reliance.

The subject of canine madness has been well handled by Mr. Gilman, in his " Dissertation on the Bite of Rabid Animals," 8vo. Mr Daniel, in his " Rural Sports," has made some good, sensible, practical observations on this subject. Mr. Thomas, in his " Shooter's Guide," is more pithy than communicative; he recommends immediate death being visited upon the victims of the disease; a very effectual mode of preventing communication, truly, but he seems not aware that premature judgments would be very likely to consign to death many good animals afflicted with other disorders than rabies. Subsequently, Mr. Johnson, in his "Shooter's companion," has printed some interesting details of occurrences within his own proper sphere. Dr. John Pinckard's "Cases of persons who have fallen victims to the bite," are well marked. But the labours of none are so much in point, as regards the horse, as the researches of my friend, Mr. John Surr, Surgeon, communicated to me, subsequently to 1810; in which year he published the result of his dissection of several horses which had died of hydrophobia, partly under his own inspection.* The substance of all my friend's observations is embodied in the foregoing pages. A writer, who adopts the signature of H. C. in addressing the publisher of the Annals of Sporting, has communicated much practical information on the subject of canine madness, that is well worthy the perusal of all persons interested in this order of created beings.

* Those papers appeared in the "Medical and Physical Journal," No. 131, and several successive numbers: Mr. Surr being accompanied on one occasion by Dr. Adams and Mr Pettigrew.

BOOK III.

CHAPTER I.

Structure and Physiology of the Foot; Mode of studying it advantageously

CERTAIN disorders of the foot owe their origin to bad structure of the limb, and the manner it is attached to the body, which influences the tread, or bearing, that the foot has upon a plane surfaced ground; others arise from accident or hard work, and a good number from the errors shoeing-smiths fall into when they neglect to adapt their work to the circumstances peculiar to each kind of horse. Furthermore, almost every individual horse has its peculiar tread, and the scientific workman should place himself in a situation to ascertain whether this be owing to such original defectiveness, or to the evil accumulations of age and hard usuage: he must not pretend to counteract, but to follow the first mentioned; the second he may endeavour to correct, to amend, and prevent its evil effects. In order to effect these objects, he should study the form and structure of well-formed limbs, learn the uses of each bone, ligament, and tendon, and ascertain how it happens that deviations from symmetry in the limb always affect the sole of the foot, sooner or later.

But so much space has been already occupied in the anatomical description of the leg, that it might properly be considered a waste of time to enter into new details to the same purpose. The reader will therefore turn back to the early sections of the first book (page 5, &c.) and he will readily perceive in what manner an originally defective limb, or the ill-adaptation of the parts to each other, or its awkward attachment to the body, may become the harbinger of one or other of the many diseases of the foot, which we come shortly to take into consideration. He will know, also, that besides this error of birth, as I call it, there are others of mismanagement: as, the employment of horses in work that is beyond their powers, or of that kind for which nature never designed them; either of which is as likely to bring on distortion of the foot, and its train of disorders, as any accident of birth to which I before alluded. Natural defects go much farther than shape or make, and the distinction between these and the inflicted, or acquired, may be aptly illustrated by the fact, that white-legged horses, whatever be their shape, are more disposed to contract "grease" than those of any other colour. This is therefore a natural predisposition to that disease; whilst the animal which is suffered to contract the "grease" entirely through mismanagement suffers an infliction as much as another, which, being put upon hard services, throws out splents, spavin, curb, &c. in consequence.

Furthermore, the shoeing-smith who should inform himself of the primary causes of badly formed feet would carry on his business with the greatest emolument to himself, and with the most satisfaction to his employers; for he would adapt his shoes to the natural defects, whilst the acquired ones he would mend by degrees until he could control the horses' heels to a healthy shape.

and thus promote the regeneration of healthy horn. He should also accustom himself to reflect on the various breeds of horses that are brought under his care, their limbs and hoofs, produced in certain situations, climates, or coun tries, as we hear them denominated, each of which requires some peculiar contrivance or adaptation. Thus, horses bred in swampy situations have long flabby limbs and large flat hoofs, to say nothing more of their long washy car cases, that predispose them to contract certain ills which come shortly to be enumerated. All those "countries" where ague prevails among mankind are unfit for breeding good horses, as is proved by the thick spongy heel and soft foot. It was to this peculiar climate I objected some years ago, when I first printed the advice given at page 18, which has since been corroborated by the opinion of M. Dupuy, as quoted before at page 130. Brittle hoof is produced by a hot, sandy breeding country, as much as by the heat and dryness of the animal's constitution. But, to whichever extreme the individual belongs that may come under consideration, mis-shapen hoof is visible from the earliest years. This increases as the animal is worked, and disease of one descrip tion or another follows, which requires the care of the shoeing-smith to modi fy, or of the doctor to cure. Thus the combination of ferrier (or iron-work er) and veterinary surgeon in the same person is not so very incongruous as at first sight may be imagined.

In the anatomical treatise that occupies the first chapter in this volume, the reader will observe (at page 11) how strenuously I insisted on the proper shape or elevation of the hoof; and he will not overlook the great service our shoeing-smith, or *ferrier* proper, may derive from duly considering this shape and ex ternal form, and of adapting his shoe to each deviation from the true form, as I shall show presently in detail. For that attendant upon and assistant to nature, who is neglectful of her deviations, or ignorant of the causes which produce them, is ill calculated for his office, whether that be ferrier or doctor; in fact, he is ever the most prizable workman of either class who is best ac quainted with those deviations, accidents, or errors that, for the most part, are inflicted we know not how. The boot and shoe maker, for example, who can best suit the bumble-footed man, is a more ingenious mechanic than he who is wholly employed in making his cordovans for perfect-footed persons only But then, the "shoe maker" of either genus should not be ignorant of well-turned feet, and the symmetry of the horse's foot should form an especial part in the education of an intelligent shoeing-smith; else, how is he to work for the preservation of the proper shape, of its restoration when time or circum stances may have effected those alterations we deplore, and strive to amend if we can not fully restore?

Let him examine nature itself in its fastnesses; let him investigate the minute parts that constitute the whole foot, to which his operations are calcu lated to afford support, or to effect alterations in its form. To aid him in his inquiries, I have annexed hereto the section of a foot of nearly perfect shape. prepared by myself, and published some time before these sheets, in order to meet and correct the blunders intelligent shoeing-smiths were every day led into by relying upon the misrepresentation of the subject contained in certain publications of the present day. I lamented this the more, because it is im possible to withhold approbation from the leading parts of the work in which the ill-conceived picture appeared, and therefore it is very likely to have diffu sed error more extensively than a less popular author could possibly inflict As an antidote to all mistakes on this interesting topic, I would recommend every one who has occasion to meddle with horses' feet, as owner, groom, or

16

shoeing-smith, to obtain a fresh hoof of a horse which has died in comparative
health, and having softened it in warm water, proceed to make a section there
of, in the same manner as I have here done.

FIG. 1. SECTION OF THE FOOT.

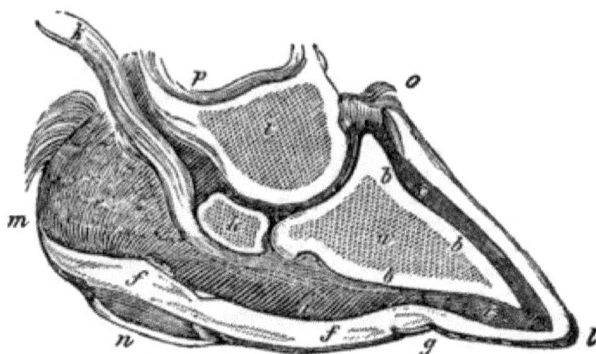

This portrait of a section of the healthy hoof was taken from a freshly
severed foot of a five year old horse, recently killed in full health.   This latter
remark veterinary readers will know how to appreciate, when comparing this
with their own preparations, which may have been derived (as generally happens) from the anatomy of diseased subjects, after the "blood" which should
have supplied fresh secretory matter has been long turned aside, or converted
to increase deformity.

To the general observer, the foot of a horse inclosed in its hoof would, indeed, seem like a corpse shut up in its coffin : and there is, certainly, no mode
of arriving at a knowledge how these act upon, and with, each other, than by
dissecting the hoof.   By this means the whole arcana of its construction are
laid open, but in no manner so intelligibly as by the section straight up and
down from the toe up to the coronet, and throughout between the clefts of the
frog and heel.   This being done, the vessels which supply the juices for renovating the wear and tear of the whole exterior are plainly bared to the view :
the ligaments, bones, and tendons, show their means and manner of action ;
and, above all, the back sinew laid flat behind the smaller pastern-bone, and
quite so at passing underneath the navicula, and at its insertion in the bottom
of the coffin-bone.   On entering the hoof it acquires the term *tendo palmaris*
among the learned, but this course only serves to puzzle the general reader.

At (*a*) on the coffin-bone, the general porosity thereof is much greater than
at any other part, being the avenue or receptacle for the blood which is diffused
throughout it, except on the surface, or border, at (*bbb*).   The shape of this
bone at the toe (*l*) is worthy of note, as being that which is best calculated to
give firmness of tread, fitting with the greatest nicety to the shape of the hoof;
or rather, perhaps, we should say, that the shape of the hoof of a healthy animal should ever partake of that which we have before us, and is evidently instructive to the shoeing-smith in his final raspings, to keep clear at the toe.
Deviations from this rule, bring the coffin-bone nearer the surface of the hoof,
as is shown in fig. 3, plate 3, where the coffin-bone (*c*) and the wall of the
hoof (*g*) are in contact; and even this representation, the picture of the College, shows that the toe of the bone is much sharper than the horn, which they
rasp away so much at (*l*), that the new shod animals go a little groggy for a
short time.

Between the hoof and the coffin-bone interpose an aggregation of secretory vessels, forming a juicy elastic substance, that prevents concussion, as would necessarily happen at every step but for this providence of nature. I have marked it (cc); but this substance, in like manner, pervades the concurrence of all other bones of the foot, only differing much in quality, and in structure a little: between the shuttle and coffin-bones it is more vascular, and the blood is still decidedly arterial. Underneath the coffin-bone at (d), it becomes more elastic, thicker, and striated, resembling pale India rubber, which qualities increase towards the heel at (e). These latter rest upon the frog (f, f), which is horny, or perforable with a point-knife, so far as (g), where it joins the toe of the hoof, more abruptly as the horse is most worked, or otherwise.

The navicula (h,) or shuttle-bone, as it has been called, moves in the midst of much elastic substance, resting upon and pressing the back sinew flat upon the strongest part of that substance, above the centre of the frog. This little bone, it will be seen, is well adapted, by its shape, to traverse the lower surface of the small pastern (i,) and the lateral edge of the coffin-bone (a,) whenever the back sinew (k,) is drawn up to lift the foot, as it does from off the ground, always returning into its place as the foot comes down. At (l) is the toe, (m) is the heel of the foot, and at (n) is the near side cleft of the insensible frog. At (o) is the coronet, or coronary ring, as at (p,) the lowest end of the large pastern bone.

At (a) when the bone is recently cut through, no difference of structure is perceivable, though upon stricter examination, it will be found at the central part more porous, than that which is adjacent to the other bones; the hardness increases towards the whole surface (b b b), where the cutting presents a perfect enamel. But the contents of the receptacle at (a), I have proved by experiment to be unequivocally the same glutinous substance (in a state of preparation) as the hoof itself. This process of nature is well explained by the old aphorism that, "arteries entering bone engender bone, those of muscle create muscle," and so on; and the blood deposited in the coffin-bone, and being dispersed over the internal part of the foot, partakes of all the qualities of bone, membrane, muscle, and skin—the whole combined becomes horn.

If my advice be worth any thing—if my earnest exhortations to investigate the subject effect their object, every man who reads these pages, whatever may be his station in or about the stable or the smithy, will not fail to make a section, or cut down the middle of a hoof at the earliest opportunity. To effect this purpose, the now industrious operator needs little more preparation than to furnish himself with a cordwainer's knife, and a butcher's saw with fine teeth: if he can add to these the use of a carpenter's vice, in which to fix his subject, he will much accelerate his labour. Having secured the foot upside downwards, he will cut down between the cleft at the heel until he comes to the bone at (b), and the wall, or horny part of the hoof (at g), where the labour of sawing is to begin. The shuttle bone (h) he will feel and hear rattle forward and backward at every stroke—the horn yields easily. As he proceeds, he will find his trouble lessened and his views of the matter in hand much enlarged by driving into the chasm his exertions have made, some two or three wedges of wood, whereby he will ascertain that the stiffest part of the horn is elastic, even though he should not have adopted the precaution of soaking his preparation, as recommended. He will thus be convinced, that the application of Bracy Clark's jointed shoe is not without its uses. But if our inquirer has soaked his horn as directed, he will find that the warm water renders it more elastic, and he will conclude that the practice of permitting their horses to stand in the kennels during the issue of hot water from brew eries, die-houses, rectifiers' premises, &c., must soften the hoofs, and indispose them for immediate concussion over the rough stones of our paved streets.

He will also thus discern why I advise, in certain cases, the enveloping the whole foot whenever the application of a poultice becomes necessary to any part of it.

On completing the section, he will discover two branches of arteries which descend into the foot at the coronet near the quarters and supplied the coffin-bone (*a*), that occupies the cavity of the horny hoof, with fine blood for its reproduction. In other words, the formation of new horn is derived from the blood, which is sent hither in good quantity, and pervades the internal part of the coffin bone in particular. In this bone the operator will perceive a cavity, or rather three hollows communicating with each other, in which the horny matter is generated. Or, probably, this is the reservoir for such particles of blood as are suited to the formation of hoof, as it may be required and called for by the process of nature, and the demands of wear and tear, of rasping and drawing inordinately, all which must subtract from its quantity, and leave the bone comparatively hollow, and less fit for resisting the hard concussions to which it is liable at every step. This fact may be ascertained by keeping a bisected foot for a few months, when the moisture having left it in great measure, in the cavity of the coffin-bone will be found a yellowish glutinous substance precisely of the same nature and colour as that which fills the space between the hoof and coffin-bone at *cc*, in the section at page 166: without odour and nearly tasteless, its uses are evidently the supply of new hoof.

Seeing this curious construction of the foot, we are compelled to allow that numerous accidents may also occur to prevent the supply of blood to the parts, to say nothing of its unfitness at times to carry on its proper purposes. The two vessels before noticed that bring this supply of new blood descend into the foot behind the small pastern bone, and pass with the back sinew (*k*) underneath the shuttle-bone (*h*), as may be noticed in the section, at page 166. Here it enters the coffin-bone at the sole, by an indentation of the bone designed for the protection of the vessels passing in and out. From the receptacle in the coffin-bone, after concoction, the blood issues forth—part of it to lubricate and nourish the shuttle-bone and its adjacent ligaments, the remainder to effect similar purposes elsewhere, but the greater part is destined to supply the horny material of the hoof.

Those "concussions" at every step, before spoken of, as affecting the action of the shuttle-bone upon the posterior point of the coffin-bone, occasion trivial injury at every step in quick motion; more harm arises as the animal is much pushed in his work; then heat and fever of the foot supervene, contractions follow, with a train of evils that have acquired different names, thirty in number, but which I have reduced by three-fourths, with a view to simplifying the subject: most of these differ only in situation. Very hard concussions, or a single injury of sufficient magnitude, produce lameness at once, which most unaccountably received the name of "strain of the coffin-joint," and under which general misconception I shall shortly give it a moment's consideration.

The student who would push his inquiries farther will next turn his attention to the muscles, ligaments, and tendons, that guide the foot; that lift it up, and suffer it again to meet the ground; that may perform these offices firm and effectively, or being relaxed, diseased, or ill-formed, they and their functions agree not with the well-being of the foot. Probably he will find it convenient to lay open this part of the arcana of progression by the horse's leg (the lower part of it) previous to severing the foot itself, seeing that the subject will then be quite fresh, and that one part may intelligibly illustrate

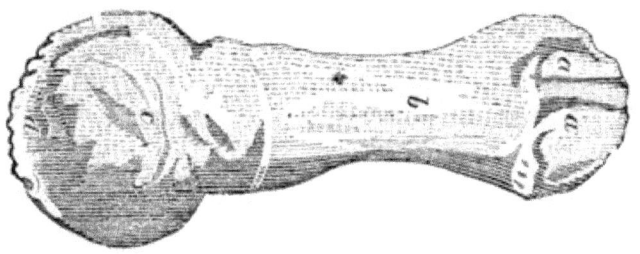

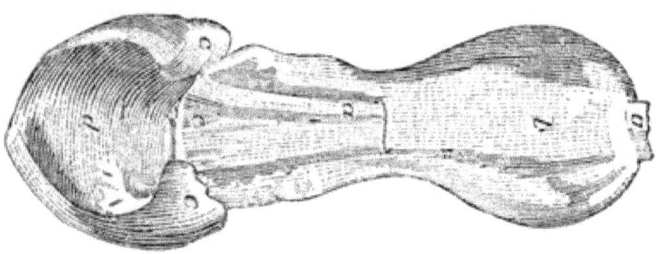

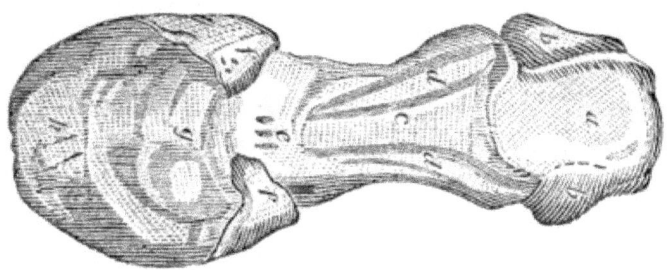

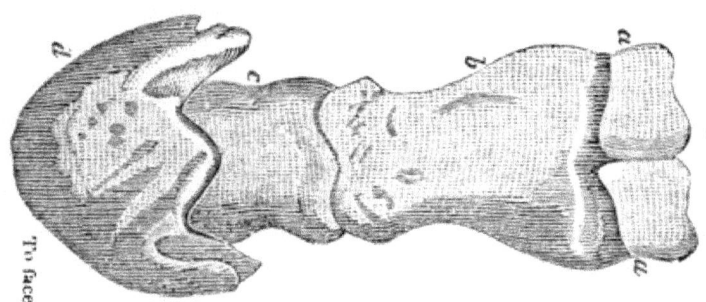

ANATOMY OF THE HORSE'S FOOT.

Plate 2

Fig. 1.   Fig. 2.   Fig. 3.   Fig. 4.

To face page 168.

16 *

Plate 3

Fig. 3.

To face page 169.

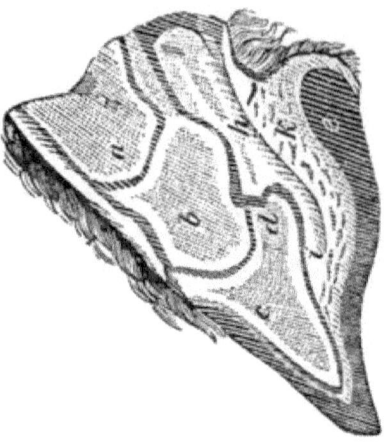

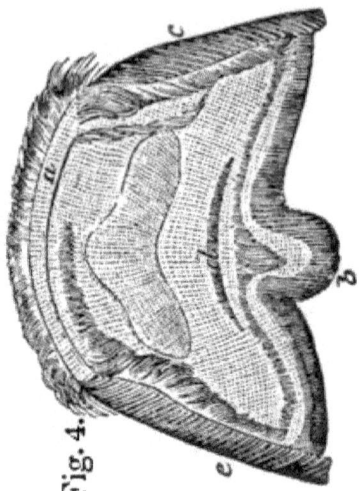

Fig. 4.

Fig. 2.

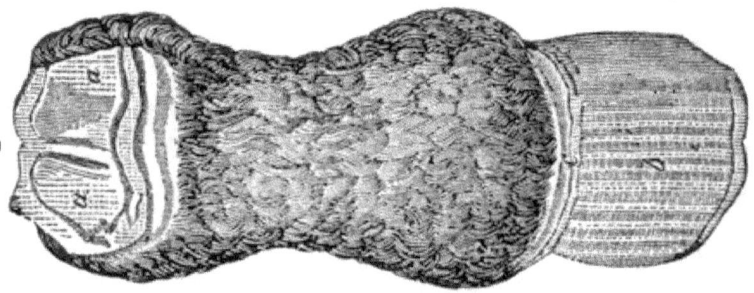

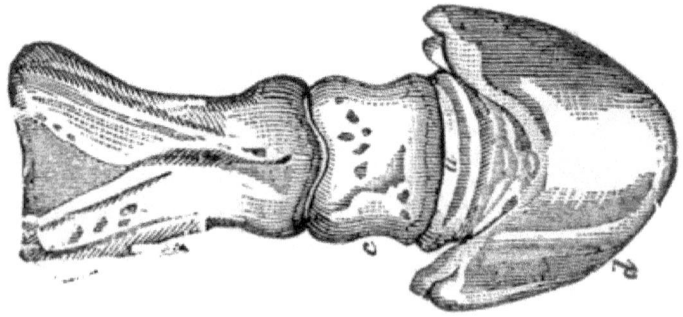

Fig. 1.

the other. This is more particularly the case with the flexor tendon, or back sinew; which he will ascertain is of great length, descending all the way from the hock, or back of the knee, behind both pastern bones, under the shuttle-bone, and is fastened to the bottom of the coffin-bone.

With the following "description" before him, he will study the figures 2 and 3 of plate 3; and after removing the remainder of the integuments, and cleansing the bones, he will then perceive the articulation of these, the manner of their working in and upon each other; and as he proceeds to repeat the investigation, he will note the difference that exists in the shape of a leg taken from a thorough-bred horse and that of a cart-horse; the one small and flat-sided, or sharp before, as best calculated for speed, the other round and heavy, as being made for heavy draught, and to support a large, muscular, and bony frame. In giving this advice, I presume he has already examined the superior part of the limb, though the lower bones and their covering come more immediately under notice in this place.

Description of Plates 2 and 3, of Anatomy of the Horse's Foot.

These figures were not designed or corrected by me, with one exception, viz. fig. 2, of plate 3; they are, however, very fair representations of the subjects studied, and depicted by members of the College. I have here a small objection to make to their mode of enlarging the coffin-bone, which they invariably draw much too big in proportion; why, I never could learn. In fig. 4 of plate 2, for example, where the whole of the integuments are supposed to be removed, the coffin-bone projects inordinately beyond the small pastern, which is not the case at all when viewed in front, or at the back. In other respects these figures speak intelligibly without further explanation.

Plate 2, fig. 1. Front view of a colt's foot, hoof, skin removed, and (*a*) the sesamoid bone, (*b*) the large pastern, (*c*) the coffin-bone, (*d*) the toe.

Fig. 2. Back view of the same—*a a* the back sinew, or flexor tendon, as it appears above its ligamentary sheath and below it, descending flat into the foot underneath the coffin-bone at (*c*); *d* the coffin-bone, having the sensible sole still adhering to it, *cc*, the lateral cartilage; *b* is the sheath in which the back sinew is enclosed, and moves at every step, but part of the sheath has been removed in order to show the course of the sinew.

Fig 3. The whole of the ligaments is here laid open by the removal of the flexor tendon, whereby is seen (at *a*) the smooth surface of the sesamoid bones over which the tendon is ordained to pass; at *bb*, part of the sheath is turned back, at *c* is the hollow part of the sheath; at *dd* the ligament that connects the small pastern to the bone above is shown, with its insertion below at *e*, whereby the large pastern is kept in position; *ff* the lateral cartilages; *g* the bottom of the coffin-bone, *h* the toe.

Fig. 4 is a front view of the same, but with all the integuments removed; *aa* the sesamoids; *b* the large pastern; *c* the small pastern; *d* the coffin-bone, but represented rather wider than ordinary.

Plate 3, fig. 1, back view of the bones, in which *a* the shuttle-bone, is seen that works loosely behind the conjunction of the small pastern, *c*, and coffin-bone, *d*; but the small pastern (*c*) has been lifted or strained upwards inordinately, as the lower part of it lies concealed, as far as the mark (*c*) in the healthy subject, behind the shuttle-bone. The shuttle bone may be seen at its middle or thickest part, in the "section of a healthy foot," at page 166; and by turning the cut sideways, the perspective will be found sacrificed to no useful purpose.

Fig. 2. View of the foot, with the hoof only removed, showing the front of the coffin-bone at *b*, and the coronary ring just above it at *c*, in which the substance is treasured up that constantly supplies the material for new horn to the foot below.   At *aa* the sesamoid bones, freshly severed at the fetlock joint.

Fig. 3. A section of a foot, agreeing essentially with my subject, at page 166, but evidently drawn from a diseased foot, the elastic process marked *cc* in that picture being wanting in this, and the shuttle-bone, *d*, having lost its function; neither do we perceive the descent of the back sinew (*k* in the preceding) to its insertion at the coffin-bone.   At *a* is the lower end of the large pastern, *b* is the small pastern, *c* the coffin, *d* the shuttle-bone, *e* the cleft of the frog, *g* the wall or hoof, *h* the situation of the sinew, *i* the sensible sole.

Fig. 4. Transverse section of the foot, from the coronet *a* to the point of the frog *b*, having the wall *ce* on each side, and showing the divided edge of the sensible sole *d*.

CHAPTER II.

Disorders of the Foot and Leg.

Introductory Observation.—ALL those derangements of the limbs which we come next to consider, I shall divide, for the reader's more ready comprehension, into—1st, those of the leg, and 2d, diseases of the foot: for it does not always happen that affections of the leg alone can be properly denominated diseases, whilst those of the foot are invariably so.   I before observed, that both, or either, may be occasioned by accident, derived from ancestry, or by the fault of misconstruction and consequent misapplication of the individual's powers.   They may be also considered as, 1st, those of the bones, 2d, of the ligaments, tendons, and muscles.   But I shall not so subdivide the heads of my treatises on the several diseases, since each will appear under the respective heads of information, besides which (as will be seen further down), whenever the bones suffer derangement, original or acquired, the integuments follow the same evil course.   Enough, however, has been said on these points in the first chapter of this volume.

Rest is the primal remedy for all acquired disorders of the limbs, whether those of hard work or of accident; but employing the animal whilst yet too young, is an universal error, which is but seldom remedied by allowing it rest when lameness once lays hold of him, much less is it capable of being cured.   The impolicy of this practice, the fruitful source of so many evils, is demonstrable by the custom of the Arabs, who never mount a lame horse, even in the desert, nor propagate from horse or mare which is permanently marked with the effects of overwork.   One remote consequence whereof is, that the foal is not entailed with a predisposition to contract readily such disorders as I come shortly to treat of; whence the superiority of the Arab breed in this respect.   At least, the fact is to be deplored, that most of our stallions of the wagon-horse breed are worked at plough and in the team at two and three years old, too much for their tender years, and permitted to cover mares at this very early age; the result of this lamentable cupidity of ownership is, that their *get* are impregnated with one or other of the maladies that I come shortly to enumerate, ere they reach maturity; but the causes and symptoms whereof I have shown are so similar, or proceed so naturally out of each other, that they differ but in name for situation, in treatment nothing.   Higher bred cattle are subjected to the same disadvantages in most breeding studs, in which the breeders prefer to derive their stock from parents which may have been successful at winning three year old stakes, or probably strained every muscle, bone, and tendon whilst yet yearlings.   We owe to the late Sir T *(*

Bunbury, of Bildeston, the introduction of this practice on a large scale, which is so evidently harmful to the rising generation—of horses.

Lameness is universally the symptom that denotes disordered limb; it is the only one perceptible for some time, until its continuance throws out some appearance on the surface; and that inquirer who can ascertain its true seat is most likely to find the cause, and to effect a cure. For instance, lameness occasioned by disordered bone, as in ring-bone and bone-spavin, is almost universally ascribed by the stable-men and humble practitioners to strain in the stifle, in the shoulder, or the whirlbone; whereby so much valuable time is lost in applying the proposed remedies at the wrong place, that those two disorders in particular make head almost irremediably before the true seat of ailment is ascertained. The same species of blunder is propagated when a disease happens to the foot, and the precise cause thereof, even when well known to those employed about the stable, is kept a secret from the owner and the doctor. " Let them find it out" is sometimes heard muttered in the distance; and in order to comply with the unfeeling permission, we pass the hand down the whole leg and foot from the top to the sole, compare the size of the lame limb with the corresponding sound one, and move the animal about. For without this examination it would be next to impossible to ascertain the precise seat of the disorder, and quite so to apply even the right remedy at the proper place. As an illustration of this position by its reverse, I may adduce the coming on of bone-spavin as that kind of attack which we can ascertain with the greatest precision of all those which lie concealed from our view and touch. It happens, too, that this is one of the few disorders of the leg that admits of cure by early applications, as it is also that which, being neglected, renders the animal wholly useless. When a horse becomes lame of a hind leg occasionally, and that after rest only, the complaint going off on taking a short exercise, we may be quite sure he labours under incipient bone-spavin, provided no other distinct cause can be adduced for his lameness; but should the lameness increase with exercise, then it does not depend upon bone-spavin, but some other malady. Further consideration of the causes, symptoms, and cure of this disorder will be found a few pages lower down.

⁎⁎ When lameness occurs to his horse unaccountably, and the inquiring reader turns to these pages for information, he had better run over once more the whole of the next six or seven heads of information; their great similarity in many respects dictates the propriety of this additional trouble, as most of the series will be found referrible to the same causes, and require much the same treatment, though differently situated.

Throughout the whole of this chapter, the reader will find great help to understanding the details, by carefully consulting the delineations of the leg and foot on plates 2 and 3, and the cut at page 166, with the description of each. References are not always made in words at length, it being presumed that he is already acquainted with the preceding pages, to which he is now referred.

QUITTOR.

Under the class of fistulous affections, I spoke of this disease at page 125 To what is there said I may here add, that as quittor is caused by sand-crack.

by a tread, or the prick of a nail, so will its situation be determined by the precise cause, on the inside of the coronet, or the outside, near the heel, or otherwise, as the cause may have been inflicted; and also, that the cure being effected by harsh means, or burning remedies, these leave the foot disposed to contract other disorders at this region, as ringbone, &c. Hence it follows, that the more moderate the means employed to get rid of this disorder, the less probability is there of the patient's contracting some other. Therefore it is advisable to try the milder remedies first, unless the quittor is of long standing and of very bad sort. The extent of each sinus, and the course it pursues, is denoted by the colour of the soft parts of the foot, being black or livid, or else scarcely tinged, according to its virulence. In order to pursue this examination more accurately, it will be necessary to stop the circulation of the blood above, by tying a ligature tight round the fetlock joint, whereby the skin of the healthy parts below will appear white, and thus more distinctly expose the nature of the sinuses. The pledgets that are to be introduced for the destruction of the pipes may thus be selected of a larger or smaller size, as the calibre of the sinus is greater or less; as also may the quality of the caustic application be made stronger or weaker, as the virulence may require.

Some hastily use the knife, and lay open the pipes freely along their whole course; and if it approach near the bottom of the foot, the coffin-bone is usually affected with rottenness (caries). This they hesitate not to scrape off, though, if the patient be of strong and vigorous constitution, exfoliation will take place without extending the operation so far. Indeed, it seldom happens that more is required than to give the disorder free vent at the coronet, whereby the necessity of operating underneath is superseded; for it will be seen that the ascent of the hoof-making particles from the sole will bring away to the orifice of the ulcer any offensive matters from below; and this process of nature effects the cure. Whenever a sinus leads towards the back tendon, or the joint, much care should be taken not to injure either with knife or caustic, for a bad-looking seam is then left behind, with lameness that terminates in anchylosis, or stiffening of the tendon, or growing together of the small pastern and the coffin joint (a) and the shuttle-bone (h), in the cut at page 166.

RINGBONE.

Causes.—At times a badly cured quittor, at others ill-shapen foot; which occasions that concussion of the hoof and small pastern bone at their conjunction, which causes the latter to swell at the coronet. Cart and wagon horses with short upright hoofs, that do not sufficiently secure the articulation of the coffin and pastern bones against injury, are most liable to this disease.

Symptoms.—Lameness is sometimes the first intimation we have of the existence of ringbone, which is at first neglected, and only ascertained by passing the hand down over the part. As usual with most diseases of the foot, the attendant commonly ascribes the lameness to a strain higher up—of the shoulder generally, as ringbones afflict the fore foot oftener than the hinder one. It consists in the ossification of the cartilage in front of the foot, which extends in time to the lateral parts also.

Remedies.—These may be applied to relieve, but no cure is to be found for ringbone. As high-heel usually accompanies the short upright hoof, the concussions of the foot may be lessened by lowering the heels. Apply blistering ointment to the seat of the disease, and firing may also be employed with advantage.

WINDGALLS.

These appear a little above the fetlock, on each side of the back sinews, and consist of small puffy swellings, that occasion no immediate inconvenience but prove that the animal has been strained in his work, unless it has been occasioned by his having been put to it too early in life. They might be occasioned by the sinus of a tumour, pointing towards the pastern joint, having been cured too harshly, whereby the joint oil issues forth upon their being pricked.

Blistering, and a run in the straw-yard, are the only remedies, though experiments are often tried (when it is found necessary to sell the animal) with preparations of muriatic acid, and muriate of ammonia diluted in water. Saturate a roller bandage herewith frequently, and partial absorption takes place.

THOROUGHPIN

Is of the same nature as the foregoing, arises from the same cause, and is equally devoid of immediate consequence to the animal's going. It consists of a soft flexible swelling on the inside of the hock joint, as well as the outside, immediately opposite each other; whence it obtains the name of thoroughpin, being supposed to go through the joint. When one of those tumours is pressed it yields, and the fluid it contains is thereby forced into that on the other side; when the pressure is removed it immediately returns to the same state as before.

This disorder has no other effect upon the animal's going, or value, than its appearance amounts to, as it conveys the information of its having been worked too hard, and too early in life, as do all these minor evils we are now considering. How this is effected, I have shown in the 15th and 16th sections of the first book, at pages 18—20. Like unto the other disorders of this class, blisters and rest are the only remedies: apply the blistering liniment composed of cantharides and spirits of wine.

SPAVIN.

Bog spavin is the more common, blood spavin but rare. Both varieties, as well as bone spavin, owe their origin to hard work in early life, in the same manner as just adduced in cases of windgall, and thoroughpin. Bog spavin is caused by the joint oil of the hough issuing into the membrane that surrounds it, and stagnating under the vein causes this to swell. The old remedy of taking up the vein by ligature should be abandoned as a long and tedious mode of cure: the circulation has then to force a new channel, in doing which irritation of the parts adjacent is the means of cure, by promoting absorption; whereas the same effect might be produced by blistering, as in the two correspondent disorders just named above.

BONE SPAVIN.

This disorder consists in a bony enlargement at the upper end of the shank-bone, inside of the hock-joint, or a little below it. It belongs to the hind leg only; and if not undertaken in time becomes incurable.

At the seat of this disorder the leg is composed of three bones, which fit together into one common cavity at the hock; and notwithstanding they appear as close together as one bone, yet possess separate motion to give elasticity to the animal's tread, and assist him in the act of progression, as may be observed in the working of his haunches when the horse is going at full speed, or making a standing leap. By mounting the colt whilst too young to bear the

17

superincumbent weight, by pushing him hard in his work, as well as by work
ing young cattle at plough, a practice some breeders of heavy horses injudiciously adopt, these bones get strained asunder, as it were, and inflammation
takes place.

As almost every one knows, by misusing young colts in the manner just described, they become cat-hammed, if they do not derive that particular mal-conformation from parentage, as before hinted (p. 161,) and is more scientifically
accounted for in the first chapter of Book I   Generally it happens that
horses so formed are good, easy goers, brisk and active ; but though well adapted for light weights, are utterly incapacitated from undertaking horseman's
weight at speed, until they are full mouthed.   If heavily mounted, or hard
driven earlier in life, they invariably throw out a curb or spavin.   The shoeing-smith frequently contributes to the contraction of bone-spavin without
knowing it, by turning up the heels of his shoes in frosty weather unevenly.
In the same manner, when the inside heel preserves its roughness longer than
the outer heel, it is clear that this last must bear lowest, and further contribute
to the evil strain that cat-hammed horses are ever liable to, about the hock.
Cow-houghed is but another name for the same mal-conformation.

Symptoms.—Inflammation is scarcely perceptible at first, or any other
symptom; and as it is vitally necessary that we should apply the remedy thus
early, we must employ the discriminating test described at page 171.   If the
existence of adhesion, which constitutes bone-spavin, be not discovered in the
manner proposed, the disorder proceeds until it may be perceived upon comparing the hocks together.   At first, the inflammation is but trivial, when the
horse is also lamest ; but when time has been allowed to unite the bone, the
heat and enlargement increase, and the spavin is incurable, but the lameness
is less.

Cure.—At the commencement only it may be effected easily, by simply
blistering the part all round the hough, in such a manner as to raise the blister
to a good extent.   Generally, in bad cases it would be advisable to repeat the
blister ; in which event let the former one be first well cleansed away with
Goulard's extract, diluted with water.   But should the duration of lameness
and degree of swelling give reason for apprehending that the adhesion is un
commonly extensive, let the part be fired previously to blistering.   Be careful
to keep the horse's head up whilst the blister is operating, and subsequently
dress with hog's lard ; but do not use any greasy applications previously to
blistering, as these only tend to harden the skin, and so obstruct the perspiration and absorption which promote the cure.

Making the shoe thin on the outside at the heel relieves the pressure when
the horse is worked : the contrary form of shoe is conducive to all diseases of
the leg bones.

CURB.

Cause.—Inflammation in the sheath of the back sinew (*b*, fig. 2, plate 2,) a
little below the point of the hock, where the sheath is attached to the muscle.
Like spavin, curb mostly affects young horses of the cow-hocked built, whose
legs stand too much under the body, and which have been worked prematurely hard, as in cases of bone-spavin.   Indeed, the two diseases bear so
much resemblance to each other, in cause and symptoms, except only as to
situation, that I feel no difficulty in referring the reader to the preceding page
for my description of these, only premising that he can not discover the coming
of a curb, by any other means than lameness, and comparing the two legs to
each other sidewise, when a diffused swelling may be seen, but very little heat
felt, by reason of the disorder being deep-seated

Cure.—It may be effectually removed at first, by blistering, as in cases of bone-spavin; but when the disease has lasted a long time, firing must be resorted to with the same precautions as those before recommended. Ease may be afforded by adding to the thickness of the heels of the shoe.

SPLENT

May be looked upon as a disorder of the fore-legs, though occurring on the hind ones, at times.

Cause.—Working of young horses before they have acquired sufficient stamina, or on labour which is much beyond their strength, as in case of spavin, curb, &c. to which the reader is referred, and the concussion which the leg receives at every step upon hard ground, stones, &c.

Symptoms.—Frequent lameness, that goes off and returns without apparent cause for either, before the splent shows itself upon the shank-bone, which it does above the knee, inside. Similarly hereto, it affects the bone of the hind leg, and then acquires the name of bone-spavin. Inflammation of the skin is soon felt, and the horse goes lame until the splent is completely thrown, and afterwards he does as well as ever, except retaining the splent mark, perhaps; but severe cases occur, that do not terminate so favourably. Such happens when the shank bone has received the concussion, that causes the enlargement and rupture, which constitutes the disease, at the hinder part of the leg, where it meets with tendons or the suspensor ligament—(See back view, plate 2, fig. 2, 3.) The lameness and the inflammation are then greatest, and the splent requires our careful attention.

Remedy.—But should not the horse throw out the splent on this last mentioned dangerous part, and become lame, he will yet suffer much in all ordinary cases; for the enlargement of the bone strains the membrane which covers it tightly, as described in Book 1. sect. 17. p. 20. For this purpose apply a warm stimulating embrocation, which affords relief in the more favourable cases; but when the splent rises under the ligament or tendon, blistering or firing must be resorted to. The latter, however, is proper only in extreme cases, and only to be adopted when blistering is found inadequate to the purpose; if the swelling is hot and tender, firing would have the effect of enlarging the whole bone of the leg, and even the blistering liquid is improper when this symptom is highly prevalent. Rather let the heat subside, or assist it in doing so by means of Goulard's extract, diluted with water, frequently applied. When this has reduced the heat, employ the following

Liquid Blister.

Cantharides pulverised, 4 drachms.

Mix with sweet oil to the consistence of treacle, and apply the same twice during the day; thus,—Let the hair be clipped off close from the part, and all round the leg, and the blister well rubbed with the hand for five or ten minutes If this does not cause further swelling and a discharge of a clammy nature, a third application of the liquid blister becomes necessary. After a day has elapsed, dress two or three days with hog's lard, and the patient may be walked about, to get rid of the stiffness. It may be proper, after this, farther to reduce the heat by more applications of the Goulard's extract, as above.

Shoeing is supposed to occasion splents sometimes, it being the practice with most smiths to make the inner heel of their shoes thinner than the outer; and the inner heel being also lower than the outer, occasions the splent bone to receive the concussion more sharply than the outer one; for, as I before observed, splents oftener occur on the inside of the leg than on any other part of it.

MALLENDERS AND SALLENDERS.

Scurvy eruptions on the bend of the knee-joints, or on the corresponding bend in the hock joint; the first mentioned term being applied to those eruptions that appear upon the fore leg, the second, sallender, is confined to those of the hinder leg.   A crack, with much soreness, accompanies both.—The cause may be found in the gross habit of body, attended by suppression of some evacuation, as stool, urine, or perspiration; therefore, to

Cure the patient, restore the defective evacuation by giving one of the two purging-balls prescribed at pages 86, 87, according to circumstances; or a urine-ball, or the emetic tartar, at page 65, or 143, in smaller doses, and the scurf decreases until it wholly disappears.

Let the hair be cut off close from the part affected, and the scurf well washed with strong soap-suds, and then rub over it daily, of the

Ointment for Scurvy Eruption.

Red precipitate powder, half an ounce.
Hog's lard, 2 ounces.   Mixed well together.

Sometimes, a poultice, in which is introduced acetated litharge, becomes necessary when the eruption is divided by a gaping crack, which the ointment may have occasioned.   The blue ointment is employed by some instead of the above ointment.

STRAINS.—LAMENESS.

These are the most deceptious class of ailments attributed to the foot of the horse; for many such are spoken of in the most confident manner which do not exist in reality, whilst others could not possibly happen to the parts indicated by the names they commonly bear; yet shall I fall into this old method of titling the various affections of the limbs, in order to make myself more generally understood.   Our neighbours, and rival veterinarians, the French, in the instructions issued to their smiths of the army, went a little farther in their complaisance to error: "All swellings of the tendons from the knee to the coronet or from the hock to the heel, show an extension or strain of the integument.   Take off the shoe and pare the foot."   In fact, their practice of giving rest in all cases of strain, which often effects a cure with very little further assistance, could not be more assuredly complied with than by thus taking off the shoes; for the Marechallerie were ill able to retain their sick horses in quarters upon urgent occasions of active service, unless they could demonstrate the fact upon the view to their superiors.   By this general mode of forming their judgment as to the cause of all swellings before or behind, we may perceive they included all "extensions" of the bone in their notions of a strain, and treated spavin, splent, curb, strain of the tendons and ligaments, all in the same manner at first.   Of these latter-mentioned we come next to consider the distinguishing symptoms and most appropriate methods of cure; and I will here candidly allow, at setting out, that our neighbours took a correct view of the general cause of all lameness: those strains which occasion inflammation of the ligaments, tendons, and muscles, always communicate fever to the foot, whence arise thrush, canker, sand crack, &c. &c.   We very improperly, as far as precision is concerned, term all lameness of the tendons

&c. a strain, though it may arise from any other cause, as frequently happens. viz. a blow given by the toe of the hind foot, in hunting over heavy lands, when the fore foot is detained too long in the ground, coming in contact with rolling stones in leaping, the kick of another horse, &c.

STRAIN OF THE BACK SINEW AND LIGAMENTS.

Cause.—Back sinew is the vulgar name for the tendon, which the reader will find depicted in a section of the foot at page 166, and marked (*k*). It ascends behind the small pastern (*i*) and large pastern, up to the knee-bend of the fore leg, or the hock joint of the hind one, respectively. In plate 2, fig. 2. at (*a*), this sinew is again shown, where it emerges out of the heel, and enters its sheath (*b*), to which it is attached in a certain degree, by means of very fine membrane, adhering from side to side, and capable of distention or relaxation. The sheath itself is attached to the two pasterns, of which it thus becomes the tendon or support; whence the back sinew and its sheath, or flexor, together obtain the plural—tendons. Within the sheath is secreted a fluid, intended for lubricating and defending the parts during the very great action to which they are liable in every effort of progression. As happens in all other secretions, this one sometimes fails to produce enough for the intended purpose, when the sinew and its sheath adhere together, or at least do not act with freedom; the consequence whereof is lameness in a greater or less degree, which may be temporary only, or become permanent, according to circumstances. If the dryness and adhesion be trivial, as happens after hard work and a night's rest, the horse upon getting warm, loses the lameness this deficiency has occasioned, for the secretion has been thereby renewed, and the lubrication is now supplied in sufficient quantity; but the horse falls lame again next day, probably, and if he can not be allowed rest, 'tis 7 to 1 that he becomes permanently lame. In this respect the French beat us hollow (as just before remarked), though they do not profess humanity so sensitively as the English; and even the Arabs, though robbers by profession, by habit, and inclination, are too sensible of what is due to a faithful animal in distress, to travel on lame horses. The same fact was before adverted to at page 170.

Symptoms.—That sort of strain which consists of relaxation of the back sinews shows itself by the horse going low upon his pasterns, in consequence of his "carrying high," or being trotted constantly in harness. Occasional lameness sometimes ensues in that fore foot which beats, or has the lead at setting out—generally the off one; inflammation of the whole foot may be felt by comparison with the heat of its fellow, which is aptly enough termed "fever of the foot" by the old farriers. This is a very puzzling kind of lameness, no other symptom than those presenting itself for us to ascertain the exact cause; and of course the less observant persons are very likely to apply the wrong medicine, and render the horse a disservice instead of doing him good. The lameness sometimes goes off without any treatment whatever but rest; it is, however, more frequently accompanied, or followed, by some disease of the sole, in consequence of the secretion of horn in the foot being obstructed. Sand-crack, thrush, corns, are among these evils, arising from supernatural heat.

Whenever it so happens, that the secretion does not restore to the entire tendons their original motion, it follows that some part adheres to another inflammation is the consequence, and the horse becomes worse and worse every day he is put to work, the lameness never leaving him altogether. When the adhesion begins extensively, the inflammation and swelling are equally so; the pain is then very great, and the lameness complete and permanent. This denotes the disorder called "strain of the back sinews." In

17 *

very bad cases, or where a slight attack has continued some time, the ligament that passes between the back sinew and the pastern bones becomes greatly diseased, and conducts the inflammation to the foot, affecting alike the sole, the coffin-bone, and the hoof, with heat.

"Fever in the foot" is that low state of the symptoms which arises from a slight attack which has been neglected; the more virulent attack must come under separate notice.

Cure.—Rest is indispensable; foment the entire foot with warm bran-water, or make the whole into a poultice sufficient to envelope the foot all over, as high as the inflammation may extend, which is sometimes as far up as the fetlock. When the heat is greatest at the sole, and the fever extends no higher than the coronet, a stuffing of cow-dung will reduce the heat considerably; it may be secured by thin splinters of wood, and changed twice the first day or two—once a day afterwards. Introduce a strong solution of nitre, and let it be strong, as you can not employ much of it. Both legs should be stuffed at the sole, though the sound one (if one only be affected) does not require changing. Let the animal have a loose stall during any stage of strain, or disorder of the limbs. Look after his evacuations, and cause them to be regular; a simple fever (or inflammation) of the foot depending very often upon nothing more than one or the other of these being stopped, which affects the whole animal system sometimes, to say nothing of a single limb.

Violent strains*, and swelling above the fetlock joint, when the lameness is very great, require strong physic; and the inflammatory symptoms, when running very high, with a quick and irregular pulse, should be lowered by bleeding. Apply fomentations of bran, or a poultice of the same, or of oatmeal in which saturnine lotion has been introduced, as much as it will bear. When the great heat of the part has caused dryness of the poultice, saturate it externally with the saturnine lotion, either by soaking cloths in it, and spreading these all over the part affected, or in a poultice as above.

After this treatment has reduced the inflammatory symptoms, but not the swelling and lameness, apply opodeldoc, which may be made as under, viz. No. 1. Embrocations. If this does not fully succeed in the course of three or four days, recourse must be had to No. 2: and if this does not prove sufficiently stimulating, apply the mild blister No. 3.

Embrocations for Strains.—No. 1.

Spirits of wine, 6 ounces,
Camphor, half an ounce,
Soap, 2 ounces.

Dissolve the camphor in half the spirits; mix the remainder with the soap, and then put both together. Rub the parts assiduously twice or three times during the day.

No. 2.

Crude sal ammoniac, 2 ounces,
Vinegar, 1 quart.

---

* The word strain, as here employed, is evidently used in the wrong sense: it should be sprain, i. e. bent or twisted out of its proper position. To strain or stretch any thing long to a greater length, as when the back sinew is strained or elongated so as to permit the pastern to slope or bend down, as in mild cases of "breaking down," would be more accurate.

Mix in a bottle. and rub the parts twice daily.   Let a long bandage, dipped in the embrocation just prescribed, be passed tightly round the parts, beginning at the bottom and making it fast above the knee, or the hock, as the case may be.   Moisten the bandage after it is on.

No. 3.

Cantharides, in powder, 1 drachm,
Spirits of wine, 2 ounces.

Mix, and rub it on the part.   Although this acts as a very mild blister, the horse's head must be tied up for a few hours while it is operating.

A course of treatment that has been followed in this manner steadily, and with due caution, seldom fails to restore the animal to a comparative soundness, if not completely so : though the swelling may remain after the lameness has ceased, it generally subsides when the convalescent animal can be permitted to walk out for a little exercise, which should take place gradually, and the use of a loose stall allowed, than which there is not in the whole catalogue of remedies a more certain adjunct to be found.   Going out too early after apparent recovery is very likely to bring on a relapse, and a relapse, as every one knows, is always more difficult to remove than the original disorder.   Time is required for the injured parts to recover their former posture and strength, if that event ever arrive.   Firing may be employed after a while, but is very often resorted to prematurely, before the tendons and ligaments have recovered position, or absorption has reduced the muscular parts to their former size, and restored their action.   When three, or four, or five months of moderate labour give reason for believing that these events have taken place, firing is likely to prove highly serviceable by bracing the whole together in a tight skin, much resembling, and greatly excelling the long bandage prescribed with embrocation No. 2, in p. 178.   The reader of discernment will please to note, that if the said artificial bracing be found to lessen the lameness in that early stage of the disorder, no less will the bracing of the natural skin by firing be found beneficial when healthy action is restored, but not perhaps the former strength.

CHAPTER III.

SHOEING.

Terms and phrases, in all matters connected with the arts of life should convey a good and most distinct notion of the thing spoken of.   This does not always happen in our day, formerly never, and proved a vast stumbling-block to the advance of science; but whoever termed the horse-shoe an "iron-defence, was a happy fellow, and deserves well at our hands, inasmuch as his appellation is goodly descriptive of the thing intended, and tells plainly what a shoe ought to be in reality.   If not made of sufficient quantity, and of a proper material, it proves inadequate to defend the hoof from injury : if made too heavy, or ill-shaped, the shoe becomes the cause of grievous offence, of pain, heat and contraction of the horn, with its train of evils.   Any workman may learn by practice, and therefore every one ought to know, at least, when too much or too little is applied.   Some feet have the wall very thick, and the shoe will require a good bearing; if very thin, it can not carry a heavy shoe, though it stand most in need of defence. Again, the horn of some horses' feet is so well-tempered and stout, that they might be permitted to go without shoes without danger, if not worked upon stony roads. Time, however, and hard work, occasion brittle hoof, and distortions, with numerous disorders that attach to the foot generally, or belong to the sole only.

When these ailments begin to show their effects, the shoeing-smith must adapt his work according to the new pattern thus cut out for him, and here begins his ingenuity: in some cases he will even have to adopt a different shaped defence for the same set of feet; but in all cases, and under every circumstance, he must fasten them on firmly to the horny wall of the foot by nailing and clenching. By paring the sole inordinately, the bones within are pressed out of position, and the wall having now no resistance in the horny sole to keep it expanded, it contracts and becomes shapeless and diseased. Partial parings overmuch produce partial accidents from without, and engender diseases within, which have received a great number of names according to the situation, but all having their origin in this or some such injury, and all producing contracted hoof and sole. The importance of avoiding this baleful practice may be deduced from the great anxiety of our ancestors to particularize, by so many different names, this single disease of the sole arising from contracted hoof. For whenever constitutional diseases fall into the foot, they never affect the sole, or any part of the bottom, unless attracted thither by accidents or contraction of the hoof, by reason of this paring and rasping away of the natural defence.

Under each of these heads of information, I shall presently place before the operative reader a few plain and intelligent precepts, accompanied by some admonitions; for most assuredly, that teacher who contents himself with telling the learner what is necessary to be done has but half performed his duty, if he leave uncorrected certain long standing errors, which he knows to exist, and to have received the sanction of ages that were confessedly working in the dark, as regards horse-shoeing above all other operations. But the method of performing this operation is avowedly not to be taught in its rudiments, upon paper. Practice is indispensable, manual labour requisite; and much of it, conducted by an intelligent mind well versed in books, is necessary towards forming the proficient shoeing smith. Hitherto, however, from the nature of the black-smith's trade, its laboriousness, and the deficiency of general education down to a late period, most of the operatives in this branch of mechanical labour were precluded from acquiring the additional information that books contain, after they had once adopted their future calling. Error and prejudice laid fast hold of our ancestors, for ages; but the prevailing national desire of acquiring the minor school endowments promises a different result at the present day, and on this occasion, when Science has been disrobed of her cloak and the niceties of Art are sought in language that all can comprehend.

The shoes affixed to the feet of their horses by the continental farriers differ materially from our own and from each other; which proves that no fixed principle is acknowledged by either of them; though the English and the French assimilate together the nearest of any, and are those, I apprehend, that approach nearest to perfection; notwithstanding the controversies and bold assumptions of superior wisdom, and the "patents" that enabled a few persons here to give themselves airs, and to set up pretensions they have miserably failed to substantiate. The jointed shoe, for instance, of Goldfinch, and of B. Clark, which is the best modification of the old semi-oval defence for healthy feet, was preceded a whole century by the French author of " Le Cheval," a folio French work, noticed by Mr. Bee in the Annals of Sporting, for 1823.

Practical Precepts.

The Shoe. In quantity or size, the common defence of the full-grown horse's foot is made nearly half an inch thick at the toe, but near the heel one fourth less; here, also, it must be made narrower by the half than at the toe, where it is an inch wide, and so continues round to the quarters, lessening away towards the heel, where it is but half an inch wide. Very near the outer edge a groove is made, not too deep, but sloping from the side next the rim, in order to throw the heads of the nails slanting, when the final hammering down takes place. A practice prevails of making this groove, called fullering, much too close to the edge; and to so great an extent does this mistaken notion prevail, that he who could so make it nearest to the edge without cracking the rim was long considered the best workman. This, however, is not the most approved method of our times; for the nail-holes that are to be punctured in this groove are thus brought too near the edge of the horn, so that the nails do not hold fast, unless driven and clenched high up on the hoof, which also is an exploded part of practice. Neither should the fullering be continued round the toe, nor to the quarters, lest you weaken the defence where its protection is most required.

Iron is the only material proper, and the toughest is the best defence, as it affords a small degree of elasticity in action, is least likely to crack, and is capable of being hardened at the wearing points at will. The toe alone is usually hardened at the time of making, unless in winter, when it may be found necessary to turn down the heels, termed frosting, when these may also be hardened, or steeled.

Some persons frost all their shoes in winter, by fullering them all over the ground surface; but this rough soon wears away, or is of little service from the first. On turning the heel down, a crippling gait is produced if the rough be long, especially with heavy horses, having low hoofs, which may be counteracted, so long as the roughing lasts, by turning down the toe also, and steeling both. But then the necessity of frequently removing the shoe, and thus impairing the wall or crust, may be remedied by making screw-holes in the ground surface of the heels, and providing a suitable supply of screws with steel heads, that may be applied and screwed on fresh every day if need be. Sizes of course would vary according to that of the horse and shoe.

Shape. For sound feet, both surfaces of the English shoe are made perfectly flat, the inner rim being thinner than the outer. The shoe extends all round the edge of the wall or crust, which it is desirable to defend, and terminates where the bar and crust join at the heel. A curve upwards, at the toe, to prevent tripping, though sanctioned by authority, and carried to an extreme by Goodwin, and others, is seldom desirable, even with heavy horses, or those which go close to the ground, and is well met by a modification of the German and French method, of forming the shoe wider than ours, and consequently less pointed at the toe. The toe being then rasped close to the shoe, no tripping takes place on that account.

The French form, or shape, differs from our English shoe, in being made wider and approaching nearer to a semicircle, and instead of being flat next to the hoof, is hammered hollow, which renders the ground surface convex; a mode of proceeding that suits admirably with their coarse footed horses, and comparatively harmless roads [meaning neir *petit chemin*, and the sides of their *grand chemins*], but is inadmissible in England, excepting perhaps with our agriculturist owners of the like ordinary cattle. They also make their shoes as thick at the heel as at the toe, which is a transgression against the general precept, at page 180, that I can not reconcile with propriety: what is more against the French, they take little heed of hardening either toe or heel

An adjusting curve upwards, which they give to the toe, could add nothing
to the security of a horse's going along safe on our roads, whatever it might
do on theirs; but their system of punching and nailing is altogether so excellent as to deserve imitation more extensively than it has hitherto been honour-
ed with, and is described with due discrimination lower down.   The horn at
the toe would of course be made of a fit shape to receive such a form of shoe
as the French; and I have reason to believe that it affects the toe of the coffin-bone in process of time, which also becomes curved upwards, precisely
after the form thus factitiously given to the horn, and doubtless gave rise to
the discrepant representations adverted to at page 166.

On finishing off fine work, let the inside of the edge or rim of the hinder
shoes be well bevelled off towards the ground, and rounded, to prevent the
possibility of coming in contact with the fore foot: with horses that are apt to
forge, the necessity of keeping the fore-shoe heel short, so that it may not
project beyond the natural heel, should never be lost sight of.   So, of the inside of the fore shoes to prevent cutting: let these also be filed off, sloping, towards the ground, though not so far as the heels.

An improved form of shoe, preferable to both the foregoing, has been recently introduced, which is an assimilation of the French and English shoes.
Instead of adhering to the old practice of fullering the ground surface of the
common English shoe, to admit of punching the nail holes therein, it is the
improved practice to hollow that surface, and leave a shoulder towards the
outer rim as a protection to the nail heads.   This is performed by a tool resembling the head of a hammer, one face whereof is well represented by the
annexed figure; the lower part at (*a*) being placed on the work, as is   *b*
usual in fullering, but nearly a quarter of an inch from the edge; the
hammer is applied at (*b*), which leaves at (*a*) the desired shoulder, and
along that surface so far as nailing is necessary, a hollowness equal to   *a*
the inner slope of the tool.   This hollow is not in fact any more than a wider
fuller, extending the width of the shoe, excepting the width of the rim left at
the outer edge; though some do further prolong the hollowing all round the
toe, and to within half an inch of the heel, whereby they leave a slight caulk-
ing that never incommodes the animal, but which may be increased in quantity and hardness towards winter.   The operation of hollowing the ground
surface just described, produces a slight convexity upon the foot surface three-
fourths of the shoe's width, leaving the outer fourth still flat to receive the
bearing of the horn.   Hereby the distance between the sole and the inner
rim is increased, and is moreover less likely to retain stones, gravel, or filth,
than is the usual flat surfaced shoe, and possesses all the advantages of the
seated shoe of old Osmer, that has been claimed by some of our moderns.
Moorcroft took great pains to recommend the seated shoe, but finding some
difficulty in getting them manufactured, he set up a machine for puncturing
out the hollows, that quite failed of success, because the power employed was
only equal to cutting soft iron, and this was found inadequate to the required
wear, we are told.   Why he did not subsequently harden, or "steel" the toe
and heels, seem surprising.

Healthy feet are those alone which I have kept in view hitherto; the bar-
shoe, concave and seated shoes being contrivances for ill-formed and diseased
feet, require separate notice.

Objections have certainly been raised against this mode of forming shoes,
that seem plausible enough at first glance: not so fast, however; for upon cool
examination they vanish.   The objectors aver, that because we can not impart the desired freedom of expansion to the whole of the foot, forsooth, we
are not to allow it at any part: if we can not get all we want, we are to reject
what is within our reach.   One of these, who is likewise the last, tells us,

' In order to admit of expansion and relaxation of the hoof by a joint, it would be necessary to make the nail-holes wide enough to allow sufficient play between the shoe and the nails [!], thus producing an effect similar to the end play of carriage springs. But even supposing (says he) this provision were made, the shoe would soon tear out the nails."

The jointed shoe.—A form of shoe was propounded for cutting the shoe into three or more several parts; then lining the foot surface with leather, and fastening on the shoe in the usual manner, with nails that were inserted into each part. But this contrivance though plausible, did not answer, inasmuch as the leather had not sufficient strength to stand the wear and tear.

Mr. Bracy Clark may have been the real inventor of the jointed shoe that bears his name, for aught I know, but he labours under the disadvantage of being preceded by about a century, by a French author; so that his battered saying that his great discovery! forms "a basis for the repose of the profession," however elegant in expression, becomes nonsense to our ears, who concede nothing to simple gentility, and less than that to self-complacent egotism. "Clark's shoe," in its various modifications, differs nothing from the generality of shoes, except in being divided at the toe, and fastened again by means of a pin, screw, or rivet. The toe would require to be made thicker than usual, let me suggest, to prevent the rivet's parting, and to secure each head of it in a counter sink; one half the thickness of the toe is to be cut away on one side at the ground surface, and from the foot surface of the other half, resembling what is termed in carpentry, a mitre; and these being brought close together, a hole should be drilled or punched through both, and let the rivet employed be the size of the hole. Whatever degree of rigidity the workman might restore to the entire shoe, it is plain that the great weight of the horse would very soon strain the rivet, so as to cause it to relax therefrom, and allow the heels to expand by so much. That this might extend over a larger part of the front of the hoof, Mr. Clark preferred nailing the shoe pretty far back towards the quarters, which I reckon among the mistaken notions of the whole class of improvers. But, mark the dissonance of our teachers! the next inventor or improver ran into the other extreme, erroneously punching and nailing up intolerably near to his rivtes or pins, for he has two of them, as per marginal cut.

Fig. 3.

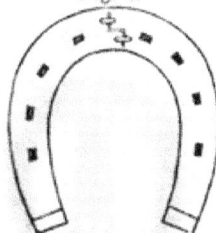

This representation of the shoe invented by Lieutenant Colonel Goldfinch exhibits a modification of "Clark's patent." Like it, the necessity of making the shoe thicker at the toe than usual with the ordinary shoes is evident. The patent was enrolled in October 1821, granting to Lieutenant Colonel Henry Goldfinch, of Hythe in Kent, an exclusive right for fourteen years; and his specification of its advantages and novelty, and the manner of making it, appears to be as follows: "The separation is to be made in any indented form, and the two parts fastened together with pins. It is further proposed to attach the shoe to the horse's hoof by driving the nails obliquely, as in the French manner of shoeing. With this view, the nail-holes are to be punched about one-third to half the width of the shoe distant from the outer edge, and tending in a slanting direction outwards." In this latter recommendation I cordially join the colonel: he was the first writer who noticed it, and is the mode of punching and nailing before alluded to, and hereafter described as the only wise course. Since 1821 it was adopted by the more intelligent smiths of the metropolis, and is hereafter minutely described. One main blunder which the colonel commits is evidently intended to correct the visible insecurity of his shoe at the joint: his holes are punched so near his patent joint as to restore the rigid immobility the patent pretends to amend.

Coleman's patent shoe for giving pressure to the frog continues in use (though in a very limited degree), notwithstanding the demonstrability of its inapplication to frogs already diseased. But, in the hands of the professor himself, and any practitioner tolerably *habile* in his profession, I was free to allow, from the very first, it might be rendered available—but not in ordinary hands;* with these it has failed of success—in some cases from the want of an assortment adapted to the various kinds of feet; a defect that may be now remedied in some measure.

Under these new circumstances, and seeing that Mr. Coleman's opinions as to pressure, and the diseases consequent upon the absence of it, are embodied in his specification, drawn up to obtain this patent (for the professor has several), he may be allowed to speak for himself on this ever interesting subject. He says, "the improvement proposed in this patent is to prevent contraction, and to relieve contracted feet, contracted frogs, flat soles, corns, sand-cracks, thrushes, canker, and quittors, and also to prevent cutting."

The patentee observes, that the "fore feet of horses in their natural state are nearly circular, but from the ordinary shoe worn in this country, which keeps the frogs from off the ground, the hoofs of horses with light fore-quarters are generally found to be more or less contracted, and this in proportion as the frogs are more elevated, and support little weight;" whence the cause of those diseases. To remedy this defect, and to afford the necessary expansion to the hoof the patentee proposed the annexed forms, observing that no specific form of shoe can be suited to all horses under all circumstances, and to every sort of road; it being necessary to alter the shoes of the same horse at different periods.

The construction of the professor's shoe will be seen in figures 4, 5, 6.

Fig. 4.	*Fig. 5.*	*Fig. 6.*
		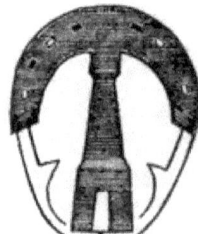

" The bar of iron down the middle of the shoe, called the frog-bar, is made broader than the frog, and welded to the shoe. This bar, when the cleft of the frog is diseased, is slit open in the middle." But all that I have seen in use are without the slit represented in the margin; and the welding on of the bar is greatly objectionable, inasmuch as the chief strain is at the junction of the bar with the shoe or tip, and I have often seen the bar break off here, or else draw the nails, and throw the patent shoe altogether.

Of preparing the hoof.—The general principle of all shoeing is to support the foot off the ground by means of the wall or crust, so that the frog shall not come in contact with the hard plain road, whilst it may be allowed to receive pressure from soft ground: the first prevents injuries and resists wear and tear,

*The subject received lengthened notice, in the Annals of Sporting, for April, 1822, p 246.

ne latter promotes the secretion of healthy horn; the proper degree of pressure being received by the heel, frog, and bars. Whatever is here said, the re foot is still kept in view, unless the hinder foot is particularly mentioned, and occasion will present itself for the distinction, as there is great difference between the two, as regards heavy draught cattle. Greater heat, fever, and affections of the lungs also cause the fore feet to contract disorders unknown at the hind feet; whilst a tardy circulation of the blood, and the consequent relaxation of the animal system, to say nothing of the evils incurred by heavy drags against the collar, produce affections peculiar to the hind feet. Something more is said of this kind of variation at page 12, Book I.

When very much flaky or rotten horn presents itself, the sole should then be pared the least, for this is a proof that great heat, or inflammation, affects the whole sensible foot, and that the hoof is then too brittle. When the flakiness is trivial, run over the whole surface with the butteris, or knife, but go no deeper than the removal of the loose flakes. La Fosse and Moorcroft were both in the right when they told us that paring the sole inconsiderately is "the chief cause of contraction," for the sole is thus rendered less capable of resisting the pressure of the wall on all sides, and of the coffin-bone within, insomuch, that were the paring carried to an extreme, this bone would protrude at the sole and come upon the ground for want of sufficient resistance. Whenever a smith applies his thumb to the sole, and then cuts again until he causes it to bend under the pressure, let him be admonished that he contributes his aid towards contraction of the foot, and some one or other disease of the sole. He has but to cut away a little more to arrive at the sensible sole, which would produce blood, and ruin the horse by a quicker mode than thumbing and cutting.

The frog seldom requires the knife; never after the removal of a shoe which has allowed it to come upon unpaved ground; for then the wearing away is carried on naturally; but if not so, the rough and rotten outside must be taken away, which some smiths effect by first tearing away the slips, or exfoliation, and then paring the mealy-looking part underneath. Hereupon the well recognised healthy horny frog makes its appearance, but is by no means to be meddled with. The cleft is to be cleared out by means of a knife having a sharp return at the extremity; but it must be evident that if the cleft has incurred no foulness, nor the frog grown luxuriantly, neither the one nor the other will require the least reduction. I will not say a word on the necessity of removing the rotten overgrown horn at the toe, and round to the quarters, so as to obtain a proper seat or bearing upon the shoe, this being an affair within every one's compass; but the rasping should always proceed with the shoe before the workman's eyes, unless when he may find it necessary to take it to the fire for the purpose of making alterations. The habit of doing this to a nicety with a single heat may be acquired without going to the fire half a dozen times, as I have seen done; least of all should the shoe be tried on hot, that the most ignorant of workmen may see where it bears most, or the least industrious lessen his labour by softening the horn. Ruinous consequences attend the application of fire to the feet, and yet I remember the period when it was the common practice to place a shovel of hot coals on brittle hoofs to ease the workman's labour!

When a foot is fitted to receive the shoe, the bottom resembles somewhat the hollow and rim of an oval dish. On being placed on a plane surface, the frog and heels bear equally; but when the shoe is applied, the frog is raised by as much as the thickness of the shoe may be at the heel. At the heels, for about an inch of its length, the rim of the shoe is to project beyond the outside of the hoof.

18

Take good heed that the inner edge of the shoe-heel bear not on the
ground more than the outer, but the contrary.

Nailing, a very important operation, requires much previous study of the
formation and functions of the internal sensible parts of the foot, many injuries being inflicted by penetrating those parts to the quick, and thereby occasioning them to fester, as we shall see presently, when treating of the diseases
incident to the sole. A good aphorism has it thus—"If it were possible to
keep the shoe in position without nailing, we should then have arrived at perfection in the art of shoeing; it follows that the less number of nails that are
driven consistent with safety, is the most commendable practice." I believe
it was Mr. Bracy Clark, in the plentitude of his many inventions, who once
proposed to fasten on the shoe by enveloping the whole hoof in an iron defence, and fastening it by screws; but the scheme failed for a most obvious
reason—its weight increased the offence adverted to elsewhere (page 179).
But I will not speculate on novelties, nor further object, simply contenting
myself with taking the evil or puncturation as one that is inevitable, though
capable of alleviation. All hands agree that the less nailing we could suffice
with, the less chance there would be of driving into the quick—hence the
firmer each nail is driven, the less liable is the shoe of loosening, and this good
never can be effected unless the nails fit the holes so nearly as to prevent shifting, and also pass through a good portion of the horn. Doubtless, a couple
of nails on each side would be sufficient to retain a light shoe for a short time,
if the work be not heavy, and allow that desirable expansion of the heel which
all agree promotes the secretion of new horn, and the health of the foot; but
we employ double the number in common work, and seven altogether in the
"improved shoe.*

As before remarked, the fullering usually practised upon the common shoe
is so near the edge, that the rim sometimes breaks off of high-tempered iron:
whilst, if it be soft, the punching inevitably drives out a bulge that the smith
seeks to reduce by hammering, which again contracts the size of the nail-
holes; the latter error occasions the nails to break off in driving; the former
leaves the heads exposed to be knocked off, or readily worn away; and by
either the security of the shoe is diminished greatly and dangerously. Besides which objection to the old method of fullering, there is a corresponding
necessity imposed upon the workman, of clenching high up on the hoof, which
increases the danger of puncturing the sensible internal parts of the foot. This
entire objection to the narrow fuller, or groove, is fully remedied by adopting
the proposed manner of punching the nail-holes farther in from the edge, thus
taking firm hold of the whole thickness of the horn, and driving out sooner
and clenching lower than is ordinarily practised—say, within 3 quarters of an
inch of the shoe in all cases.

The nail commonly used is much too long in the shank for any kind of shoe,
and too thin near the head; but should be of the same thickness throughout
from the head so far as the pointing takes place. The material must be of the
toughest quality, equal to Swedish, insomuch as the nail may bear bending
forward and backward half a score times without breaking; it should have a
counter-sink head to match with the second punch-holes, and the hammering
which the head receives before, at, and after driving, sufficiently hardens it to
resist the immediate effects of wear. Do not point the nails too much, lest
they splinter in the driving, nor make two or three punctures before you drive
home each nail: both these practices proclaim the clumsy workman.

Punching.—As before intimated, the nail-holes should be punched as far
from the outer edge of the shoe as the web is thick. A small punch of the
size of the nail's shank is to be first driven smartly and visibly through, but
not so deep as to raise a burr on the hoof side. Then, open the hole with a

pritchel; and a large counter-sink punch, the size of the nail-head, is then to be employed, but not driven so deep as the small punch; the first being of the size of the nail-shank, the second is to receive a small part of the head. This mode is, of course, best adapted to the "improved shoe" recommended at page 182, where a shoulder and groove supply the place of fullering. But in every form of shoe, and every modification of nailing, the manifest advantage of admitting the nail-head to a rest or protection from rude concussions against the ground, must be evident, when the counter-sink part of the head is allowed to lie deep in the shoe.

The number of nail-holes has hitherto been eight, but a better practice prevails in some forges of driving seven nails only, three on the inside, four outside; whereby the fourth nail outside is thrown so much farther back than the third nail inside. More play is thus allowed for expansion at the quarters: and if the punching and driving be performed effectively, the hold thus obtained will be found fully adequate to any service to which the greater number of nails is applied. The safety of the shoe depends more upon the nails' passing through good sound horn, and filling up the punch-holes in the shoe, than upon their number. A good workman can hear when the nails thus *tell*, by the sound of driving. After punching, the smith must not apply heat or a hammer to the shoe, with a view to reduce any bulge, or burr, which the punch may have occasioned; for this exploded practice spoils the shape and size of the holes, upon the fitness whereof wholly depends the security of the shoe. Indeed, good and proper iron does not readily incur either of those objectionable forms, nor will it break or chip off at the fuller-edge (when such a plan is adopted) like ordinary metal.

Driving the nails home properly includes no small share of skill. Formerly, he who could drive highest into the crust without occasioning lameness was reckoned the best workman, whilst the French method of driving both into sole and crust is an error in the contrary extreme, and argues no little slovenliness and disregard of the construction of the sensible part of the foot. As may be seen and accounted for by reference to the section at page 166, immediate lameness is not always likely to succeed the pricking of the sensible part at *cc*, but matter may form underneath, and lameness ensue at a future day, unless upon removal of the shoe it issue forth at once in the shape of blood. The hoof, which may have lost the elastic substance of this sensible part through age or infirmities, as represented at (g) fig. 3, plate 3, is usually "pricked to the quick" at once, and flinches, or goes crippling away from the smithy.

According to the most improved modern mode of punching and nailing, the nail should enter at the conjunction, nearly, of the sole and crust, so as to penetrate almost the whole thickness of the crust,* and be driven slanting outwards, so that the clinch be little more than half the usual distance above the shoe. If the nail-holes be punched too near each other, and the driving be performed by a workman who drives and draws his nails, and then peers into the punch-holes, then points his nail and drives again—however well his work may appear when put out of hand, he will but have prepared the hoof for fresh injuries at the next shoeing: after this treatment portions of the hoof are apt to come away, and the smith is thus compelled to fasten on his *defence* by the toe, or at the quarters, and so produce fresh *offence* and incurable lameness.

Do not nick the hoof, as is too commonly practised, previous to turning the clenches; as most feet can not afford to lose so much of their natural support, and even the stoutest foot ought not to be subjected to the loss of so much of its main strength. Neither rasp off the clinch, by way of finish, for the same reason, but hammer it down like the head of a rivet.

* As shown in the figure of Goldfinch's shoe, at page 183.

Mr R. B. Teast recommends a construction of the foot surface, that seems very well calculated to attain his object, the preventing contraction and amending the several evils arising therefrom, by raising a ridge along the whole ex tent of the shoe so far as the heels, exclusively, thus making an inclined plane outwards of the thickness of the wall or crust of the hoof. The punching and nailing takes place at this ridge, thus affording secure driving for the nails, and a safe hold upon the iron. Withinside, the shoe is convex on both surfaces, but admits of modification, at will, on the ground surface. The hoof must be prepared to receive this form by paring away the horn lower on the inner part than on the outer side, or external edge of the wall; in fine, so as to correspond with the form of the shoe. At least, this is what I understand in the course of reading a series of very obscurely written "Practical Observations" on the subject; for, although the plan seems admirable, none of my connexions have seen it in actual use, notwithstanding I called at his forge for that purpose. This was in 1821, soon after the promulgation of Mr. Teast's plan.

By the means proposed, the hoof is spread outwards at every step, so as to afford expansion to the heels, and avoid pressure upon the sole; an advantage for such horses as are weak or thin-soled that is at once obvious and gratifying. His idea of giving pressure to the convex sole, by making his shoe with the whole foot surface inclining outwards, is more vague, since all that is requisite is attained by the first method.

Notwithstanding the French method of punching has been spoken of in terms of approbation, and their mode of driving and clenching low is recommended to imitation, let it not be supposed that in other respects they make the best shoeing-smiths in the world, but the contrary. Their finest shoeing is sad, slovenly work to look at; and this very excellence of theirs is more attributable to laziness than to design or plan. As one instance of this undesirable quality, they assign two men to placing the shoe, a lacquey holding the foot and bringing the tools, whilst *le marechal* himself hammers it on with much pomp. In Portugal they employ three, which includes the *gallegos*, or porter.

CHAPTER IV.

DISEASES OF THE FOOT.

When these can not be traced distinctly to any specific cause, they are fairly attributable to ailment of the whole system dropping into the legs, and "fever in the feet" decidedly so, in my opinion, when both are so afflicted. Therefore it was that I noticed this disease along with "strain of the tendons," to which I attribute its origin, as much as to other causes of general heat of the foot. Indeed the whole structure of the foot of the horse is so peculiarly curious that it almost deserves a separate study, but we must always keep in mind, whilst considering its ailments, that the great irritation kept up by its extreme action is readily communicable from the one to the other, so that we can not intelligibly separate the leg from the foot, when speaking of the ailments of either, notwithstanding I have thought proper to begin this chapter with the disorders that are situated higher up, and mean to close it with such as only make their appearance below.

But there remains still another distinction that may as well be drawn here, before I enter into other particulars, as to fore foot and hind foot. In all the little dissertations which I have ventured upon in this book, and elsewhere, as to the structure of the foot, and all the dissections I have made from time to time, I have taken the fore leg and foot only, with one unimportant exception. I know not why this preference was first made; I believe it to be general, but is of very little importance. For the hind foot, though a little smaller, and somewhat more upright in form, corresponds exactly in all its parts with the fore one, until age and deformity comes on; the back sinew descending from the hough behind the pastern bones, until its insertion underneath the coffin bone of the hind leg, in the same manner as before described, as pertaining to the knee and bones of the fore leg. Further, I believe the name given by the learned to this sinew in the hind leg differs from that given to it in the fore-leg, that being *tendo plantaris*, this one the *tendo palmaris*; a distinction that became necessary, perhaps, that they might be enabled to make themselves understood by each other, when speaking of this important tendon as belonging to the one or the other leg.

When we reflect upon this strict accordance between the structure of the fore foot and the hind, and then look over and lament the numerous disorders that the first is liable to, whilst the hinder one is comparatively free, it gives us reason to pause. But without entering upon an elaborate investigation of this difference as to health, I come to the conclusion that we ought to attribute diseases of the feet, as I have already those of the body, to excessive heat of the vascular system, promoted by the great exertions the animal is put to, and the rude concussions the fore feet in particular endure at every step; thus creating heat and attracting hither any evil humours that may afflict the body generally

FRUSH, OR THRUSH.

A running of matter at the cleft of the frog was formerly called "a running frush;" the moderns, however, write it "thrush." But, *to frush*, being old English for, to break, or crack, or crush, like the cracking of walnuts, I prefer that term before thrush.

Cause.—Depraved habit of body and disordered pulse always accompanying the appearance of a frush, I have no hesitation in ascribing its origin to that remote cause, especially as it is proved to be a deep-seated morbid accumulation; aided more immediately, perhaps, by an injury received whilst travelling, either by the bruise of a stone, or the insertion of gravel at the parts. This latter, however, is not a necessary cause of frush, though the gravel and dirt work into the ulcer as soon as it opens; for the lurking approach of the disease towards this consummate symptom may be ascertained by turning up the hoof and pressing the cleft, which will give pain and occasion the animal to flinch : inflammation has already begun at the insertion of the back sinew in the bottom of the coffin-bone, where the branches of crural artery also enter the bone, at the bottom whereof is the sensible sole which separates it from the horny sole. See this structure of the foot described at page 166, &c. Filthy stables promote frush, and, when the cure may be nearly effected, cause relapse in nine cases out of ten.

Symptoms.—The earliest, as just said, is denoted by tenderness at the cleft, accompanied by sharp, quickened and irregular pulse, as usual in all cases of local inflammation, being at the same time both cause and effect. Of course it follows, that as the disorder in the sensible frog proceeds towards maturity, the blood feels and tells of that fact by increased disorder of the pulsation. These timely indications being neglected, as usually happens, if the animal be then put along over stony or newly-dressed roads, the first discovered symp-

18 *

tom will then be his tumbling down through acute **pain**. The cleft opens, and an issue of a most offensive kind presents itself.

Cure.—If not speedily taken in hand, canker will be the consequence of a neglected frush. But, as scarcely one in ten will take the precaution to ascertain, from the state of his pulse, when the horse is likely to acquire this or any other inflammatory disease, it may be deemed impertinent in me to say, that the preventive of frush in its worse state may be found in purging physic and a cooling regimen, as prescribed for general inflammatory and febrile complaints, set down at the commencement of book 2, pages 59, 63, &c.; for this disease frequently depends on some untimely suppressed evacuation, as the urine, stool, or perspiration; then let these be restored by giving the diaphoretic powders, purging or urine balls, according to circumstances. A very much hurried pulse would of course point out the necessity of immediate bleeding; for the animal so suffering in the vital function must necessarily contract disease of some kind or other; and that particular organ or member which may be least able to bear it is sure to feel its effects soonest. This is as likely to happen to a horse with defective frog, as to its size, texture, or shape, as to any other part of him. Then, let the careful owner examine and find out the least perfect part of his horse, let him watch it closely in all its weakness, and endeavour to detect the first symptoms of illness, that he may aid nature and restore her functions, before these run riot beyond the help of art.

When the frog has been pared away, and the filth of the sore removed, wash it with a solution of vitriolated copper, and apply a pledget dipped in tar or turpentine at the opening. If the case be a bad one, the wash may be made stronger by the addition of a few drops of vitriolic acid to the solution; and the tar may be poured into the opening whilst warm. Place dry tow, and keep it in position by means of splints. Repeat this tar dressing every other day, until the injured parts slough off. Purging physic will be necessary to complete the cure.

CANKER.

Evidently a corruption of the word canker, as applied to a running sore in human ailments, it is yet well silently to permit the innovation, the better to keep the two practices separate, in small as well as more extended affairs.

Causes.—Precisely the same as those which produce the frush, only making its appearance at various parts of the sole, frog, &c. Sometimes the canker is but an aggravated frush a very bad or neglected cure becoming in my estimation a canker, and next to incurable; whereas a frush, taken in time, is easily cured. Our French neighbours write of the two under the same head, of cancer, let them be seated wheresoever they may.

The symptoms are those of frush, extended also to the bars of the frog, the heel, the sole, &c.; and so is the

Cure; with this addition, that the paring must be carried on to the extremity, baring all the diseased parts, though these extend over the whole bottom of the foot Cut away the proud flesh to the quick, and when it has bled a little, apply

The Powder.

Sulphate of copper, 1 ounce,
Corr. sublimate, 4 drachms,
Prepared chalk, 1 ounce.

Mix and sprinkle it over the exposed surface. If the disease makes a hollow between the hoof and the coffin-bone, the powder must be introduced there by means of a spatula, or flat piece of wood, with a bit of tow on it; but do not leave the tow behind, as that might produce a fresh disorder. Butter of antimony is preferred by some to the foregoing powder, because it is a liquid and acts more generally; but it operates only for a short time before its effects cease, being killed by the moisture of the disease it was meant to destroy. Bind up the foot until the following day, when the application must be repeated, after wiping away roughly as much as possible of the diseased parts.

As it is found of some importance to the cure, that the foot should be kept as much as possible from wet and filth, and seeing that the mode of tying on a great bundle of tow in cloth, in the manner now in vogue, often fails, a light shoe, adapted to the present shape of the foot, should be put on, for the purpose of sustaining the dressings, &c. which may be found necessary to put on. The shoe has another advantage over the tying fashion, inasmuch as it allows of the animal to place his foot fairly on the ground, a position that mainly conduces to the cure by promoting the secretions, especially when at length he can move about. Let the shoe be narrow-webbed, with a groove on the inside edge, so as to admit of a tin slider being shoved in and drawn out, when you desire to examine the under surface of the foot to change the dressings, &c. Such a shoe will obviate the complaints usually raised by our stable attendants, that they can not keep on the dressings, nor preserve the foot from damp, which always retards the cure; for they are most of them bunglers at bandaging, owing to the very little practice which falls to the share of any one person among the whole fraternity. Splents of wood may supply the place of tin, when this latter may not be at hand.

Whenever the cankered parts slough off, and leave a more healthy appearance, the powder need no longer be applied at those particular places. Upon these lay on a dressing of tar, in which has been introduced about a tenth part of blue stone, powdered. Let as much pressure be given to the sole as can be contrived, to prevent the granulation of new flesh coming on too luxuriantly, which is otherwise very likely to happen, on the edges of the wound particularly. In this respect, the grooved shoe will be found effectually serviceable. If, notwithstanding all your care, the edges will grow too fast, touch that part with lunar caustic; and in case the horny substance grows over the still cankered parts, it must be again pared away and laid bare. Perhaps the animal is young and vigorous in other respects, and his system probably would promote the secretion of new horn quicker than an older, or less healthy horse; this difference should teach us to employ some digestive for the dressings, which has less tendency to promote the growth of new horn than tar has, which would be found more proper for old horses. For the younger animals, let turpentine be substituted, into which has been mixed a small portion of vitriolated copper.

From what has been said, the reader will perceive the dressings require changing with some degree of judgment and discrimination, and that they should not be passed over or delayed, as he values the horse; for, upon this marked attention alone depends the cure, and such a cure as shall prevent a relapse. Of course he will not fail to take care of the evacuations, as in case of frush; nor that the earliest exercise the animal takes be proportioned to the amount of disease he has undergone in an inverse ratio.

Prevention.—As we have seen that inflammation is the immediate cause of all disorders of this class, and seeing that the irritation which produces this has been brought on by distress of the parts for the want of due pressure on the frog, any one whose eyes are open may see the necessity of paring down the heels so that the frog may have a bearing, when the horse is walked over field or turf for example. For hereby it will be seen, on turning to the brief description I thought proper to give at the beginning of this book, pages 167 169, &c. of the internal conformation of the foot, that the healthy action of the parts upon each other is only to be kept up by the pressure of the sensible frog.

" When the frog is not sufficiently pressed upon, (says Mr. Coleman) it becomes soft from the accumulation of the fluid which it naturally secretes in great abundance from the fatty [elastic] substance, which lies immediately under the tendon." This view of the process tallies tolerably well with my own examination of the subject, at the pages just referred to, and elsewhere.

SAND-CRACK.

When suffered to continue, the cure is attended with great difficulty, and the disorder may therefore be divided into two stages or degrees, like many other affections of the horse. The name of sand-crack is derived from the worst of these states, when sand, gravel, or dirt, has got into the crack, which constitutes the disease.

Cause.—Brittle hoof will occasion sand-crack of a very bad sort, but the accompanying cause is the cessation of the function of supplying matter for forming new horn in the vessels leading from the coronet. This may arise from an external injury at the coronet, or severe treatment for some other disorder of the foot, as a running frush. As the hoof is always hot, one main cause of sand-crack is referred to heated roads, to travelling in deep, hot, sandy countries; scantiness of water, and removal out of a cold to a very hot climate, as from England to India (East and West), are all known to cause the heat and brittleness which accompany sand-crack.

Symptoms.—A split or crack in the hoof, on the inside quarter of the fore foot, for the most part, but often on the front of it, down towards the toe, and occasionally on the outside, and also near the heel. Sometimes it appears on the hind foot, on the front of it, and prevails with us generally in hot weather. Sand-crack is either superficial and easily remedied, or deep and extensive, requiring much attention, and an operation or two in its different stages.

Cure.—It will be seen that a slight crack may, by working the animal, become one of the worst species. Pare away the rotten parts, if such be found, and make a transverse incision across the upper part of the crack ; wash out the sand or dirt, apply daily tincture of tar, with a pledget of tow, and give the horse rest. Bind round the hoof tight with listing, and stop up the sole with cow-dung, and this treatment will suffice in ordinary cases. But when the crack extends so high, that there is no room left for making this incision across, to stop the progress of the crack, the disorder has assumed its worst aspect : the edges of the crack internally now press upon the sensible part, or laminated substance that holds the coffin-bone and hoof together, and inflammation succeeds, if blood does not issue forth. If the crack affects the coronet, you may draw one side of it down to the quick about an inch with good effect, but no farther, as that would occasion the hoof to divide more readily. Rest, however, will restore adhesion to the upper part of the crack, and when this has taken place extensively, the operation of cutting across, or of firing it across with one line only, may be performed with every prospect of success ; for as the hoof grows down, which it does from the coronet, this transverse artificial crack you have made intercepts the material for forming horn, on its inside, and thus contributes greatly to fill up the chasm below. To increase this supply of the horny material, let the coronet be anointed with a solution of tar and tallow and hog's lard daily, which should be extended to the horny part of the hoof.

Stopping has been mentioned as necessary to be adopted on the first or mildest attack, being very conducive to recovery of the lost function of secreting the proper horny material. This, of course, will be attended to in every other stage of the disorder, whereby moisture is applied to the dry brittle hoof, and conveyed to its most sensible part internally. On the uses and advantages of this simple remedy I took occasion to say a few words when speaking of frush at page 190, and recommend the application of a web-shoe grooved, as an excellent auxiliary to the cure, by keeping on such dressings as might be found necessary; and by allowing of speedy removal, it will also save much time and labour to that description of persons who are seldom inclined to bestow too much of either. A number of other contrivances may be adopted to apply the same remedy—for affording cool moisture to the hoof, the readiest of which is the leading him forth to a shaded place, and there tethering him up, where he might stand upon the natural sod, grass, clay, or soft ground, without a chance of running about to make the case worse, as would inevitably happen. In default of this convenience, a good substitute is the sponge boot, with bran poultice to cover the whole surface of the foot from toe to heel. In all cases remember to keep out the sand, gravel, or dirt, which is ever likely to insinuate itself and protract the cure, and, if not otherwise come-at-able, an opening must be made for that purpose by cutting down one side of the crack, as before recommended. Proud flesh will be found at this part of the opening, which must be dressed with a solution of blue vitriol.

If the crack be near the heel, merely thinning the horn and taking off that part which bears upon the shoe, will assist the cure ; and whatever shoe may be put on, care should be taken to prevent the crack from bearing on it : the bar shoe for heavy horses is esteemed indispensable, and some farriers apply it invariably. Others again fire the foot, from the upper half of the hoof, above the crack, to the fetlock joint. This is certainly decisive practice, as the hoof is renewed with new horny matter, being so supplied as to thrust off the old one; but all violent remedies should be avoided until the milder ones have been tried and failed, which will not be the case with those means I have recommended above.

The bearing of our English shoe being commonly on the outer surface of the hoof, promotes the crackling and chipping off of the wall. Mr. Teast's shoe (see page 188) is admirably calculated to prevent this disorder, but is nowise calculated to cure it, when once the ruin has fairly commenced.

THE CORN.

Causes.—I shall not repeat what is already said on the two first-mentioned appearances a few pages higher up, but merely add, that distortion and undue pressure on the sensible sole occasions that irritation which brings on inflammation of its edge, where the shuttle-bone, or heel-bone, presses down upon it at every step, and causes the utmost bending that the minute elasticity of the hoof allows of; but contraction of the heel, which accompanies hot, brittle, and inelastic hoof, prevents its bending duly and truly, and lateral pressure upon the quarters follows. The sole being thus unduly pent up, the circulation is obstructed in its passage to and from the cavity of the coffin-bone, and a deposite of blood, which soon becomes offensive matter, is the consequence. Bad shoeing, whereby the heels are pinched, also when the ragged hoof is left, which may have contained particles of sand, will cause irritation, and end in corn, or figg.

Symptoms.—The mischief thus commenced within shows itself between the bar and the crust, or wall of the hoof, in a dirty-red tumour, with greatly increased heat. Lameness, in a degree proportioned to the badness of the corn, is usually the first symptom that directs our attention to the sole   *Figg* is but another name for the same kind of corn when situated close to the bar of the frog, a little farther back in the hollow of the sole. Pain, very acute on the touch; or, when the horse treads on a hard substance, he issues a moan, or grunt: it is that sound in which his voice is aptly likened to the complaint of the human sufferer.

Cure.—Although oftentimes very troublesome, returning again and again when the farrier apprehends he has cured it radically, yet no affection is easier of a partial remedy, or effected by more ordinary means. Deceived by the name, perhaps, resembling the hard excrescence called a corn, on the human foot, they proceed at once to "pare the corn out to the quick, till the blood starts;" but they heedlessly put on the same shoe upon the same thick heel and hard hoof which first brought about the malady, and the lameness returns. Let the heel of the shoe be cut off on the side that is afflicted, or if both sides have corns, a bar shoe is recommended as giving pressure to the frog. The heels are then to be rasped away free from any contact with the shoe; if they are thick and hard, this will give them play—if thin and tender, they will thus be freed from pressure. The thick heel is most commonly affected, and should be softened by an extensive poultice that is to cover the whole foot, after the corn has been pared and treated with butter of antimony. Tar is then a very desirable application, or Friar's balsam; and if inflammation is again discovered, poultice the foot once more. Fire is applied by some, but the hoof is permanently injured by the actual cautery; and whatever good is achieved is thus counterbalanced by the evil. Vitriolic acid mixed, carefully, with tar, in the proportion of one-tenth of the former to nine-tenths of the latter, will promote the absorption upon which the cure depends.

But in some desperate bad cases, the matter has already formed within, most offensively, and discharges at the coronet by means of that curious process of nature which I described at a preceding page, as affording the coronet the material for forming new horn to supply the wear and tear of the hoof. Upon paring away the horny sole, which now becomes necessary, the offensive matter will be found to have spread itself underneath the sensible sole, which will ooze forth and give immediate relief to the coronet. Let so much of the horny sole as lies loose from the sensible sole be pared away, and a dressing of tar, or of Friar's balsam, be applied as before directed; and if inflammation is again discovered, apply a poultice; should the growth of horn be found too luxuriant, discontinue the tar.

CHAPTER V.

Of Strains Generally.

STRAIN OF THE COFFIN-JOINT.

Cause.—As previously observed, lameness of the foot does not consist in a strain of the joint within the hoof, but is referrible to general concussion of all the parts, and is rather a strain of the back sinew at its conjunction with the bottom of the coffin-bone. Inflammation and accelerated circulation follow, and numbness of the foot succeeds: these, if not remedied betimes, are followed by ossification of the tendon, of the ligament of the small pastern, and the cartilaginous process at (f), fig. 3, plate 2, also becomes bone. That the joint of the coffin-bone with the pastern-bone may be strained, is very probable; but no injury, blow, or concussion, can affect it, which does not at the

same time affect all the component parts of the foot.   Thin hoof and sole are
most liable to this injury.

Symptoms.—Sudden lameness, that is always increasing, and has scarcely
an intermission, without any appearance to account for it on the limb; and
the persons who permit the horse to incur this disorder by their carelessness
seldom have the candour to acknowledge that they know the cause to be a
tread, a rolling stone, or a stumble, and the doctor is left to "find it out."
A most every one imagines the lameness to reside higher up, as in the shoulder or the hip joint.   Great heat and tenderness of the part soon come on;
the latter symptom may be ascertained by striking the hoof in front with a
key or small stone, when the animal will flinch considerably more than when
the corresponding foot is struck in the same manner.   When the horse would
stand at ease, he usually does so with his toe pointing forwards, so as to keep
the pastern in a straight line with the leg, and thereby take off the tension or
pressure upon the back sinew and ligaments: the inflammation shortly after
reaches the upper part of the sinew, as may be ascertained by passing the hand
down over it when the patient flinches.

Cure.—Blistering at the coronet and fetlock repeatedly will reduce the inflammation within.   A poultice covering the whole foot also tends to the same
effect, which will be further assisted by paring the sole, if it be not already too
thin: reduce the frog also, and do the same for the corresponding foot.

Formerly they pared the toe tolerably close, and bled it there, by making a
longitudinal incision: the usual application of tar, &c. then completed the
cure.   But this is an operation that is seldom performed with sufficient exact-
ness, the incision being too often made unwisely deep, so that other diseases
were thus generated at some future day.   Others, again, passed a seton through
the heel to the hollow of the frog, taking care not to touch the sensible sole.
A third set apply the actual cautery, which comes least recommended of either
of the remedies just described; especially when we consider that the actual
disease is very often mistaken for some other; a remark that implies how much
caution should be used in first ascertaining the exact seat of the lameness, its
cause and symptoms, ere we set about the cure by such violent means.

STRAIN OF THE SHOULDER.

Horses that are weak before, and low footed, with an unsteady tread, are
most liable to contract this disorder, which consists in a twist or sprain of the
strong muscles that attach the shoulder to the body.   I think the horse is very
liable to incur this disaster, in a petty degree, whenever his progression is accelerated to the utmost of his powers; but we must guard ourselves against
placing entire reliance upon the hastily-delivered opinions of empirics, who
boldly pronounce when they hope to deceive, and expect belief from the cre-
dulous.

Cause.—Much the same as those which occasion concussions, blows, and
numbness of the foot, with their consequences, which we consent to call
"strain of the coffin-joint," without the most distant possibility of knowing
whether this misfortune ever has happened.   When the horse is subjected to
any rude accident, as a kick, or being thrown down, or slipping on pavement,
ice, &c., or treading on a loose stone, he is very likely to incur strain of the
muscles of the shoulder.   See conformation of the shoulder, in chapter 1, p.
10.

Symptoms.—Decided by swelling upon the chest, or at the top of the shoulder; but we think it desirable to ascertain whether the accident has taken
place before this symptom becomes apparent.   Lameness immediately succeeding any or either of those accidents, which may be distinguished from a

strain lower down by the animal's drawing his toe along the ground, from in ability in the part to lift it off the ground; but when he throws out the foot in a semicircle, described by the segment at page 11, this shows that the hurt is chiefly confined to the lower part of the shoulder near the elbow. Taking up the foot and bending the limb will further prove the existence of strain in the shoulder, if the animal evince pain; whereas, if it lie in the foot, and not in the shoulder, the lame leg can be moved as supple as the sound one. The difficulty of ascertaining the real seat of lameness is sometimes so very great, being entirely invisible, as to put us upon all sorts of expedients to find out the real seat of the disorder. For this purpose, hold up his head high, and after comparing and finding no difference in the shape of his two shoulders, let go the head, when he will be observed to flinch upon bringing it towards the affected side. Let a person rattle some corn in a sieve at a distance behind, now on this side, now on that, and he will be observed to evince pain at turning the neck so as to strain the affected side; not so if the pain be in the foot, of course. As the horse will step short, and also throw out his leg somewhat in a semicircle, when he has received a prick in shoeing, this latter sign is not to be taken as finally indicative of "strain in the shoulder," until the foot has been examined, and the shoeing-smith questioned as to his skill and carefulness.

Cure.—If the injury be considerable, as when the horse has been thrown down, he should be bled at once, in the plate vein when it is local, but in the neck when the injury has been more general. A laxative ball, or a purgative, must follow as a matter of course, proportioned to the actual state of his body at the time. A fomentation of camomile flowers, or of scalded bran, should be applied largely and assiduously at the chest and inside the elbow, and these remedies, with rest from all labour and exercise, generally perfect the cure.

When the swelling is great, but not extensive, as in the case of a kick, spirits of wine, in which a fourth of its weight of camphor has been dissolved, should be rubbed in. This will supersede the necessity of walking the horse too early, with the hope of recovering the "use of his limb," by promoting the lymphatic absorption. A rowel is sometimes employed, when the heat and swelling are very high, with good effect; but the old system of previously boring and blowing, and laying on "a charge," is exploded as barbarous and inutile. On the symptoms abating, let the convalescent horse have a loose stall, and in proportion as his action may be free from lameness, so should be regulated his return to walking, to exercise, and to work. Before he can be fit for his former occupation, it generally is found best to give the horse a run at grass; but previously he may try his powers in a contracted plot of ground in the homestead, to prevent his indulging too freely in exercise, seeing that he is very liable to a relapse, which is generally more difficult to overcome than the original attack.

Swimming the horse "for strain in the shoulder joint," was a favourite remedy formerly, but is deservedly exploded, although we could be certain that the joint intended to be cured were the elbow. This accident, however, does not happen often: I never saw a marked case, and merely deem it possible; yet has the practice still its advocates. And it may be serviceable in other respects, as the muscles are thus brought into play, and the whole limb employed in quite different kind of action to that of walking on terra firma. Some persons submit their horses to bathing, by entire submersion; an operation that was performed most adroitly, about thirty-five years ago, by a stableman named Denis Lawler, in the bay of Dublin. His manner was to ride his horse to a convenient depth of water, and then jumping forward suddenly on the animal's head, thus souse it head foremost to the bottom. The feat caused great marvel at the time; but not so the total disappearance, upon one occa sion of the performer: poor Denis is supposed to have received a kick to the bottom, and his body drifted out to sea, as his Howth friends heard no more of him after that, though "New Harbour" underwent thorough repair since Lawler's last kick.

STRAIN OF THE WHIRL-BONE (HIP-JOINT).

A supposititious disorder, that is more frequently found to be a tardy attack of bone spavin, that is slow in coming forward, and upon which M. la Fosse has thought proper to be very facetious: "a horse has the spavin, or he has it not," says he; "for it is not like a jack-in-a-box, that waits to make his appearance when you pull the string." Either spavin or strain of the whirl-bone, he concludes, must be the disorder of the hind leg, when the animal draws its toe along the road, as described just above as being a symptom of strain in the fore leg. When the animal has received injury in the region of the hip, the camphorated spirits recommended in a preceding page (196) should be applied; but if the heat, swelling, and tension, do not abate by this treatment, blister the parts with the mild blister, No. 3, at page 179. This application, with rest, is adequate to any ordinary case of hip-joint accident.

STRAIN OF THE STIFLE.

Simple "lameness" would better designate this so-called strain. The same remedies as those prescribed for whirl-bone strain will apply to this part; also fomentations, physic, and if the case be inveterate, a rowel, &c. &c. one after another. Camphorated spirits, or ultimately a mild blister, are useful and proper, according to the circumstances just set down.

STRING-HALT.

A catching up of the hind leg at every step the horse takes, constitutes what is termed string-halt. It is one of the incurables; but this consideration shall not deter me from observing, that this over-action of the hind leg may be brought about by art, or rather the ingenuity of man operating upon a known function of nature.

The cause, naturally arising, is very obscure; but the horse-exhibitioners, having occasion for much show at their amphi-theatric courses, sought to bring on this "high show" by puncturation. To them let the secret belong; it is barbarous and unseemly.

STRAIN OF THE BACK

When the immediate covering of the bones, described in Book I. at page 19, become relaxed, and thereby fail to hold the joints together sufficiently firm, the consequence naturally arising from this circumstance is, that they bend a little out of place, at every movement the animal makes, and the least accident confirms the strain, or sprain. Merely straining or stretching will effect this evil at times, though that be no greater than an effort to relieve nature by a motion.

Cause.—Mostly affecting draught cattle of the heavy kind, and principally incident to cities and towns, where dray and cart-horses are obliged to turn short upon slippery stones, we may ascribe this disorder to what is called a wrench, or twist in the human practice. The steady pull, unattended by a turn, is not likely to occasion hurt of the back, be it ever so hollow originally; because the effort that is made to pull a great weight causes the joints to press straight against each other, every capsule being than filled with its next corresponding convex bone.

Symptoms.—A kind of separate motion for the hind quarter, compared to the fore one, of which the exact perceptible division is the seat of the injury. Sometimes it appears as far back as the loins, but when farther forward than the twelfth spinous process, (at G. 24) on the frontispiece, it affects the respiration, and with it other vital functions, and the animal suffers in his general health. It may be muscular or ligamentary, or compounded of both, in which cases the parts adapt themselves to the derangement that has taken place, by thickening their substance, and the first lameness decreases greatly: in this event the horse's condition is not at all affected. While staling, his efforts are somewhat ludicrous; as are also all attempts to make a trot of it when out of harness. These symptoms have deceived some persons into the belief, that the one or the other exertion has caused the strain; whereas it is only the earliest demonstration of it to the observer.

Remedies applied early may assist nature, but the lameness never can be cured completely. If the wrench or sprain has been of a violent sort, as in case of "strain in the shoulder," let the animal be bled to a good extent, i. e. from three to five quarts, according to the quantity or degree of violence the animal has sustained; for it usually happens that it has been strained all over, in various parts. Two dray-horses, which were employed in pulling beer butts from the cellar of a public house, being backed too close to the steps, fell in, the weight of the hinder horse dragging in the fore one upon him. Much contusion was the consequence, as well of the accident itself as in dragging them out: they were in fact strained all over, so that they could scarcely stand for a while. Bleeding copiously, however, to the amount of six quarts, reduced the tendency to inflammation; and although they might be pronounced hurt all over, and the hind horse in particular, both did well after physicking, and a few days of light work.

I have found a fomentation of hot vinegar of very great service, in a well marked case of recent strain; the plan recommended by White, of administering it by means of a woollen cloth or rug, steeped and loosely wrung out, being followed. A fresh sheep-skin, just flayed, was applied immediately, and the lameness sensibly lessened, after two days, applying the fomentation four or five times.

FOUNDER

Is a disorder, or rather a complication of disorders, of the fore feet. Some controversy has crept into our books of farriery latterly, as to what really is founder: and whilst some would confine their consideration of the subject to the foot only, others follow the fashion of grooms, and ascribe the incurable lameness that has no visible specific cause, to an affection of the chest. Hence "chest founder" of the stables, and the "body founder" of White. "Shoulder-shook" is a provincialism of the smithy, when the farrier can perceive "nothing amiss" with the feet—so far as he can see, feel, or understand. Surbating was another name given to the symptom we now recognise as founder, at a time when it was the practice to divide and subdivide every disorder under many, useless, and unmeaning appellations.

Cause.—Hard work, bad shoeing, age and ill-usage, either of which produce so many other disorders pertaining to the horse in his domesticated state, precede founder; for, we never meet with it unless the animal has been so treated or kept, and I look upon it rather as a complication or effect of several diseases of the foot. Some of these, we have seen, are liable to be mistaken for others; therefore do they get maltreated, imperfectly cured, or retain the

seeds of future disease; and founder is the name given to that which is otherwise inscrutable, has no other origin, and is badly defined by all writers and
talkers upon the subject. Out of this dilemma I do not at present attempt to
rescue it: I care not for terms, unless insomuch as they can assist us to un-
ravel the character of a disorder. Contracted heel is the slow cause of most
cases of founder, whereby the quarters press on the coffin and shuttle bone
and thus prevent the action of the latter, which is very great at every step,
and is mainly conducive to the proper secretion of the horny material before
spoken of pretty much at large. To "a chill" is generally attributed the immediate cause of founder; and indeed the poor animal which has suffered
severely at the hands (or spurs) of his master is most open to acquire any ill
which chill or cold may inflict. When this chill takes place, the attack is sudden and usually violent.

Inflammation always attends the first symptom of founder, if it be not an
immediate cause thereof, arising, I have no doubt, from the waste or destruction
of the secretion marked (c) (c) in the cut at page 166. To this conclusion I
am come the more positively, by reason of the absence of those secretory vessels in the feet of old, foundered or otherwise diseased horses; which secretions were designed to furnish the material for forming new horn and giving
elasticity to the tread. Fig. 3, plate 3, at (g) shows the progress of incipient
founder, where those vessels are represented as nearly dried up, and adhesion
has begun of the inner surface of the hoof and the coffin-bone. What must
follow, but brittle hoof, battered feet, or surbating, want of elasticity in the
sensible frog and tendon, accompanied by inflammation, which is a cause, if
not caused by founder?

But young horses sometimes, while breaking in, by the violence that is
deemed necessary, are foundered by the rough rider, through the rupture or
forcing asunder the connexion between the hoof and coffin-bone, just spoken
of. In such cases, the animal being vigorous and the foot replete with juices,
the coronet is greatly affected by oozing out there, in its blood and lymph state.
If youth and general good health should bring the animal through his sufferings, its feet will ever after bear external marks of the internal injury.

Symptoms.—Curved, wrinkled, or striated hoofs, ever attend those animals
which have been so over strained in youth, appearing as if the horn had been
carved or indented; which arises from the coronet furnishing the horny material too luxuriantly, before it has received sufficient concoction within the coffin-bone, as before described, at p. 168. Lameness in one or both fore-feet,
with evident pain, and great heat in the whole foot, attend founder in every
case. At the first attack of acute or violent founder, the horse is observed very
restless in his fore feet, which he endeavours to ease, by alternately changing
position, and lying down when he should be feeding. He brings his hind
legs far under his belly for the same purpose, and if he is roused by hunger or
mandate he lies down again. Considerable alteration takes place in the pulse,
which indicates fever, and the patient breathes short with pain. The progress of those symptoms is very rapid, seldom occupying more than a day or
two.

The slower or chronic founder begins with apparently rheumatic pains and
awkwardness of going, for which he usually receives the whip. After a while,
flattening sometimes appears on the front of the hoof, and the heels contract:
the older animals have now short, brittle, shining hoofs, with the small pastern
bone deeper sunk than heretofore; the hollow of the sole is converted into the
convex, or pumice foot, so that the animal can scarce find foot-hold on the
ground, but will slip and slide about. He is then considered groggy, that is to
say, "like a drunkard," and may last many years: this is chest founder, and
indeed the whole limb is usually affected up to the very chest.

Whether the attack be of the acute or the chronic kind, it dies if not relieved; for the coffin-bone becomes rotten, and the hoof is cast off without the possibility of ever being renovated. In some constitutions, nature lends its aid in critically raising a tumour at the coronet, the breaking whereof and the discharge of offensive matter effects a cure. The same sort of critical tumour as denoting the crisis of general fever, or inflammation, was noticed at page 114, "Critical Abscess."

Remedy.—As soon as discovered take off the shoe, note well the condition of the sole, the heat, and other symptoms, for according as these vary, so must the remedies be changed. Draw the soles a little with the buttress if found too thick, not otherwise; rasp the heels and quarters, which will ease the pain occasioned by the binding of the hoof, and give room for the action of the foot; a fact that may be ascertained by bending it at the pastern, forwards and backwards, before the operation, and trying the same experiment afterwards. Apply a bran poultice warm to the whole foot daily, but do not add to it any greasy or oily substances as is too often practised. The sponge boot may be employed with advantage, made large. After three or four days, that the horn has recovered its former consistency, put on the shoe gently, and walk the patient, to try in how much he is now lame; and if the attack has been a slight one, he may recover with very little more treatment than a turn out in a meadow will afford. Otherwise the feet must be stopped, and kept moist and cool, as directed in case of Canker, at page 191.

In all cases, (except where the foot is pumiced, or the sole is very thin), the jointed shoe of B. Clark, or of H. Goldfinch (page 183), will be found serviceable, as being well calculated for giving play or action to the parts of the foot, which produce the secretion that is so salutary to the renovation of new hoof; but which the disorder we call "binding of the hoof" has sadly perverted into an offensive and harmful matter.

The proper secretion of the juicy elastic substance, for the formation of new hoof, being essential to the restoration of the horse, and as the lameness will not wholly subside unless this process goes on healthily, resort must be had to blistering, provided he still goes lame any. This should extend from the coronet and quarters to the knee, and be repeated, taking care to keep the heels open and the sole stopped. The good effects to the sole that will be found to result from blistering, shows the connexion or companionship that exists between the legs and feet, as I took occasion to observe at a former page.

But, as to drawing the sole, as before recommended generally, there is one exception: if the lameness and other symptoms come on after an inflammatory fever of the whole system, then we ought to look upon it as an effect of the fever seeking to throw off its dregs thus critically; and a swelling and discharge at the coronet may be expected soon to take place that should be encouraged, and treated as simple abscess, not fistulous. When this is the case the bar-shoe is better adapted to keep the parts in position, that the discharge may proceed temperately.

In default of sending the sick horse to a meadow, he may be allowed to stand on a clay-made floor in an outhouse by day, or any slip of soft ground; but by no means adopt the plan of putting the patient upon litter that is damp, and is therefore half rotten and heating. A number of contrivances for affording coolness and natural pressure to the sole and frog, besides the foregoing, have been resorted to, and among these the admixture of vinegar, alegar, verjuice, or solution of nitre with the clay, with the stopping, &c. are well calculated to answer the purposes intended. Rubbing the knees with turpentine is also serviceable.

Physic would not of course be neglected at the earliest stages of lameness, adapted to the previous state of the patient's bodily health, and calculated to lower the access of inflammation, which so much pain must naturally produce. Either of the three evacuations being suppressed, or imperfectly performed, must be restored, and a purgative, a urine-ball, or a diaphoretic powder be administered as occasion requires, and opportunity presents itself: of course, neither of those will be given while the animal is out of doors.

PRECAUTIONS NECESSARY TO BE OBSERVED ON BUYING A HORSE.

Much as hath been said of the make, shape, and proportion of the various breeds of horses in Book I., some few precautionary hints, still more familiar, seem desirable in this place. Of the several points of inquiry to which purchasers apply themselves, the age of the animal is ever considered the most deserving of attention ; the state of its legs, bodily health, and eye sight, coming next in order, if general appearance does not precede every other. On each of those heads I offer a few words of advice, most of which are tolerably well recognised, though seldom in print, as most of those who deal in horse-flesh acquire their knowledge from experience rather than books. Before all things, the new horse-dealer should guard against imposition, and not "look at a horse" where he has got to withstand two or three masked advisers. To be sure, no one desirous of a nag would submit to the imposition of a cart-horse instead ; but, next to this kind of gross attempt, the thorough-paced dealers practise deceit of every species, and throw obstacles in the way of cool examination, especially when we come to investigate the seat of any actual defect.

General appearance: an idea of a good horse.—And first, that we may make no blunders, and the younger portion of readers be thinking of one part of him, whilst I am talking of another, let the annexed plate of " Terms commonly made use of to denote the external parts of the horse," be kept constantly in sight, so that there be no mistake of that sort.

Previous to stating our own old English notions, it may not be useless to quote the instructions with which the purchasers of cavalry for the French military service travelled (as I believe) over that country. Its coincidence with our own opinions and practices is at least curious, though on such a topic no Englishman whatever requires instructions from a foreigner, if his own assertions are to be taken for genuine. " The persons sent to purchase horses should not only keep in mind the colour, height, and price of horses for which he is to treat; but also the usual defects of the country, that he may guard against them ; these are, faulty sight, flat hoof, too brittle, or too soft, and affections of the lungs.

" Those things being well thought on, the purchaser will look at the horse sideways at a tolerable distance : he will choose him as nearly as possible one tenth longer than he is high, measuring from the breast to the quarter, and from the withers to the ground, so that if the horse be five feet high, his length should be five feet and a half.

" Preserving the same situation, he will see 1st, If the horse has a small head, not too fleshy, perfectly free from tumours, and well placed, neither carrying it too low nor too high ; 2nd, If he has not an ill-shaped neck, with his windpipe hanging too low, or bending ; 3rd, Whether his withers be either too sharp or too large, with fleshy shoulders ; 4th, Whether he is not hollow backed ; 5th, Whether his chest be well formed, neither too round nor too flat, 6th, Whether he be low-bellied, with a small sheath ; 7th, Whether he be touched in the wind ; 8th, If his fore legs are not too slender, or his hock do not bend forwards too much ; 9th, If the tendons or back sinews be not ailing, i. e. either sore to the touch, or else stiff-jointed ; 10th, Whether the animal be not either long-jointed, or short-jointed ; 11th, Whether he be strained in

19 *

tne pastern joint, going low; 12th, Whether flat-hoofed, with low heels; 13th,
Whether he be not narrow at the hind quarters; 14th, Whether he has not
spavin, windgall, or curb, ring-bone, or thorough-pin, or is likely to cut. Examine his sole and heel for thrush, canker, or corn, and if contraction has taken
place."

Having thus before us the Frenchman's precautions, we come to the Englishman's long accepted description of a good horse, and nothing else. His
head ought to be lean, of good size, and long; his jowls thin and open; his
ears small and pricked; or, if they be somewhat long, provided they stand
upright like those of the fox, it is usually a sign of mettle and toughness. His
forehead long and broad: not marefaced, but rising in the middle like that of
a hare, the feather being placed above the top of his eye, the contrary being
thought by some to betoken approaching blindness. His eye full, large, and
bright; his nostrils wide, and red within; for an open nostril betokens good
wind. His mouth large, deep in the wykes, and hairy. His windpipe big,
unconfined, and straight when he is reined in by the bridle, for, if it bends like
a bow (or cock-throttled), it very much hinders the passage of his wind. His
head must be so set upon his neck, that there should be a space felt between
the neck and the jowl; for, to be bull-necked is uncomely to sight and preju-
dicial to the horse's wind. His crest should be firm, thin, and well-risen; his
neck long and straight, yet not loose and pliant, which our north countrymen
term withy cragged; his breast strong and broad; his chest deep at the girth,
his body of good size and close ribbed up to the stifle; his ribs round like a
barrel, his fillets large, his quarters rather oval than broad, reaching well down
to the gaskins. His hock bone upright, not bending; which some do term
sickle-hougled, and think it denotes fastness and a laster. His legs should
be clean, flat, and straight; his joints short, well knit, and upright, especially
at the pastern and hoofs, with but little hair at his fetlock; his hoofs black,
strong, and hollow, and rather long and narrow than big and flat. His mane
and tail should be long and thin rather than very thick, which some think a
mark of dullness.

Some do affect a small head at all hazards, thinking none other belongs to
a good horse, but much will depend upon how it is set on; if that be upon a
crane-neck, as usually happens when very small, he will carry unsteady, with
tail up as a counterbalance; and if large head arise from thickness of the jowl,
this will also be a real deformity and interfere with his safe going: hard mouthed usually accompanies the great big head at the jowl. Expanded forehead
is quite a different thing, and belongs to neither of those objections, but on the
contrary is a redeeming sign of good breeding for any kind of faulty head, or
long or short, or thick or thin. The crest being slightly curved is always accompanied by distinctly marked windpipe. No horse with a bad shoulder can
carry his rider with ease and pleasure on the road, though a large one be requisite for harness, or a very oblique one belong to a speedy horse; because it
is the hind legs that send the animal along, as was eminently the case with
Eclipse. See pages 5, 9, 10, of Book I., where many other points to our present purpose are discussed.

As to bodily health, also, the reader will not have far to look to enable himself to judge how any animal is affected which he may desire to purchase.
The whole volume now in his hands is devoted to a description of the functions of animal life, and of their derangement.

AGE.

General appearance bespeaks the age of every animal, to those who have
much practice in ascertaining that point, and whose interest may be said to

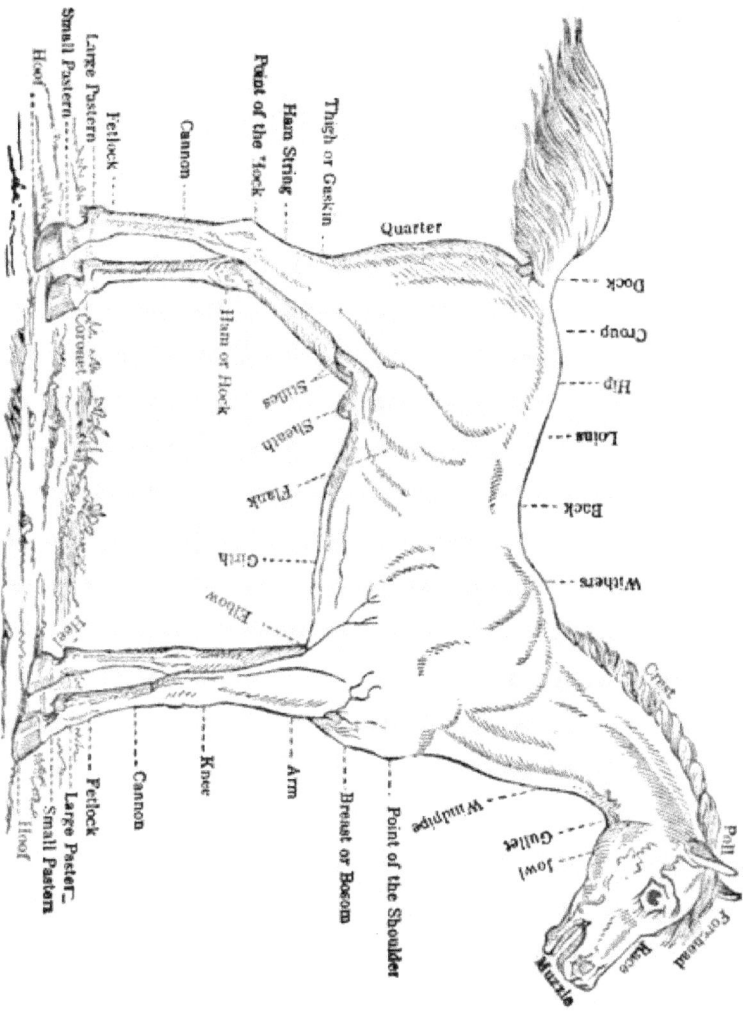

TERMS COMMONLY MADE USE OF TO DENOTE THE EXTERNAL PARTS OF THE HORSE.

Hoof
Small Pastern
Large Pastern
Fetlock
Cannon
Point of the Hock
Ham String
Thigh or Gaskin
Quarter
Dock
Croup
Hip
Loins
Back
Withers
Crest
Poll
Forehead
Face
Muzzle
Jowl
Gullet
Windpipe
Point of the Shoulder
Breast or Bosom
Arm
Knee
Cannon
Large Pastern
Small Pastern
Hoof
Fetlock
Coronet
Heel
Elbow
Girth
Flank
Sheath
Stifle
Ham or Hock

sharpen their judgment: in the horse we are enabled to make a fair estimate of his years from the birth, to ten or twelve, by means of its teeth, but then we should guard ourselves against a number of deceptious tricks that are practised on the unwary.

A certain juvenility of countenance and springiness of action, legs long compared to the carcass, or filling up, large at the knees and other joints, wide jowl, rough coat, and intractability, denote the foal and colt in succession; all which indications vanish gradually as it advances towards maturity, and becomes full mouthed.　Heavy cattle assume premature age and sometimes deceive us upon the first view; nor do such decay when aged so fast as the more spirited, fretful, and lighter breeds; and as no one would purchase a horse for use before it be fit for his purpose, nor take to one that is worn out, the vendor hesitates not to stretch a year or so, one way or the other, as may best suit his own interest and his customer's wants.　To aid their nefarious designs, they are said to file the marks of age in colts' teeth, and to bishop the aged, for confirmation of their falsehoods.　But we never rely wholly upon those marks, but turn our attention to the curve of the tushes in the horse's mouth, and the sloping forward of the corner teeth in both sexes, to detect the imposture.

When we open the mouth of a full grown, or four year old horse, we perceive twelve nipper teeth in front and twenty-four grinders behind: between the two sets, above and below, a space is seen on the gum, designed by nature to receive the bit, and termed the bars of the upper or lower jaw, as the case may be.　About an inch behind the last of the front teeth, the male has tushes at this age, which seldom occurs with mares.　The tushes coming up in the lower jaw sometimes occasion soreness at the bars, when these are to be lanced and the tushes appear: this the dealers effect prematurely at times; and having also drawn out the two front sucking teeth, this causes the "horse teeth" to come up soon, so that the animal may appear four years old before its time.　Pursuing the same species of deception, they proceed to draw the remaining sucking teeth, that the animal may assume the appearance of a five year old.　Jockies have then a pass word for this operation, which they term "all up!"

In examining the mouth to ascertain the age, we leave entirely out of consideration the grinding teeth, and chiefly rely upon the under jaw; though when deception may be suspected, the buyer should refer to the upper teeth also, as these follow the same course of nature as the lower, but do not decay so fast in old age.

At fifteen days old the fore teeth (two above and two below), appear above the gums, the outside shell first, having muscular substance in the middle of the two shells, which fills gradually up, till about the end of the first year, when the surface becomes smooth, and a small ring is observable towards the root of each.

Meantime, when the foal is a month old, the next two teeth (one on each side) above and below, appear in like manner; and at thirteen months the fleshy cavities of these fill up, and a ring is observable as in the former.

At four months old the corner teeth come up, and the filling up is similarly effected at sixteen or seventeen months old.　After this period the whole six teeth wear even, and so continue smooth and unmarked until two years and a half, the corner teeth being still the least perfect, the front ones largest.

During this state of the mouth, if the unprincipled dealer would give his animal's teeth the mark of three or four years old, he is said

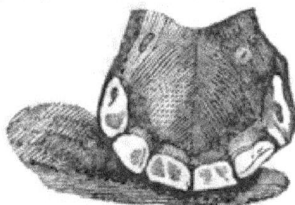

2 1-2 years old

to "file" two or four front teeth hollow in the middle; though in fact it is burnt in with an acid that is capable of destroying the hardest substances. But this falsification may be detected, 1st, by comparing the upper with the lower jaw which they omit "to file;" 2d, by noticing whether the marked teeth have the ring before described, as pertaining to the sucking teeth, but does not belong to the horse teeth—lastly, these latter are larger, of a brownish yellow tinge, and soon acquire tartar, very unlike the fine whiteness of the sucking teeth

On rising three years old, the two fore teeth (below, and two above) fall out, and are replaced by two horse teeth, having the hollow mark in the middle, as shown in the annexed cut. As just said, they are also larger and of a darker colour than the sucking teeth. But between the third and fourth years, two further colt's teeth (as well above as below) shed, and are replaced by "horse teeth," i. e. larger and browner than the sucking teeth, with the black mark; the tushes also push forth, and the horse is now fully mouthed, as represented by the cut in the margin.

Only the corners now remain unchanged from colts' to horses' teeth. These differ from the others in being shorter, smaller, and of a shell-like appearance, until the middle of the fifth year, when these also are displaced by horse teeth, in shape much like the former, and their marks but just perceptible within the upper surface. But, toward the completion of five years of age, they become larger, are more strongly marked, and are grooved on the inside, which groove denotes the age to be five with precision; no deception can be practised on this point, nor as regards the tushes, which are now curved, having grooves inside, that may be felt with the finger, and seen as represented in the figure annexed. At this age the two front teeth give proof of being worn, principally on the outer edge · the wearing away goes on, and at six years the surface is level, or as they say, "the mark is gone," whilst the next two teeth also begin to wear. Now, also, the grooves just spoken of in the corner teeth fill up; the curve in the tushes is diminished, at seven years their grooves fill up in like manner, and become convex in another year or two. Up to this age only the two corner teeth retain the mark, and that but slightly; when the horse acquires the term "aged," and these two likewise soon after become smooth.

This is the state of the lower jaw at seven years old, but the teeth of the up[per]

Rising 3 years.

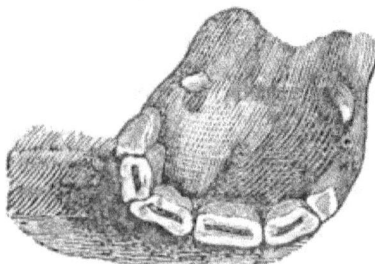

4 years old.

5 years.

7 years.

jaw do not fill up so fast by two years; so that a tolerably shrewd guess at the age of a horse may be formed until it is twelve. The marks in the two front teeth of the upper jaw are not obliterated until eight years old, and the next two become smooth only at the tenth year of its age; being each two years later than happens to the corresponding teeth of the lower jaw; whilst the two corner teeth above do not lose their marks until the twelfth year.

The tushes of old horses, then, have neither curve nor groove; they wear away at the points as if they had been broken off and polished again; the corner teeth appear long and leaning forward; the upper teeth project over the lower, and all lose their oblong shape, whilst the the gum recedes and leaves their roots bare, so that the teeth seem as if grown longer. When the teeth do not so meet evenly, certain dealers file away the projecting teeth; for this denotation of old age, which may be attributed to a strong mouth, sometimes happens prematurely without any other corresponding sign. In two or three other respects we notice similar deviations from the general rule, that none know how to account for, unless it be that such animals were got by old parents, when the hollowness over the eyes will be found to disfigure young colts of the most tender age. Some, again, lose the mark in all their teeth except the corners, as early as five years old; others have hard mouths and the bars almost callous; but all these have the hollow just spoken of. This defect dealers endeavour to rectify by puncturing the skin and blowing it up. On the other hand, some horses are so strong in the mouth, or rather healthy that the marks of five years old are retained by them until six or seven.

Extreme old age may be further ascertained by the mouth, with moderate accuracy. Up to ten or eleven years old, the teeth generally retain their oblong figure and touch each other. From this period the teeth contract in size, become roundish, and leave a small space between them; which space increases up to the fourteenth or fifteenth year, when each tooth assumes an angular shape, and projects forward, irregularly. In another year or two the under lip hangs down, the jaw becomes heaped and contracted, the gum recedes considerably from the roots, and the shape of the teeth is then of an oblong, but directly contrary to the first.

Moreover, the eyes of a horse approaching twenty years wax yellowish, he winks much, and the inner skin of his mouth turns outward. If naturally of a gray or roan colour, the darker spots turn rusty, and he is then what they term "flea bitten:" gradually he turns gray, beginning with the head and finishing with the legs.

LAMENESS

Is not easy of detection, when the horse comes from the hands of a dealer; who of course makes the best of him, and endeavours to inveigle our judgment, and to throw obstacles in the way of examination. Perhaps, when a horse walks queerly, or unaccountably odd, this should be sufficient cause for rejection; but if he suits the purchaser's purpose in other respects, we are usually induced to look at him a little further, and this is the reason that the seller always puts his tit upon the pace he can perform best, commonly the gallop. All paces are natural ones, except backing and cantering, and are all modifications of the walk, trot, and gallop. The walk is made in four equal steps; the trot in two, and the gallop likewise in two, except at setting off. When the steps are not made in equal time, then is the horse lame. This is observable when he walks, is more apparent when he trots, but is scarcely perceptible when he gallops: therefore to judge whether a horse be lame or not, he should be put upon the short trot, because at the long trot it is more difficult to be discovered by seeing him go, than by hearing the difference if sound in each alternate step. Hence, it will be seen, we must examine whether a horse be lame by a gentle trot. To judge whether the lameness oe before, let him come towards you, then the fore leg which falls to ground the quickest is the sound leg, and the contrary one is faulty; but to find out whether the inequality of his paces proceeds from defect in the hind leg, make him trot from you, and that leg which is longest in coming to the ground is affected in some manner or other; and in either case the faulty leg is to be closely examined, according to the instructions before set down, page 171 Even then, unless the person has great experience, he is liable to be imposed on, as the poor animal is often lamed of a fore and a hind leg, at the same time: an occurrence that may have been inflicted on one leg in order to counteract the first appearance of actual lameness in the other. To detect this cruel imposition, it becomes necessary to examine every leg, to turn up all the soles, and to ascertain whether the horse has not been pegged between the shoe and the sole, or his "heels opened," by the shoeing-smith for the purpose of sale.

BLOOD-LETTING.

EVERY one, almost, can bleed a horse in some way or other, and it is often found extremely desirable that the operation should be performed without delay. But, like many other excellent remedies in the hands of unreflecting persons, this one is frequently employed imperfectly, as well as too often. Each has its peculiar notions, either as to the fit part whence the blood should be taken, the time when it becomes necessary, or the quantity proper to be taken; the latter being the more common error, as it is also the more excusable, inasmuch as they can plead "authority" either way for what they do, is nevertheless demonstrably ruled by wrong principles. The practice of bleeding at given periods, be the quantity taken ever so small, is most injudicious, to say no worse of it; for, why should we employ a curative when there is nothing to cure? especially when we thereby substract from vitality itself.

"Only bleed in cases of inflammation," say the French farriers, and then they take a large quantity, under the impression that taking a quart, or a little more, from a mass of two hundred and twenty quarts, which is fairly calculated to reside in a middling-sized horse, "is but trifling with the disorder." The existence of inflammation, or fever, is to be ascertained by the state of the pulse, upon which I was tolerably minute whilst speaking of fever, &c. (pages 61—64) the number of beats, and the kind of vibrations, being well considered, previously to taking the fleam in hand, when the quantity drawn should be commensurate to the extent of the disorder. Keep in mind, however, the exceeding danger of mistaking one series of febrile symptoms for another, as may be judged of by turning to those of "low fever," at page 67, when bleeding would destroy the horse. See also pages 68, 69. Without question, if the operator entertains a doubt about the symptoms as indicated by the pulse, the least quantity he takes is likely to perpetrate the smallest amount of harm; whilst, if he be correct in his observations, and has witnessed the good effects of bleeding in strictly similar cases, its inadequacy can effect no good whatever, nor repetition amend the matter one tittle, but the contrary. So that he must be wrong either way.

For, as I proved at the pages before referred to, the disorders for which bleeding is found serviceable depend less upon the quantity of blood that may be in the system at the time of the attack, than upon the construction or "state of the blood,"* and the degree of irritation that may exist in the ves-

_____
* That is to say, the proportion of its then component parts, which is mainly affected by the disease.

sels that contain it; both which affections, or causes of disease, are more frequently to be moved by the manner of taking any given quantity than by the actual weight, or rather the measure thereof. If the blood, for example, be drawn from a small orifice, no matter how rightly judged the quantity may be, however consonant to the proportions I have prescribed at page 63, yet the irritation of the blood-vessels, known by the rigid feel of the artery, will not be reduced, nor the animal recover. "He has been bled," is thrown in the face of the doctor, "and is no better: we have even preserved the blood." But the thing has not been performed with requisite skill. Among other absurdities, the operator will perceive the impropriety of permitting the blood to escape upon the ground, and then guessing at the quantity drawn; than which no practice can be more slovenly and fallacious.

A measure should be provided, marked with graduated circular lines, and numbered from the bottom by pints each. Glass forms the neatest vessel; but pewter offers a less brittle material in horse-medicine. The blood should be preserved awhile in the vessel, that the form it assumes in coagulating may be noted and remarked upon; as commonly happens most indiscreetly by all bystanders, whether it be caught or not; for very few can pronounce accurately, upon the view, the quantity of disease the blood indicates, particularly when it is on the ground; nor yet when in a vessel, unless it be caught properly.

Let the vessel be presented so as to catch the blood fairly, and not trickle down the sides, whereby the manner of its coagulation is affected. Blood that is drawn from a healthy horse, soon congeals in nearly one uniform mass, about one fifth of water only remaining at the top; from the residue you may wash away the red or colouring particles, and leave a pale thick coagulum or lymph. In a pound of such blood will be found these proportions—viz. 8 ounces of thick lymph, 5 ounces of the red or colouring particles, 3 ounces watery. If the operator keeps stirring the blood until it cools, the water does not separate, but the whole forms one homogeneous mass. In cases of great inflammation or fever, the watery proportion is much less, and the blood is then consequently more viscid or thick; which proves that this viscidity is an accompanying symptom of the disorder, as maintained in various parts of this volume; but, as the fever goes on, the animal loses appetite, and he makes no more new blood; the blood then becomes thinner in consequence of the deposite of lymph made in its circulation, and the red part predominates. On the contrary, in low fever and all languishing disorders of a tardy circulation, in cases of œdematous tumour, the watery part is found in the greatest proportion, and the red part is then almost extinct; in inflammatory fever the red particles predominate, the water is nearly dried up, and the lymph greatly decreases.

Instruments.—The fleam and blood-stick have been attacked as remnants of the old school, but were unjustly stigmatized as a rude method of obtaining blood. In the hands of judicious persons, the fleam has been found equal to every purpose that was required, and when used adroitly no other means of blood-letting, probably, ever will supersede it. But during the rage for improvements and new inventions, that prevailed a few years since, they sought to avoid a certain clumsiness of its application by introducing the lancet to general use. True it is, that the awkward method of making two or three aims with the stick, before striking at the fleam, occasions the horse to shy, especially whilst every vessel of the head is swelling with blood, in consequence of the application of the ligature round the neck; and equally true, that careless operators frequently cut through the vein, so as to cause subsequent disorders; though others, again, dangerously wounded the carotid artery

20

that passes under the vein; yet are there insurmountable obstacles to the general use of the lancet, that can never be overcome.

Of those, I need mention but one objection, viz. the time occupied in making the opening—seldom less than four or five seconds, which causes the animal to move its head, and thus to defeat the intention of making a sufficient orifice, whereby the adipose muscle of fleshy animals is allowed to interpose, and the blood trickles down the neck, and part of it gets underneath the skin. By the way, this happens when the operator does not bleed sufficiently high up the neck, the skin and muscle being much thicker lower down. Upon large animals, likewise, the lancet is wholly incompetent to its purpose, owing to the very thick teguments it has to pass through, leaving entirely out of consideration the substance of the vein itself. To remedy those objections, the spring fleam is more advisedly employed by less practised hands, and is found to combine the advantages assumed for the lancet, whilst it secures the requisite orifice punctured by the fleam.

Bleeding is now performed without previously applying a ligature, as it became apparent that the blood which was thus detained in both veins, distended also the capillary vessels in the head, which pressed upon the brain. Hence it frequently happened that vertigo came over the animal, filling it with the apprehension of danger. Sometimes it fell down through compression of the brain, and plunged; whereupon the disconcerted operator was known to give it up for a bad job, at the moment it became more than ever necessary, charging the fault to account of the horse's restiveness, with an expressed intention of resuming the attempt at some more favourable opportunity. But this was a promise he was seldom able to redeem cleverly; the alarm excited by striking the fleam again and again scarcely ever subsiding, for the tension of the vein would but increase with the continuance of the ligature, and caused it to slip aside more certainly. Apoplexy and death has ensued from the same cause, namely, the application of a ligature, and the consequent bursting of the fine blood-vessels of the brain.

A large vein is more desirable to take blood from, as an evacuation that is to relieve the whole system, than a small one, and the jugular or neck vein, within a hand of the jowl, is ever preferable; because the small do not conveniently admit of making so large an orifice, for the quick escape of the blood, upon which so much benefit depends; nor for the same reason allow of drawing a sufficient quantity at one time, to effect any good upon the spasmodic tendency or irritability of the vessels.

Local bleeding, in the plate vein for example, for a bruise in that region, does not enter exactly into my present view of the subject of blood-letting; though as much service to the part affected may be derived from drawing off from the circulation at the neck vein, as spraying a vein immediately at the seat of the evil. Bleeding in the foot is the only exception I should make; unless the practice of incising the bars of the mouth when the animal will not take his corn, be another, or at least not of importance sufficient to be mentioned at all, even as an exception.

The jugular vein being sought for where it is largest and nearest the surface, this will be found upon pressing it with the finger, a hand's breadth from the setting on of the head, a very little below the place where a branch comes from the lower jaw, and joins another from the upper part. The Frenchman instructs his *marechal* thus pithily on this topic, as on several others—"Do not bleed your horse in the head, but as near to it as possible." Its situation being thus found take the fleam between the fore-finger and thumb of the left hand, and pressing gently upon the vein below with the other fingers, the vein will rise; then strike, with stick or spring, as the case may be, and continue the pressure until the proper quantity of blood is drawn off. If this latter necessary attendance is found inconvenient, the ligature may now be applied without danger, but with no additional advantage.

Pinning up the orifice is the final part of blood-letting that is frequently overdone; that is to say, too much of the skin is drawn up over the orifice of the vein, so that the blood will flow underneath the skin, which causes a swelling; and a fistulous tumour is the consequence, that is very troublesome to cure. Where the quantity of blood taken has been small, leaving a redundancy in the system, this latter misfortune is likeliest to happen; but when the quantity taken has been large, and the horse rests quietly after it, the pinning up may be dispensed with, for the blood ceasing to flow of itself, the parts being brought together will adhere almost naturally, by holding the finger at the orifice for a few seconds. But when you must use a pin, be careful it does not prick the orifice of the vein.

Is it necessary to add, that the fleam should be clean, and otherwise in good order?

Rules. 1. Always give purging physic after letting blood. 2. Never bleed immediately after a run; nor at the moment pretend to pass judgment on the pulse, as it is then flurried. 3. You may bleed after a fall, or a contused wound; though the pulse be not quick, it will then be irregular: incised wounds do not require bleeding, since enough escapes at the wound. 4. If the blood in the measure be very hard, with buff at the top, the animal may be bled again: it indicates high fever. 5. If the blood scarcely coagulates, the poor creature ought not to have been blooded at all.

ON ACUTE FOUNDER.

Founder, as a general subject, is one of great importance; and when it is considered as probable, that if it does not destroy, it at least renders useless more horses than all other diseases put together, its importance can hardly be rated too high. To a proper consideration of it, however, it must be regarded as consisting of two kinds, and these essentially differing from each other. The one is an acute attack, dependent on diffused inflammation or fever, like the inflammations of any other important organs: the other, a chronic, occasioned by local inflammation, sometimes dependent on constitutional liability, but much more frequently on outward occasional causes.

But as an acute founder appears to be the most general disease in this country, I shall confine myself entirely to a consideration of it.

ACUTE FOUNDER appears to have two origins, in one case being a true metastasis of primary fever, or translation of disease from one part to another; in other instances the attack appears to be made more directly on the feet themselves. In a great many instances it can be directly traced to the effect of obstructed perspiration; or at least of the sudden alternations of temperature, operating in the production of general febrile affections, whose translation to the feet is sometimes perhaps accidental, and at others may be produced by some cause which has already weakened them. In this latter way it often occurs after very severe exertions; as very hard riding or driving, with previous, present, or subsequent exposure to wet or cold, particularly of the feet, as washing them immediately after the horse arrives; or the tendency may perhaps be increased by first exposing the feet to cold and afterwards suddenly removing them into a warm stable; the vessels of the feet not being able to bear this sudden alteration, distend and fall into inflammation It may in many of these cases occur prior to general fever, which will then be symptomatic; or it may be consequent to it, when the founder itself is the effect of translation; and both are frequently occasioned, as before stated, by re-peated and long continued exertions with subsequent exposure to cold, espe-cially by the custom of washing the feet and legs when hot.

Founder very frequently proceeds from cold too suddenly applied to the body from a current of cold air acting upon it when in an over-heated state, or from drinking freely of cold water. The symptoms are at first these : when the horse begins to cool, he appears very stiff and feeble in his fore quarters, and, when forced to move forwards, he collects his body, as it were, into a heap, and brings his hind feet as far forward under him as he can, in order to remove the pressure of the weight of his body from the fore legs a d feet ; at the same time he sets his fore feet to the ground with great pain ; his fore parts are extremely hot, and sometimes his legs are considerably swollen, and evidently painful to the animal when touched.

As soon as the complaint has risen to any height, the feet will be found in-intensely hot, and the pastern arteries pulsating very strongly; there is some-times some little tumefaction round the fetlocks, and when one foot is held up for examination, it gives so much pain to the other that the horse is in danger of falling. The poor beast groans and breaks out into profuse sweats at one time, and at others is cold ; his eyes are moist and red, and his whole appear-ance betokens that he is labouring under a most painful inflammatory affec-tion.

In this state, the complaint shows itself the first three or four days, after which its effects are various. In excessively bad cases, when the symptoms stated have raged a few days, a slight separation of the hoof at the coronet may be observed, from which a small quantity of thin matter may be pressed ; the sensible laminæ of the foot, now losing their connexion with the insensible laminæ by the effects of the inflammation, the hoofs gradually separate, and at last drop off. At other times the effects are not quite so violent : still how-ever the termination is sufficiently unfortunate; for coagulable lymph is thrown out, which equally forces off the hoofs; but not until the parts underneath have acquired some solidity, nor till the germ of a new hoof appears, which if suffered to grow never proves perfect ; on the contrary, the horse usually remains permanently lame. In other cases the laminæ, losing their elasticity and power, yield to the weight of the coffin-bone, which becomes pushed back-wards, and in its passage draws with it the front of the hoof, which falls in ; the pressure also of the coffin-bone destroys the concavity of the horny sole, which becomes convex, or pumiced, leaving a large space hollow towards the toe, which very frequently turns up.

But when the attack is not commenced with that violence which has been detailed, or when an early and judicious plan of treatment is adopted, the ter-mination will be more fortunate ; the horse will stand longer upon his feet, the pulse, which at the onset of the disease is very high, will gradually fall, these favourable appearances will increase daily, and in the end the animal will re-cover the use of his feet.

As soon as the disease is discovered, take away blood from the neck to the amount of four, five, or six quarts, as circumstances may require, or size and condition will permit; back-rake and throw up clysters, but unless there be much costiveness present, do not give strong purgative medicine, as the high state of irritative fever which is generally present, forbids such practice. Mild laxatives should be given twice or three times a day until the bowels are moderately opened, together with the fever ball, recommended some pages further back, twice a day, until the inflammatory symptoms have sub-sided.

The feet should be attended to after the general bleeding, &c. In the first place let the shoes be taken off, and the soles pared a little ; the hoof should be rasped as thin as is prudent, which will greatly relieve the internal sensible parts, which are tender and swollen, by removing the pressure of the sole and hoof from them; let the feet be immersed in warm water or apply poultices to them, or if preferred wet cloths may be kept round them; if the general febrile symptoms still continue repeat the bleeding and the medicine. As soon as amendment becomes apparent, feed mildly, and allow the horse to rest; do not proceed to exercise until the feet have gained some strength, nor must it be forgotten that feet once foundered, require great caution in their future management, as they are very liable to become again affected on any considerable exertion.

20 *

INDEX.

A.

K.

L.

M.

21

SUPPLEMENT

TO

MASON AND HIND'S

POPULAR SYSTEM OF

FARRIERY:

COMPRISING

AN ESSAY ON DOMESTIC ANIMALS,

ESPECIALLY THE HORSE;

WITH

REMARKS ON TREATMENT AND BREEDING;

TOGETHER WITH

TROTTING AND RACING TABLES,

SHOWING

THE BEST TIME ON RECORD, AT ONE, TWO, THREE.
AND FOUR MILE HEATS;

PEDIGREES OF WINNING HORSES, SINCE 1839;

AND OF THE MOST

CELEBRATED STALLIONS AND MARES;

WITH

USEFUL CALVING AND LAMBING TABLES, ETC., ETC.

BY J. S. SKINNER,

Editor now of the Farmers' Library, New York; Founder of the American Farmer, in 1819
and of the Turf Register and Sporting Magazine, in 1829; being the first Agricultural and the first Sporting Periodicals established in the United States.

PHILADELPHIA:

J. B. LIPPINCOTT & CO.

1867.

DEDICATION.

WITHOUT going through the formality of asking leave to say " by his gracious permission," which, if sought, might have been withheld, this SUPPLEMENT TO MASON AND HIND'S POPULAR SYSTEMS OF FARRIERY is respectfully dedicated to COL. BALIE PEYTON.

It is not that a contribution so inconsiderable is deemed worthy of him, or the subject so interesting; but that the Author would fain embrace any fair occasion to manifest to him, and through him to their common friends at New Orleans, his grateful remembrance of their kindness when among them.

There would be, moreover, an essential propriety in dedicating to Col. P. a more adequate offering of this sort; as he is known to be a breeder and warm amateur of the high-bred horse; and, in his own spirit and character, exemplary of what is best bred and most excellent among men.

J. S. S.

NEW YEAR'S DAY, 1848.

For the nonce at Annapolis, Md

PREFACE.

THOUGH, under ever fluctuating but sometimes propitious circumstances, the very climax of equestrian power may have been reached in a few cases in the United States, as in the country from which we derived our skill and material, is it not still worthy of all consideration how we may contrive to *belay*, as the sailors say, what we have gained in that important branch of Rural Industry—not only as a means of individual enjoyment, but as a prolific, indispensable source of National power and wealth?

However serious and apparently insurmountable may be the difficulties that stand in the way of farther improvement of domestic animals, and especially the HORSE —either in the general absence of the necessary means and appliances, and of adequate encouragement for the care and expense attendant on the production of Horses of high qualities, there ought, surely, among well-informed men, to be no obstacle arising from *ignorance of the art of breeding*. Hence it is that in sending forth the *Ninth Edition* of this popular work on Farriery, while nothing seemed to be needed in the way of description or treatment of the diseases of domestic animals, and while the author of this Supplement was only called on to extend the *stud-book* in a manner to embrace the pedigrees to which breeders and dealers might have occasion to refer, he could not forego the opportunity to offer some such additional matter as, to him at least, seems to be of sufficient value to render it acceptable and useful.

In the introductory remarks on the relations existing between Man and the animals destined for his use and amusement, and the obligations these relations impose, the writer has but expressed the sentiments he has ever entertained, of duty on our part to respect the feelings and comfort of the humblest among them; and has endeavoured to encourage continued exertions for their melioration by showing how successful and progressive such efforts have been, even up to the present time.

To these observations of his own are appended those of writers of acknowledged judgment and authority — accompanied by such *notes* as appeared to be apposite and well-founded; and to these, again, have been superadded a few tables and other items which might not elsewhere be conveniently met with. His undertaking, kind reader, 'hath this extent, no more." All, then, that the author of the " SUPPLEMENT TO MASON'S FARRIER" has to ask of you is that you will bear in mind that there has been no engagement to *write* anything—much less a *Book on Farriery:* for that there was no call or necessity. With this intimation, the reader will please accept for what it is worth and with all due allowances, the little that has been volunteered—by one who may claim to have been all his life an amateur if not a connoisseur of the Horse.

J. S. S.

Edit. Farmers' Library

CONTENTS.

SUPPLEMENT, ETC.

ON THE RELATIONS BETWEEN MAN AND THE DO-MESTIC ANIMALS — ESPECIALLY THE HORSE — AND THE OBLIGATIONS THEY IMPOSE.

"La connaissance de la conformation exterieure du cheval est beau-coup moins répandu qu'on ne le pense vulgairement : elle repose sur des etudes d'anatomie de physiologie, de mecanique, et d'histoire naturelle dont peu de personnes se font une juste idée."

IF animals were classified by naturalists in the order of their intelligence, docility and usefulness, the HORSE and the DOG would occupy, in relation to Man, the jux-taposition they have assigned—on the ground of physical structure—to the impracticable baboon and the grotesque and chattering monkey; and in lieu of groping in the darkness of antiquity for the period when they are sup-posed to have been entrapped or subdued, by fraud or violence, we should the rather conclude that Nature placed all the domestic animals where we have ever found them — in close association with Man, administering to his pleasures and wants; lightening his toils and sharing his dangers, and constantly advancing, like Man him-self, under the improving influence of civilization and the arts that belong to it.

In contemplating the whole animal kingdom, does not Man—standing preëminently at the head of it, surrounded by the domestic races — present everywhere the most lustrous spot on the varied map of living creation? From the everlasting snows of the north to the burning sands of tropical deserts, his faithful dog follows at his foot; the horse is at his side — submissive to his will; — the patient ox bows his neck to the yoke; and the sheep and the hog are present to supply his clothing and his food. Far otherwise is it with untameable and predatory birds and beasts. Restricted to particular regions by an all-wise Providence, the absence of food and climate congenial to their nature forbids them to roam beyond limits comparatively circumscribed. And do not these arrangements for our benefit, and which give us " dominion over all the earth and every creeping thing that creepeth upon the earth," enjoin on us the duty of studying their habits, their economy, and all the laws of their existence —with a view to their improvement for our advantage, in every way consistent with kindness to them and with gratitude to HIM,

> " Who in his sovereign wisdom made them all ?"

And while these considerations teach us to be merciful ourselves, do they not convey the admonition

> " Ye therefore who love mercy, teach your sons
>   To love it too!"

The very fact that to them has been denied the power of speech, and the necessity of uncomplaining submission under every hardship, ought to put us constantly on our guard against practising, or permitting to be practised, any, the smallest measure of abuse or ill treatment. Thus every man of common humanity will study their comfort in all things, consistently with the purposes for which they were designed, and will never even mount his faithful horse without seeing that whatever is needed has been done to give an easy *set* to his saddle—and, still more, that all is *right about his feet !*

Doctor Rush, in a beautiful and benevolent eulogy on
the Horse, in one of his lectures, related a touching anec-
dote of a highly intelligent and successful Pennsylvania
farmer, who, stricken down suddenly with apoplexy in
his barn-yard, expired on the instant—with this last direction to his herdsman on his lips: *" Take care of the
creatures !"* And the biographer of an eminent English
Chancellor relates, as from himself, how his beloved son
had preferred to him, in his very last moments, a petition
in favour of his faithful terrier; *" And Father, you'll
take care of poor Pilcher, won't you ?"* Nevertheless,
after all the care that can be taken, we should probably
be amazed if we could know the amount of pain unwit-
tingly inflicted on animals dedicated to our service, and
some of whose bodies are at last consumed to afford us—
as some would contend—superfluous nourishment, referring back as they do to that golden age when

> " Man walked with beast—joint tenant of the shade ;
> The same his table and the same his bed—
> No murder clothed him, and no murder fed."

Even all unnecessary harshness of reproof should be
avoided—for it is well known that some animals are even
more susceptible of painful and violent emotions, from
various causes, than some men, whose hardened nature
and familiarity with vice, render them as insensible to
the reproaches of others as to the stings of their own
conscience. Those, for instance, who have studied the
character and affections of the horse—with a view to his
diseases and moral susceptibilities—need not be told that
while sharp and threatening words will so disturb him as
to quicken his pulse some ten beats or more in a minute,*

---

* The natural constitution of different varieties of the same class of
animals is worthy of close attention. In small and thorough-bred horses,
for instance, the pulsations of the heart are about 40 to 42—while in
the larger, cold-blooded cart-horse, they do not amount to more than 36.
But when ill-treated, as before suggested, their pulsations are increased.

ne has in very memorable cases been known to fall dead
under the excitement of the sexual and other passions.
That he is sometimes animated by the strongest spirit of
rivalry, and a noble ambition to excel, has been occasionally evinced by violent attacks on his passing rivals
on the turf—and very recently the case occurred with a
noble animal which fell dead at the very winning-post,
in vainly struggling for victory, on the Pharsalia course at
Natchez.   The contest which had this melancholy issue
was between Col. Minor's JENNY LIND and Col. Bingaman's BLACK DICK :

" Dick was the favourite at odds.   Some even bets
were made that he would win at three heats—and some,
if the heats were broken, would not win.   Jenny drew
the track, and after some little manœuvring, they got off
together, but Dick outfooted her and took the track on
the turn ; at the half-mile post she had got her head to
his hips, and they ran locked round the upper turn ; at
the head of the front stretch she began to draw clear of
him, and spurs were applied.   ' Then burst his mighty
heart,' for he soon was seen to reel, but he still struggled
on ; his jockey Mat, leaped unharmed from his back, and
the noble animal fell dead within ten feet of the winning-
post, which he had left not two minutes before in perfect
health and the finest condition.   No shout of triumph
hailed the winner: all was sympathy and regret.   Two

---

say, ten in a minute.   The natural circulation of the sheep is about 70
per minute.   The average pulse of a full-grown ox, in a state of health,
in England, is about 40—but this increases in a climate of higher temperature.   Doctor James Smith (Journal of Agriculture, vol. ii. p. 92,)
finds that in the climate of Louisiana the pulse of the ox, in its natural
state, is from 68 to 75—rising on the slightest excitement to 80.   Every
one knows how destructive is the moral influence of *fright* to a flock of
sheep—when, for instance, they have been badly scared by dogs.   It
often happens that they never recover from its effects.

For all farmers who have occasion to *fatten* animals, we must take
room for three words—*warmth, cleanliness*, and *quietude*.   They are
the *veni-vidi-vici*, in *their* fields of action.

of our most talented medical gentlemen immediately made a post-mortem examination, and came to the conclusion that the death of the horse was produced by apoplexy, caused by congestion of the heart, brought on by over-excitement and violent exertion."

The annals of domestic animals abound in cases to show how liable they are to acute affections and suffering, far beyond the apprehension of the most considerate and humane.

Thus much, good reader, have we gladly seized the opportunity, and even gone a little beyond the requirements of our publishers, to say in the way of appeal in behalf of speechless creatures, as alive to pain as to a sense of gratitude for generous treatment; and having already adverted to the obligation we are under to study the laws of their existence, and the means of their melioration, it may now, even be insisted that in the whole range of the occupations and interests of breeders of their own stock, there are few things that demand more consideration and skill than does this very branch of rural industry.

The study and the pride of every one should be, not merely to maintain them at a point of excellence already acquired, but to have them progressively improving in whatever constitutes economy and value; for why should any man indolently conclude that his stock has already attained the *ne plus ultra* in the way of amelioration, however superior it may be? Such is not the fact, nor, it may safely be affirmed, would it be consonant with the orders of Providence, or even with our own interests, that it should be so. To man has been given dominion over the beasts of the field—that, like the earth itself, he should cultivate and improve them; and for that, among other purposes, was he endowed with the great, distinguishing, and godlike power to prosecute intellectua

investigations into every department of nature and industry. Doubtless our ancestors, more than a century
ago, were ready to believe—what indolence is ever ready
to whisper — that the several races of domestic animals
most immediately under their care, had then already been
carried up to the maximum of improvability; yet which
of them has not been vastly bettered in the meantime, in
all their valuable points—and that, too, not by any sudden or accidental accession of one or more good qualities, but constantly and progressively; by a closer study
and a better knowledge of the laws of animal and
vegetable physiology, and by the application of other
appropriate sciences. In the plain English of the motto
chosen for these reflections what is there said of the
Horse may apply to other animals :

" The knowledge of the external conformation of the
horse is much less extended than is generally supposed.
It reposes on the study of anatomy, of physiology, of
mechanics, and of natural history, in a manner of which
few persons have a just conception."

In 1710, by the estimate of Dr. Davenant, — a writer
of unquestioned candour and authority,—the weight of
" black cattle" (so called, because, at that day, most
cattle were of that colour) averaged but 370 pounds; the
weight of the calf was estimated at 50 pounds; and the
average of sheep and lambs, taken promiscuously in the
London market, was only 28 pounds. After the lapse
of 120 years, — with far less of science applied to the
subject than at this time,—M'Culloch, in his dictionary,
so highly characterized by the accuracy of its statements,
puts the average of cattle at 556 ; sheep and lambs at
50; and calves at 105. But the late accomplished Professor Youatt, in his able work on cattle, estimates the
average weight now at Smithfield at 656 ; that of sheep
and lambs at 90 ; and calves at 144 ; — the weight of

each having doubled in 130 years; and that, as before said, not by any accidental importation from abroad, or fortunate cross at home, but by a course of careful, systematic, and sagacious attention to the laws and principles of breeding and feeding. The horse, standing at the head of the list,—sharing and supporting man in all his most pleasurable as well as toilsome and dangerous enterprises,—naturally engaged his earliest attention and most assiduous care, to cherish and improve to the highest pitch, his noble faculties of strength, speed, and endurance; and thus may have been already brought to the zenith of his capabilities, if indeed he has not passed the culminating point; but see what must have been achieved by the stimulus of the turf, and art in the breeding-stud, to raise the bred horse of England to a height of perfection, even above the wonderful capacity of his south-eastern ancestry, — the very "drinkers of the wind" themselves!—for we have the high authority of Nimrod, the crack writer of England on all field-sports, for saying that, on the best Indian authorities, "the best Arab, on his own ground, has not a shadow of a chance against an imported English racer, in anything like a good form." The celebrated race on the Calcutta Course, between *Pyramus* and *Recruit*, — the former the best Arab of his year; the latter a second-rate English race-horse, by Whalebone, the property of the Marquis of Exeter,—settled this point, inasmuch as allowance was made for the comparatively diminutive size of the Arab,— it being what is termed a give-and take match, or weight for inches; in which Recruit carried 10 stone 12 (152) pounds; and Pyramus only 8 stone 3 (115) pounds, an extra allowance of 7 pounds having been given to him as an Arab.

Pyramus, says the reporter of this race, is as good

an Arab (he had previously beaten all the best Arabs in
Calcutta for the gold cup) as has appeared for many years.
His condition was undeniable; the distance was all in
his favour, and he was ridden with superior judgment—
so that the result of his match with Recruit may be considered to have established this an axiom : that no allowance of weight, within the bounds of moderation, can
bring the best Arab—even in a climate most congenial
to him—upon a par with an English thorough-bred horse
of moderate goodness. In addition to all these circumstances in favour of Pyramus, it must be remembered
that Recruit only landed on the 28th May, (the race was
run in January), after a voyage of five months."

In England, where the progress of improvement was
greatly accelerated by a seasonable infusion of Arabian
and *barb*aric blood, the *bred*-horse—standing, in respect
of the equine race, as the capital on the Corinthian pil-
lar—has reached a point of perfection that, if it can be
kept up, we can hardly dare hope will ever be excelled
In that country, four-mile races are nearly abolished, and
it has been said with every show of reason, that early
training, light weights and short distances, are impairing
the stoutness of the English race-horse and hunter, and
their capacities to stand up and go the pace as in the
palmy days of the English turf. In our own country,
the annals of the course show, that our climate is highly
congenial to the constitution and physical development
of the horse — and that whenever the sport has been
fashionable and the rewards adequate, he has ever been
ready to meet all reasonable expectations—rather advancing than falling oack.

When Floretta won her race in Washington—winning
the 2d heat in 7.52, against such nags as Oscar, Topgallant and First Consul, it was deemed a mar·ellous

performance ;* and sportsmen thought that the acme of
speed and bottom had been reached in our country in
the days of Sir Charles and Eclipse, yet have not their
best achievements been eclipsed by two illustrious and
yet living rivals of each other — *Boston* and *Fashion?*
But what have we not to apprehend should what seems
to be threatened come soon to pass, and the turf— the
only sure test of speed and stoutness, be *allowed to go
down?* We remember once at a dinner-party at the
British Minister's in Washington, to have inquired of the
late John Randolph of Roanoke, whether the Old Dominion maintained, unimpaired, her claim to a superior race
of horses? "No, Sir; no, Sir," was his shrill-toned
prompt reply; "Since we gave up horse-racing and fox-
hunting, and turned up the whites of our eyes, our horses
as well as our men have sadly degenerated."

Finally—justice, truth, and a sense of obligation for
the assistance derived from his labours, in the small contribution we are here making to the breeders and amateurs
of the Horse, demand of us to say, at the least, that if
the American Turf should decline, it will not be for want

---

* This was one of the most memorable contests that ever came off
on the Washington Course. Horses were horses, and men were *men*,
in those days. Fair-top boots, powdered heads, and golden "guineas"
were all the go—and for fairness and honour, a "stain was felt like a
wound."

The horses were thus placed :

Dr. Edelin's c. m. *Floretta*, by Spread Eagle, 6 years old, 5 1 1
Gen. Ridgely's b. h. *Oscar*, by Gabriel, 6 yrs. old, 2 2 2
J. B. Bond's b. h. *First Consul*, by Flag of Truce, aged 4 3 3
Col. Tayloe's b. h. *Top-gallant*, by old Diomed, 6 yrs. old, 1 4 4
M. Brown's b. m. *Nancy*, by Spread Eagle, 6 years old. 3 dr.

In this race Floretta was closely run by Oscar and First Consul —
each heat was run under 8 minutes, and the *second* in 7.52. Each horse
made play from the score, and the time was better than had been made
on that Course even up to 1829. Has such a field of men and horses
come to that post since ?

In another pace—the trot—it was deemed marvellous that "old Top"
should go his mile with 150 pounds weight in 2.45. But Lady Suffolk — well dashed with the old Messenger blood — has done hers in
2.28½, and is yet in full if not improving vigour.

of an able, industrious, and tasteful advocate and illus-
trator of its advantages and uses, as long as W. T. Porter shall continue to animate and guide the " Spirit of
the Times." Extensive acquaintance and coëxtensive
popularity—the just fruits of accomplished manners and
an obliging temper—have made him the focus of a most
varied and recherché correspondence: while his own
tact, scholarship and nice appreciation of what is good
in the literary and the sporting world, enable him to turn
all his rich resources to the best account, for the enjoyment of his numerous and refined readers—for the most
part, gentlemen of *blood and mettle.*

ON THE FORM OF ANIMALS,

BY HENRY CLINE, ESQ. SURGEON.

WITH NOTES BY J. S. SKINNER.

The form of domestic animals has been greatly improved by selecting with much care, the best formed for
breeding—but the theory of improvement has not been
so well understood, that rules could be laid down for
directing the practice. There is one point particularly,
respecting which the opinions of breeders have much
varied, which is, whether crossing the breed be essential
to improvement.

It is the intention of this communication to ascertain
in what instances crossing is proper, and in what pre-
judicial; and the principles upon which the propriety
of it depends.

It has been generally supposed that the breed of animals is improved by the largest males. This opinion
has done considerable mischief, and would have done
more injury had it not been counteracted by the desire
of selecting animals of the best form and proportions,
which are rarely to be met with, in those of the largest size

Experience has proved that crossing has only succeeded in an eminent degree, in those instances in which the females were larger than in the usual proportion of females to males; and that it has generally failed when the males are disproportionally large.

The external form of domestic animals has been much studied, and the proportions are well ascertained. But the external form is an indication only of internal structure. The principles of improving it must therefore be founded on the knowledge of the structure and use of internal parts.

The lungs are of the first importance. It is on their size and soundness that the health of an animal principally depends. The power of converting food into nourishment, is in proportion to their size. An animal with large lungs, is capable of converting a given quantity of food into more nourishment than one with smaller lungs, and therefore has a greater aptitude to fatten.*

The Chest.

The external indication of the size of the lungs is the form and size of the chest; the form of which should

---

* [In farther explanation of this principle, it may be added, from an author who had evidently read and relied on this able Essay of Surgeon Cline, that muscular exertion facilitates the return of venous blood to the right side of the heart, and in long continued and violent exertion, the respiration being quickened, the lungs—if small—are unable to *arterialize* and get rid of the blood as fast as it is pumped into them; consequently, if there is not room for the blood, congestion takes place, and the horse becomes what is termed "blown"—the lungs being gorged with blood, and sometimes the animal is destroyed by it. In England it is said to be "well understood that a majority of horses that perish under a hard press 'across the country,' are *narrow-chested!*" The conical form, not of the body, but of the *chest*, as laid down in the next paragraph, is very observable in the best paintings of Fashion. There, and in her quarters and hocks, appear to us to lie the great sources of her yet in this country unequalled speed and stoutness.— J. S. S.]

have the figure of a cone, having its apex situated between the shoulders, and its base towards the loins.

The capacity of the chest depends upon its form more than on the extent of the circumference; for, where the girth is equal in two animals, one may have much larger lungs than the other. A deep chest therefore is not capacious unless it is proportionally broad.

The Pelvis.

The pelvis is the cavity formed by the junction of the haunch bones with the bones of the rump. It is essential that this cavity should be large in the female, that she may be enabled to bring forth her young with less diffi-culty. When this cavity is small, the life of the mother and of her offspring is endangered.

The size of the pelvis is chiefly indicated by the width of the hips and the breadth of the *twist*, which is the space between the thighs.

The breadth of the loins is always in proportion to that of the chest and pelvis.

The Head.

The head should be small, by which the birth is facil-itated. Its smallness affords other advantages, and gen-erally indicates that the animal is of a good breed.

Horns are useless to domestic animals. It is not dif-ficult to breed animals without them. The breeders of horned cattle and horned sheep, sustain a loss more extensive than they may conceive; for it is not the horns alone, but also much more bone in the skulls of such animals to support their horns; besides there is an addi-tional quantity of ligament and muscle in the neck which is of small value.

The skull of a ram with its horns, weighed five times more than another skull which was hornless. Both these skulls were taken from sheep of the same age, each be'ng four years old. The great difference in weight depended chiefly on the horns; for the lower jaws were nearly equal, one weighing seven ounces, and the other six ounces and three quarters; which proves that the natural size of the head was nearly the same in both, independent of the horns and the thickness of the bone which supports them.*

In a horned animal, the skull is extremely thick. In a hornless animal it is much thinner; especially in that part where the horns usually grow.

To those who have not reflected on the subject, it may appear of little consequence whether sheep and cattle have horns—but on a very moderate calculation it will be found, that the loss in farming stock, and also in the diminution of animal food, is very considerable, from the production of horns and their appendages. A mode of breeding which would prevent the production of these, would afford a considerable profit in an increase of meat and wool, and other valuable parts.

The length of the neck should be proportioned to the height of the animal, that it may collect its food with ease.

The Muscles.

The muscles and tendons, which are their appendages, should be large; by which an animal is enabled to travel with greater facility.

---

* [It is matter of surprise that among the varieties of cattle imported, no one should bring the celebrated *Suffolk* polled or hornless cattle. Besides the advantage here enumerated, valuable animals are sometimes killed by being *gored*. In respect of this breed, Youatt speaks very highly. He says they sometimes give 32 quarts of milk, and 24 is not uncommon, in a day—and adds:—"There are few short-horn cows; although far superior in size to the Suffolks, and consuming nearly double the quantity of food; that will yield more milk than is usually obtained from the smaller polled breed." Formerly the Suffolk polled cattle were generally of a *dun* colour, and thence commonly called Suffolk *duns*, but that colour has of late been repudiated.—J. S. S.]

The Bones.

The strength of an animal does not depend upon the size of the bones, but on that of the muscles—Many animals with large bones are weak, their muscles being small. Animals that were imperfectly nourished during growth, have their bones disproportionately large. If such deficiency of nourishment originated from a constitutional defect, which is the most frequent cause, they remain weak during life. Large bones, therefore, generally indicate an imperfection in the organs of nutrition.

On the improvement of Form.

To obtain the most approved form, two modes of breeding have been practised—one, by the selection of individuals of the same family—called breeding *in-and-in*. The other by selecting males and females from different varieties of the same species; which is called *crossing the breed.*

When a particular variety approaches perfection in form, breeding in-and-in may be the better practice—especially for those not well acquainted with the principles on which improvement depends. *

---

*[Professor Youatt says, on this subject [breeding in-and-in]: "It is the fact, however some may deny it, that strict confinement to one breed, however valuable or perfect, produces deterioration." By what he afterward says, as will be seen, he must have meant confinement to *one family or strain* of the same breed. The rule should be this: that valuable qualities being once established, which it is desirable to keep up, should thereafter be preserved by occasional crosses with the best animal to be had of the same breed, but of a different family, This is the secret which has maintained the bred Horse in his great superiority— for although, as Nimrod avers, the immediate descendants of eastern horses have, almost without an exception, proved so deficient of late years that breeders will no more have recourse to them than the farmer would go for immediate improvement to the natural or original oat ; yet the breeder is glad to cross his stock with one of another strain or family of the same blood, taking care never to depart from the blood of the south-eastern courser which flows in the heart of all families of Horses of th. highest capabilities.

When the male is much larger than the female, the offspring is generally of an imperfect form. If the female be proportionally larger, the offspring is of an improved form. For instance, if a well-formed large ram be put to ewes proportionally smaller, the lambs will not be so well shaped as their parents; but if a small ram be put to larger ewes, the lambs will be of an improved form.

It is here worthy of remark that Nicholas Hankey Smith, who resided a long time among the Arabs, in a work entitled "Observations on Breeding for the Turf," gives as his opinion that colts bred in-and-in show more blood in their heads, are of better form, and fit to start with fewer sweats than the English turf-horse; but when the incestuous intercourse has continued a few generations, he says, the animal degenerates.

This plan of breeding in-and-in, says Youatt farther, when speaking of cattle: "has many advantages to a *certain extent*. It may be pursued until the excellent form and qualities of the breed are developed and established. It was the source whence sprung the cattle and the sheep of Bakewell, and the superior cattle of Colling—and to it must be traced the speedy degeneracy, the absolute disappearance, of the new Leicester or Bakewell cattle; and in the hands of many an agriculturist, the impairment of constitution and decreased value of the new Leicester sheep and the Short-Horn beasts. It has therefore become a kind of principle with the agriculturist to effect some change in his stock every second or third year—and that change is most conveniently effected by introducing a new bull or ram. These should be as nearly as possible of the same sort coming from a similar pasturage and climate, but possessing no relationship, or at most a very distant one, to the stock to which he is introduced"—and these remarks "apply to all descriptions of live-stock," says Professor Johnston, author of the Farmer's Cyclopedia.

This is the secret whereby Mr. GEORGE PATTERSON, of Maryland, has not only kept up but improved the size and beauty of his North Devons. Every "two or three years," a new bull the best to be had in England, is introduced to his cows. The neglect of this precaution, and breeding in-and-in too closely, are the true reasons why we so rarely see the descendants of imported stock in this country equal to the originals. Too close breeding tells in Man as well as in beast; hence the famous lines of Lord Byron when speaking of the nobility:

"————They breed in-and-in as might be known,
"Marrying their cousins, nay, their aunts and nieces,
"Which always spoils the breed, if it increases."

But, after all, we must look closely to the *form* of the parents as well in Horses as cattle—for, let the world dispute as it may, whether "blood is everything," or "blood is nothing,"—be the blood what it may, who has ever seen, as Apperley asks, an instance of a misshapen horse and ill-formed mare producing winners?—J. S. S.]

The proper method of improving the form of animals, consists in selecting a well-formed female, proportionally larger than the male. The improvement depends on this principle, that the power of the female to supply her offspring with nourishment is in proportion to her size, and to the power of nourishing herself from the excellence of her constitution.

The size of the fœtus is generally in proportion to that of the male parent; and therefore, when the female parent is disproportionately small, the quantity of nourishment is deficient, and her offspring has all the disproportions of a starveling. But when the female, from her size and good constitution, is more than adequate to the nourishment of a fœtus of a smaller male than herself, the growth must be proportionately greater. The larger female has also a greater quantity of milk, and her offspring is more abundantly supplied with nourishment after birth.

To produce the most perfect formed animal, abundant nourishment is necessary from the earliest period of its existence, until its growth is complete.

It has been observed, in the beginning of this paper, that the power to prepare the greatest quantity of nourishment, from a given quantity of food, depends principally upon the magnitude of the lungs, to which the organs of digestion are subservient.

To obtain animals with large lungs, crossing is the most expeditious method; because well-formed females may be selected from a variety of a large size, to be put to a well-formed male of a variety that is rather smaller.

By such a method of crossing, the lungs and heart become proportionately larger, in consequence of a peculiarity in the circulation of the fœtus, which causes a larger proportion of the blood, under such circumstances, to be distributed to the lungs than to the other parts of the body; and as the shape and size of the chest depend upon that of the lungs, hence arises the remarkably large chest, which is produced by crossing with females that are larger than the males.

The practice according to this principle of improvement, however, ought to be limited; for, it may be carried to such an extent, that the bulk of the body might be so disproportioned to the size of the limbs as to prevent the animal from moving with sufficient facility.

In animals where activity is required, this practice should not be extended so far as in those which are required for the food of man.

On the Character of Animals.

By character in animals is here meant, those external appearances by which the varieties of the same species are distinguished.

The characters of both parents are observed in their offspring; but that of the male more frequently predominates.[*]

*[To the contrary of this, as to Horses, T. B. Johnson, author of the Shooter's Companion, and a writer of high authority, says : " although it is a maxim universally admitted, that an equal degree of precaution should be used in respect to the Horse, it is doubly and trebly necessary with the mare—because strict observation has demonstrated that nearly or full two out of every three foals, display in their appearance *more of the dam than the sire ;* and that there are more fillies than colts fallen every year will not admit of a doubt."

This positively asserted predominance of females over males, may be accounted for on the principle established by very numerous experiments in France with sheep, if not with other animals—on the results of which the experimenter, whose name is not remembered, based and confidently asserted his theory, that the *sex* of the offspring, in all cases, depends much on the *comparative vigour of the parents.* By putting old ewes to young rams in the prime of life, he never failed to get a *large proportion* of ram lambs ; and, *vice versa,* when young ewes in their prime were put to a ram lamb, which had not yet attained his full growth and development, or to old ones far gone in the down-hill of life, then a very large proportion were females. A great number of experiments were given corroborative of the doctrine. Is it not reasonable to suppose that an influence sufficient to control the sex, would have an effect on exter-

This may be illustrated in the breeding of horned ani-mals; among which there are many varieties of sheep, and some of cattle, that are hornless.

If a hornless ram be put to a horned ewe, almost all the lambs will be hornless; partaking of the character of the male rather than of the female parent.

In some countries, as Norfolk, Wiltshire, Dorsetshire, most of the sheep have horns. In Norfolk the horns may be got rid of by crossing with the Ryeland rams; which would also improve the form of the chest and the quality of the wool. In Wiltshire and Dorsetshire, the same improvements might be made by crossing the sheep with South Down rams.

An offspring without horns might be obtained from the Devonshire cattle, by crossing with hornless bulls of the Galloway breed; which would also improve the form of the chest, in which, the Devonshire cattle are often de-ficient.

Examples of the good effects of crossing the breeds.

The great improvement of the breed of horses in Eng-land arose from crossing with those diminutive Stallions, Barbs, and Arabians; and the introduction of Flanders mares into this country was the source of improvement in the breed of cart-horses.

The form of the swine has also been greatly improved, by crossing with the small Chinese boar.

Examples of the bad effects of crossing the breeds.

When it became the fashion in London to drive large bay horses, the farmers in Yorkshire put their mares to

nal form and colour? It may be a reason why some of our very popu-lar stallions, being overtasked, have had so few of their get to rival them in power and fame. Every reader may cast about for himself, for in-stances, to see how far and to *what* other animals the principle applies. After all, in an economico-agricultural view, it is much more important that the stallion should be all right because it is *his* blood that is to be diffused far and wide.—J. S. S.]

much larger stallions than usual, and thus, did infinite mischief to their breed, by producing a race of small chested, long legged, large boned worthless animals.*

A similar project was adopted in Normandy, to enlarge the breed of Horses there by the use of stallions from Holstein; and, in consequence, the best breed of Horses in France would have been spoiled, had not the farmers discovered their mistake in time, by observing the offspring much inferior in form, to that of the native stallions.

Some graziers in the Island of Sheppey, conceived that they could improve their sheep by large Lincolnshire rams, the produce of which, however, was much inferior in the shape of the carcase, and the quality of the wool; and their flocks were greatly injured by this attempt to improve them.

Attempts to improve the native animals of a country, by any plan of crossing, should be made with the greatest caution; for, by a mistaken practice extensively pursued, irreparable injury may be done.

---

* [This was the effect experienced in Maryland, by the use of *Exile*, a Cleveland bay, of the highest breeding of his sort in England, imported by the late Robert Patterson about the year 1820. At three years old, he was advertised for sale, and stated to be then upward of 16 hands high.

They may do very well, with their long legs, long backs and long tails, for the heavy, lumbering slow coaches of millionaires, to drive to church, and occasionally to make a swell in town, but they are not fitted for the country — and especially not for this country. True, for the coach-horse we want substance, but we want that substance well placed, deep, well-proportioned body, rising in the withers, and slanting shoulders, short back well ribbed home, and broad loins; sound, flat, short legs, with plenty of bone under the knee; and sound, open, tough feet. " In fact, coach-horses should be nothing more than large hackneys, varying in height from 15 hands 1 inch to 16 hands 1 inch." Such horses, of good colour, and well matched, will always command a high figure from the swelled heads in our large cities—men who have grown rich as the conduits of exchange, between the producer and the consumer of Agriculture and Manufactures. — '. S. S.]

In any country where a particular race of animals has continued for centuries, it may be presumed that their constitution is adapted to the food and climate.

The pliancy of the animal economy is such, as that an animal will gradually accommodate itself to great vicissitudes in climate and alterations in food; and by degrees undergo great changes in constitution; but these changes can be affected only by degrees, and may often require a greater number of successive generations for their accomplishment.

It may be proper to improve the form of a native race, out at the same time it may be very injudicious to attempt to enlarge their size.

The size of animals is commonly adapted to the soil which they inhabit; where produce is nutritive and abundant, the animals are large, having grown proportionally to the quantity of food which for generations they have been accustomed to obtain. Where the produce is scanty, the animals are small, being proportioned to the quantity of food which they were able to procure. Of these contrasts the sheep of Lincolnshire and of Wales are examples. The sheep of Lincolnshire would starve on the mountains of Wales.

Crossing the breed of animals may be attended with bad effects in various ways; and that, even when adopted in the beginning on a good principle; for instance, suppose some larger ewes than those of the native breed were taken to the mountains of Wales and put to the rams of that country; if these foreign ewes were fed in proportion to their size, their lambs would be of an improved form and larger in size than the native animals; but the males produced by this cross, though of a good form, would be disproportionate in size to the native ewes; and therefore, if permitted to mix with them, would be productive of a standing ill-formed progeny

Thus a cross which, at first, was an improvement, would, by giving occasion to a contrary cross, ultimately prejudice the breed.

The general mistake in crossing has arisen from an attempt to increase the size of a native race of animals; being a fruitless effort to counteract the laws of nature.

The Arabian Horses are, in general, the most perfect in the world; which probably has arisen from great care in selection, and also from being unmixed with any variety of the same species, the males have therefore never been disproportioned in size to the females.

The native Horses of India are small, but well proportioned, and good of their kind. With the intention of increasing their size, the India company have adopted a plan of sending large stallions to India. If these stallions should be extensively used, a disproportioned race must be the result, and a valuable breed of Horses be irretrievably spoiled.

From theory, from practice, and from extensive observation, which is more to be depended upon than either, it is reasonable to form this conclusion, that it is wrong to enlarge a native breed of animals; for in proportion to their increase of size, they become worse in form, less hardy, and more liable to disease.*

---

* [For this plain reason, our farmers should have recourse to well-formed bulls of a smaller or middling size, rather than to those of a larger breed than the average size of their own cattle, and also why it is far better to employ compact, short-backed, well-formed, thorough-bred stallions, than cold-blooded stallions of larger size.

Essential difference has been found, by analysis in France, between the *blood* of the ordinary Horse and that of the aristocratic race descended from the south-eastern courser. It is stated to be less *serous* than that of the common Horse. One cannot but admire the ardour with which, in France, they are now applying the sciences to enlighten all branches of agriculture, as it has been so much more and more successfully applied to other industries. A society of the first men of that country is devoted to the *melioration of the Horse*, and they undertake to predict the time not distant when " *la science du cheval*," the science of the anatomy and physiology of the Horse—will be as well understood and agreed upon as any principles in Geometry.

The reason that, in our country, agriculture has benefited so much less by the application of the sciences, is that the policy of the government has a tendency to *disperse them*, while it concentrates other classes Instead of compelling the consumer — the shoemaker, the tailor, the wheelwright, and all manufacturing consumers to come from abroad as well as at home, and settle down nearest to them, the agriculturists foster a policy which compels them—over bad roads—to expend half the produce in carrying it to the fashioner and consumer. — J. S. S.]

ON THE IMPORTANCE OF MORE ATTENTION TO THE
PRINCIPLES OF BREEDING — THE STALLION AND
THE BROOD MARE.

To every lover of the Horse, possessed of a knowledge
of his fine points and capabilities, it must be lamentable
to perceive how miserably ignorant and careless the mass
of breeders of that noble animal appear to be, as to all
the precautions which are indispensable to maintain him
at the point of excellence which is known to be attainable
—much less by well-digested and rational systems of
breeding and rearing throughout the country, to meliorate
his form and invigorate his constitution ; and on no one
point is there, seemingly, more pernicious indifference
displayed than in regard to the *condition* of the stallions
they employ, as set forth in the Essay which these remarks are intended to introduce.

Well has it been said, in the introduction to the *"An-
nales des Haras et de l'Agriculture,"* that if the importance of a question is to be measured by the number of
those who are occupied with it, that of the multiplicat.
and of the amelioration of the Horse ought to hold the
first rank in Political and Rural economy. The traditions
of antiquity—those of nations, whether barbarous or
enlightened—writings the most ancient as well as the
most modern—prove to us the estimation which Man,
in all times, has attached to this his *most noble conquest,*
to use the expression of BUFFON. The Horse, as there
alleged, is in truth the most fruitful source of the riches
of States, by his indispensable instrumentality in the
cultivation of the soil. He is one of the most direct
agents of their power by the use that is made of him in
armies, whether in peace or in war ; and has contributed
much more than is generally considered, to the civiliza

tion of communities, by facilitating intercourse between
them and the individuals of whom they are composed.

It is not, then, astonishing that in the abstract, so much
importance should be attached to the multiplication and
improvement of an animal so useful ; but is it not
amazing that this universal admission of his value, and
the general interest of society in cultivating his finest
qualities, should give rise to no association or system in
our country, based on reason, and guided by scientific
principles? On the contrary, everything is left to chance,
to ignorance, and to narrow and sordid calculations of
economy. True, we have societies that group the Horse
with every other animal and thing, and offer petty pre
miums for the mere exhibition of the best that may
happen to be convenient to, or purchased for the show ·
but should not an object so important be made the sub
ject of special associations, and of legislative encouragement, directed to a thorough investigation of the princi
ples to be followed in all enlarged and judicious plans
for the melioration of the whole race? Look at the
amount of capital involved in the whole Union—4,365,669
horses. Value these at an average of $50, and we have
a capital of $218,283,450, which, with anything like
judgment or system, might be brought to an average
improvement of at least twenty per cent. in a few years.
What is the number lost by exposure to sudden vicissitudes of weather—to bad shoeing—in short, to ill
treatment and ignorance of the management and the
remedies *prescribed in this work*, no one can venture to
estimate. Youatt sets down the loss of *cattle* by disease
annually in England at $50,000,000!—and the loss of
sheep at one-tenth of the whole number; and though
there the veterinary art is taught as a science in the endowed colleges, and regular professors practise it
throughout the kingdom, he says it is difficult to **say**

wl n u is the greater source of this immense loss to th. agriculture of the country—" the ignorance and obstinacy of the servant and the cow-leech, or the *ignorance and supineness of the owner*." The Horse, in a state of nature, even the colt—until subjected to ignorant hand- ling and cruel management, is much healthier than after he comes under the hands of him who ought to be his kindest friend.

If such be the immense mortality in England, what nust it be among Horses in this country, where not one farmer in a hundred knows how to tell the colic from the botts, or the thrush from the scratches — ignorant alike of symptoms and of treatment?

Properly appreciating the importance of a constant supply of Horses for their cavalry, as one of the most efficient arms of her military power—the French Govern- ment takes it upon itself to supply its thirty-six thousand communes with stallions, whose services are put at the lowest rate, the average being set down at 5 or 10 francs, (one or two dollars,) and these stallions are required to be not under a certain age—four at the least—nor under a certain standard of height, according as they are tho- rough-bred, half-bred, or slow draft: 1 m. 49 centimes, or a fraction over 14.2 for thorough-bred; 1 m. 55 c. for half-bred ; and 1 m. 55 c. for heavy draft stallions— and undergo every year rigid inspection, to guard not only against palpable deformity of shape, but against any latent or transmissible diseases. Opposed as is the genius of our political institutions to regulations, too minute, of individual industry and concerns, yet it is hard to say why a planter's tobacco or his butter should be subjected to rigid inspection, and condemned and taken from him for bad quality or short weight, and yet that any fat, lazy, lounging rapscallion should be allowed to *set up a public stallion* without spirit or action, and

too often tainted with some hereditary disorder or defect
of body or temper—to deform and poison everything he
.s allowed to touch.   The Arabians, after having brought
their breed of Horses to the highest degree of perfection
of which they consider them capable, are said to have
preserved their splendid qualities of great endurance
with highly organized matter and natural soundness of
limb, by prohibiting the use of stallions until *approved
by a public inspector.*   " Breeders of all kinds of Horses,"
says Nimrod, " but of the race-horse above all others,
scarcely require to be cautioned against purchasing or
breeding from mares, or putting them to stallions, constitutionally inferior.   By constitutionally inferior is
chiefly implied, having a tendency to fail in the legs and
feet during their training, which too many of our present
racing breed are given to—although the severity of training is not equal to what it was some years back.   It
would be invidious to particularize individual sorts ; but,
says he, we could name stallions and mares from which
the greatest expectations were raised, whose progeny
have sacrificed thousands of their owners' money, entirely from this cause."   After instancing numerous cases
to show the heritableness of diseases—glanders among
others — of horses, sheep, and cattle, " these conside-
1ations," continues an eminent French writer, Professor
Dupuy, on the Veterinary art, " are to us of the greatest
moment, since we have it in our power by coupling and
crossing well-known breeds, to lessen the number of animals predisposed to these diseases.   Acting up to these
ideas, our line of conduct is marked out.   We must
banish from our establishments, designed to improve the
breed, such animals as show any signs of tuberculous
disease or any analogous affection."

Thus much have we felt called upon to say, introductory of the following able dissertation on the *condition*

of the stallion — anonymously written by some gentleman who has evidently observed the precaution too
often neglected; to *understand his subject*, before he
oegan to speak upon it. It is taken from the " FARMERS'
LIBRARY," for which it was written, and where, it
may be needless to say, *such* writers will always be truly
welcome. As against the assertion of Surgeon Cline,
with whom the author of this Essay agrees as to the predominant influence of the male in characterizing the
progeny, we have, in another place, arrayed the opinion
of Mr. Johnson, it is but fair here to adduce, in support
of the affirmative side of the proposition, the all-power-
ful testimony of Mr. Apperly, who says: " Virgil, in
his excellent remarks on breeding Horses, tells those of
his readers who wish to gain prizes to look at the dam;
and until of very late years, it was the prevailing opinion
of Englishmen that in breeding a racer the mare is more
essential than the Horse, in the production of him in his
highest form; and we know it to have been the notion
entertained by the late Earl of Grosvenor—the most extensive though not perhaps the most successful breeder
of thorough-bred stock that England ever saw. The
truth of this supposition, however, has *not been confirmed*
by the experience of the last half century, and much
more dependence is now placed *on the stallion than on
the mare*. The racing calendar, indeed, clearly proves
the fact.

" Notwithstanding the prodigious number of very highly
bred and equally good mares that are every year put to
the horse, it is from such as are put to our very best
stallions that the great winners are produced. This can
in no other way be accounted for than by such horses
having the faculty of imparting to their progeny the
peculiar external and internal formation absolutely essential to the first-rate race-horse; or, if the term 'blood'

be insisted on, that certain innate but not preternatural virtue peculiarly belonging to some horses, but not to others, which, when it meets with no opposition from the mare—or, in the language of the stable, where 'the cross nicks' by the mare admit of a junction of good shapes— seldom fails in producing a race-horse in his very best form."

After all, when the reader shall have carefully perused the following disquisition, he will, we think, be apt to concur with us in the belief that incalculable loss and deterioration ensue from an almost universal want of attention to the *condition* of the stallion, and from igno- rance in what true condition consists. The maxim of the *feeder* of the ox may be embraced in the words *warmth, cleanliness* and *quiet.* Not so with the grazier of *stock*-cattle—for they may be kept too warm; nor with the owner of a Stallion; yet too generally they manage him as if he had nothing to do but to *eat, drink,* and *sleep*—except when suddenly aroused to go through violent agitation to the opposite extreme.

—On the subject of the comparative agency of the male and female parent in the modification of the progeny in form and character, as sir Roger expressed it " much may be said on both sides." There needs no citation of instances to show the influence of the male progenitor in modifying the exterior form and colour, of the off spring, and may we not infer it in regard to its internal structure, its temper and character? Neither can we deny the share of the female parent in the same influences —see how often the calf, in its marks, exhibits an exact copy of its dam. But there are cases of what is called *superfœtation,* which go to show some extraordinary power of the male in *transmitting* his influence even to the second and third generation on the fruits of subse- quent conceptions from sexual intercourse between the same dam and other males. No fact in Natural History need to be better proved; and circumstances lead us to believe, though we are not aware that the question has occurred to naturalists, that this always occurs with the first or virgin conception; and if so, it admonishes the breeder to be especially particular in the selection of the male to which is granted the high privilege of the first access. Out of many cases that might be referred to, the reader's memory may be here refreshed as to two that are somewhat familiar.

Twenty-six years ago, in the London Farmer's Journal was recorded the case which had then lately appeared in the Philosophical Transactions, on the authority of Earl Moreton, stating that his lordship possessed a male animal called *Quagga* by the Hottentots — in whose mountains they abound. It closely resembles the Zebra, but of a smaller size. He determined on obtaining a foal by this animal, from a chestnut-coloured mare of seven-eighths blood, which had never been bred from. This gross prostitution—as we should call it—took place, and accordingly a female hybrid progeny was produced, which bore, in form and colour, decided indications of mixed blood, but proved incapable of breeding—as is almost universally the case with mules; but not *quite*, as the writer has proved in his edition of Youatt on the Horse, (Lea & Blanchard,) on the most unquestionable testimony.

This mare of seven-eighth Arabian blood was soon after sold to Sir Gore Ousley, who afterward bred from her, by a very fine black Arabian stallion, two colts. These Lord Moreton went to see and examine,—the one a two-year old filly; the other a yearling colt—both of which were as strongly characterized by Arabian blood as might be expected where there was fifteen-sixteenths of it present—but both in their colour and hair of their manes, they showed a *striking resemblance to the quagga*

The whole statement was fully verifie to the Society by Doctor Woolaston, a member of it, who examined both the filly and colt, and who was " distinguished for his very extensive knowledge."

Following the communication of Lord Moreton in the Transactions, is one from Dr. Woolaston, relating the case of a black and white sow, of Mr. Western's celebrated breed of hogs (she being the property of a Mr. Giles) which was put to a wild boar, of a deep chestnut colour, that was soon after by accident drowned. The pigs produced, which were the sow's *first* litter—partook in appearance of both boar and sow, but in some the chestnut colour of the boar strongly prevailed. This sow was afterward put to a boar of Mr. Western's breed. The pigs produced were some of them stained and clearly marked with the chestnut colour which had prevailed in the former litter. Her next litter, by a boar of Mr. Western's spotted, black and white breed, were also stained with marks of the wild boar — although in no other instance, with any other sow, had the least tinge of the chestnut colour been observed.

Another very striking instance of the transmissible influence which survives the f. st and impresses itself on subsequent conceptions, occ red under the observation of the writer of this, and wa‘ it is believed, related in a small volume scribbled and ublished under the title of " The Sportsman and his Dog." The case was that of a beautiful coach-dog bitch, Annette, presented to him by that earnest and efficient promoter of agricultural improvement, GORHAM PARSONS of Massachusetts, along with her full brother, Lubin. Though closely watched for the first signs of sexual appetite, with a view to a litter of the genuine breed for the great pleasure of giving them to friends to whom they were promised, a stray dog, of large size, of white colour, except his *black ears* contrived to steal the first access to the bitch, and in all subsequent litters, by Lubin, one pup always appeared to attest the indelible impression made in the enthusiasm of a first embrace. It may gratify curiosity to note such facts, and may serve, beyond all dispute, to show how cautious every breeder should be in the choice of the male—especially the one first employed.

But how vain to endeavour to account for these things! Nature invites us to study her ways, and science is most efficiently applied to every art and every industry, when it most closely conforms to her laws : but she has certain arcana of her own, which she keeps in reserve, and which defy the scrutiny of the most curious and importunate inquirer. We see enough to know that her laws are enacted by an All-Wise and Overruling Power ; and can never be too grateful for the faculties that enable us, so much above other created beings, to study and understand them, and yet more for that hopeful thirst for knowledge which is leading us on from one discovery to another, until, in view of what science is revealing from year to year, who shall say how near we may be permitted to approach the Supreme Intelligence ? Oh that our love of peace and of each other, may keep way with our progress in knowledge!—for of those to whom much is given, much shall be required—else, has it been well asked,

> ————" why was Man thus eminently raised
> Amid the vast creation ?   Why empowered,
> Through life and death, to cast his watchful eye
> With thought beyond the limits of his frame—
> But that the Omnipotent might send him forth
> In sight of angels and approving worlds :
> Might send him forth the sovereign good to learn ;
> To chase each meaner passion from his breast,
> And through the storms of passion and of sense
> To hold straight on, with constant heart, and eye
> Still fixed upon Man's everlasting palm,
> The approving smile of Heaven."

There is, as elsewhere intimated, if we consult Nature, always acting for the best,—reason to conclude it was intended, with domestic animals, that the male should exert the greater influence over the form and qualities of the progeny. Were it not so, how slow and ineffectual would be all attempts at amelioration, for it is through *one* male that blood and form and qualities are imparted to great numbers—while, with the female, but a solitary effect or result can be accomplished during a whole period of gestation. In herds of wild Horses, Nature allows troops of mares to be engrossed by the stallion of most courage and strength, thus guarding against the inevitable degeneracy of promiscuous intercourse—and he again, after a season or two, is supplanted by some rebellious young rival, stronger if not braver than he, before time enough has elapsed to stamp the whole race by that degeneracy which follows incestuous intercourse long continued. Here again we are invited to follow, and, as art may always do, improve upon, if we *do* follow, the laws of Nature. But, alas, of breeders of animals it may be said, "they have sought out many inventions" that violate her laws, and the consequence is, a miserable race of ill-formed, decrepit garrans, fit neither for harness nor saddle, for the road or the chase, for peace nor for war, nor for anything but—dog's meat.

AN ESSAY ON THE CONDITION OF A STALLION

The word *condition* is used by horsemen in a different sense from that in which it is understood as applied to cattle by the mass of farmers. By *condition* the farmer often means a high state of fatness; the horseman, on the contrary, makes use of the word to indicate the greatest health and strength produced by *reducing* all superfluous fat, bringing the mere flesh into clean, hard and powerful muscle, and invigorating the lungs and other internal organs, so that they may promptly discharge their respective functions, and suffer no damage from uncommon stress—the whole in order to the animal's performing labours and sustaining a continuance of action to which he would not be adequate without such especial preparation.

By the *Condition of a Stallion* is meant the state of the system in which the male horse should be kept, in order to deriving from him the greatest excellence in the progeny.

Too many persons are content to breed their mares to a horse whose figure suits them, without regard to his *condition*. The mention of one prominent instance alone will be sufficient to show that good condition is essential to the production of a valuable progeny. A remarkable case occurred in England some years since, in so high a quarter as to attract public attention, and consequently the fact of the account's obtaining currency without contradiction is a fair evidence of its correctness. The Prince of Wales, who afterwards became George the Fourth, owned, and was in the habit of riding as a hunter, an entire horse of unequalled excellence. In consequence of this horse's superior qualities, His Royal Highness caused a few of his own mares to be bred to nim in the spring, after he had been kept in the highest condition as a hunter throughout the winter, and the produce, on growing up, proved every way worthy of their sire. When His Royal Highness, as Prince Regent, became seriously engaged in the cares of Government, and therefore relinquished the pleasures of the chase, being desirous to perpetuate the fine qualities of this stock, he ordered the horse to be kept at Windsor for public covering, provided the mares should be of the first quality; and in order to insure a sufficient number of these, directed the head groom to keep him exclusively for such, and to make no charge, with the exception of the customary groom's-fee of half a guinea each. The groom, anxious to pocket as many half guineas as possible, published His Royal Highness's liberality, and vaunted the qualities of the horse, in order .o persuade all he could to avail themselves of the benefit. The result was, the horse being kept without his accustomed exercise and in a state of repletion, and serving upward of a hundred mares yearly, that the stock, although tolerably promising in their early age, shot up into lank, weakly, awkward, leggy, good-for-nothing creatures, to the entire ruin of the horse's character as a sire—until some gentleman, aware of the cause, took pains to explain it, proving the correctness of their statements by reference to the first of the horse's get, produced under a proper system of breeding, and which were then in their prime, and among the best horses in England.

Almost every observing farmer in this country has remarked that whenever, within his knowledge, an ordinary work-horse has, by chance, covered a tolerably good mare, the foal thus produced has, at maturity, almost invariably become a better animal than it was expected to be, and in many cases proved quite superior to the get of the high-priced and highly pampered stal-

lions of the neighbourhood. What was the cause of
this? Condition. The work-horse, by constant and
severe exercise, was brought into health and strength,
and his stock partook of the state of his system at the
time of copulation. Why is it that many experienced
farmers, after having tried the best stallion within their
knowledge, frequently resort to the keeping of one of
their own colts or farm-horses entire, for the service of
their mares, and actually obtain as large and as good
and saleable stock from such a one, as that from the
public stallions of far superior size, form, blood, and all
other qualities, except this indispensable *condition*?

It may be stated that, generally, whenever the get of
a stallion has proved, at maturity, to be of remarkable
excellence comparatively with the sire, such horse has
been, at and previously to the time of getting such valuable stock, kept without pampering, without excessive
sexual service, and with a good share of exercise or
labour.

To show the effect of a peculiar state of the system in
the parents at the time of copulation, instances may be
cited from various sources. We will content ourselves
with two—and first take a lamentable case in the human
species as given in the valuable work on " The Constitution of Man," by George Combe :

" In the summer of 1827, the practitioner alluded to
was called upon to visit professionally a young woman
in the immediate neighbourhood, who was safely deliv-
ered of a male child. As the parties appeared to be
respectable, he made some inquiries regarding the absence of the child's father, when the old woman told him
that her daughter was still unmarried; that the child's
father belonged to a regiment in Ireland ; that last autumn
ne had obtained leave of absence to visit his friends in
this part of the country, and that, on the eve of his de-

parture to join his regiment, an entertainment was given, at which her daughter attended. During the whole evening she and the soldier danced and sang together; when heated by the toddy and the dance, they left the cottage and after the lapse of an hour were found together in a glen, in a state of utter insensibility, from the effects of their former festivity; and the consequence of this interview was the birth of an idiot. He is now nearly six years of age, and his mother does not believe that he is able to recognise either herself or any other individual. He is quite incapable of making signs whereby his wants can be made known, with this exception, that when hungry he gives a wild shriek. This is a case upon which it would be painful to dwell, and I shall only remark that the parents are both intelligent, and that the fatal result cannot otherwise be accounted for than by the almost total prostration or eclipse of the intellect of both parties from intoxication."

For another instance of a peculiar constitution derived from a parent at the time of copulation, and owing to a temporary excitement of the animal, a respectable farmer related to the writer of this Essay that he witnessed the effect of pain and nervous agitation on a stallion just before the moment of covering, in the production of a wild, timid, violent and worthless colt. The sire was in repute as one of the best horses ever kept in the dis trict; and his stock afterward justified the opinion. The groom became angry and beat him in his stall in a cruel manner, and then led him out and allowed him to cover the mare, which was one of a perfectly quiet and orderly temper. The consequence was the production of an animal totally valueless, as above mentioned.

That the doctrine here held is no " new thing under the sun" is evident from many venerated authors. Plutarch says " The advice which I am now about to give, is indeed no other than what hath been given by those who have undertaken this argument before me. You will ask me what is that? 'Tis this, that no man keep company with his wife for issue sake, but when he is sober—as not having before either drunk any wine, or, at least, not to such a quantity as to distemper him; for they usually prove wine-bibbers and drunkards whose parents begot them when they were drunk; wherefore, Diogenes said to a stripling somewhat crack-brained and half-witted, 'Surely, young man, thy father begot thee when he was drunk?'"

Shakspeare intimates the same belief in making a hero insult his enemies with the taunt

"For ye were *got* in fear."

On no other known principle than this *condition*, or a peculiar state of the system at and before the time of copulation, can be explained the important fact which forms at once a criterion of skill in the scientific breeder, and a stumbling-block to the ignorant and unreasonable one, who would expect success without giving himself the trouble of investigating the natural laws which govern the subject of his operation : such a person is too apt to argue within himself that because the same parents at different times produce offspring of opposite character-istics, there can be no certain rules by which to create determinate qualities in the progeny: such a one would maintain that, because all the children of one married couple are usually somewhat different in characteristics from each other, there can be no means of predicting, with an approach to certainty, the qualities to be pro-duced in the offspring by a particular sexual intercourse. Now this *law of condition* accounts for the difference between individuals produced at several births from the same parents. The case of twins, in the human species, serves to strengthen this argument, inasmuch as the two

persons produced at one birth usually bear a close resem
blance to each other, in all respects.

It is known that ideal impressions on the femal
parent, subsequent to conception, frequently take permanent effect on the offspring.   That such causes do not
usually give the leading characteristics to the progeny,
is evident from these considerations:

1st. The consequences of such impressions on the
female, are usually somewhat of an unnatural or monstrous order, being different from the traits of either
parent, and from the common nature of the variety to
which the animals belong.

2d. It is a settled point with breeders that the pro
geny is more strongly characterized by the traits of the
male, than by those of the female parent.   This fact is
well known ; and indeed it can hardly be expected otherwise than that the sex which bears so much the stronge.
impress of character, should impart the more visible resemblance to the offspring.

3d. It is an ascertained law of Nature, that peculiar-
ities of climate, food, occupation and most other circum
stances affecting the well-being of an animal, produce in
its constitution a change such as is necessary for the wel-
fare of the species; and that this proceeds throughout
many generations, until the animal becomes completely
adapted to the circumstances of its existence.   [The
same thing occurs in the vegetable kingdom.]

This last consideration, of the gradually altered state
of an animal through successive generations, is a strong
instance of the effect of *condition ;* and it is by a regard
to this invariable law of Nature, of self-adaptation to circumstances, that the cultivation or improvement of any
breed is to be effected.   " Hence the most acid and
worthless grape is by skilful culture rendered sweet and
luscious , flowers without attraction are gradually nurtured

into beauty and fragrance; the cat may be made to present all the rich colours of the tortoise-shell, and the pigeon may be ‘ bred to a feather.’ ”

Let us now endeavour to deduce a useful, practical conclusion from the foregoing arguments. If our doctrine be correct, the horse-breeder will depend upon the *condition* of the stallion, in order to the producing of valuable stock from him, as well as upon his other qualities of pedigree, speed, action, bottom, wind, temper, spirit, form, style, size, colour, &c.

The next practical question is, how this condition is to be attained, and how the animal is to be kept at the required standard in this respect. The requisite *condition* is only to be attained by *training* for health and strength in a great measure according to the system of training for races : supplying an abundant nourishment of the best quality, allowing sufficient periods of repose for digestion, and giving regular and strong exercise, the whole with such variations as only experience and close observation, under constant practice, can dictate.

The aptitude of an animal to benefit by training is often inherited, like other qualities, from its parentage ; and judicious breeding alone can insure a continuance of the desirable quality, or create a propensity for it by proper crossing, when it does not exist in the parents.

The age at which the horse is best adapted to undergo a course of training, is just at the close of his most rapid period of growth, while the system is in its greatest freshness and vigour. This period is at about five years old. The powers of a horse will augment by suitable treatment in this respect until about the age of nine years : and, in order to obtaining the most valuable stock, a stallion should not be put to service before attaining a full development of his powers, nor kept at it after his form or energies appear to be affected for the

worse. He should be, then, between five and fifteen
years of age, if of an ordinary constitution ; but if of remarkable energy and endurance, and exhibiting no symptom of debility, may be continued until past twenty.

Trainers find their endeavours to produce the highest
state of strength, in an animal, greatly impeded by any
excitement of the sexual appetite. It is then the more
necessary to keep the horse in a state of training throughout the year, impressing most forcibly a tone of health
and strength upon his system at the time when his nerves
are liable to the least distraction; and continuing the
course carefully thoroughout the season of copulation ;
never allowing such excess of service, or of the excitement of sexual appetite, as to induce a disturbance of
spirit or temper, or a relapse from the most thoroughly
strong, healthy and regular tone of the system.

G. B.

TABLES.

TdE following Tables may be so often useful to the classes of persons for whom this work is intended, that it has been thought expedient to give them a place.

The list of medicines embraces such as ought to be kept constantly on hand, not only in every training and livery stable, but by every farmer and breeder who aspires to good management, and to deserve the praise of all men who happen to visit his establishment, and who know, as the French say, what is *comme il faut.* Some other medicines might well be added, but it is thought best not to leave any excuse to the indolent and improvident to say that too much is required—but we will begin with

WEIGHTS AND MEASURES.

Apothecaries' or Troy weight is most usually employed in medicine. In this, a pound contains twelve ounces:

1 lb. is........................	5760 Troy grains.
9 oz. or three-quarters of a lb.	4320 " "
6 " " a half lb.	2880 " "
3 " " one-fourth of a lb.	1440 " "
1 " " 8 drachms	480 " "
7 drachms	420 " "
6 "	360 " "
5 "	300 " "
4 " or a half oz.	240 " "
3 "	180 " "
2 "	120 " "
1 "	60 " "
1 scruple	20 " "

APOTHECARIES' WEIGHT.

Twenty grains	one scruple
Three scruples	one drachm
Eight drachms	one ounce.
Twelve ounces	one pound.

MEASURE OF FLUIDS.

Sixty drops	one fluid drachm.
Eight fluid drachms	one fluid ounce.
Four fluid ounces	a measure or nagg's.
Sixteen fluid ounces	one fluid pint.
Eight fluid pints	one gallon.

LIST OF MEDICINES,

And other articles which ought to be at hand about every training and livery stable, and every Farmer's and Breeder's establishment:

MEDICINES.

Aloes, Barbadoes,
Alum,
Arrow Root,
Basilicon, yellow,
Camphor,
Castile Soap,
Goulard's Extract,
Honey,
Hog's Lard,
Linseed Meal,
Nitre,
Oil of Caraway,
Oil, Castor,
Oil of Cloves,
Oil of Olives,
Hartshorn,

Resin,
Spanish Flies,
Sweet Spirit of Nitre
Spirit of Turpentine,
Salt, common,
Soft Soap,
Tar,
Tartar Emetic,
Tincture of Myrrh,
Venus Turpentine,
Vinegar,
Vitriol, Blue,
Verdigris,
Wax,
White Lead.

Apparatus for Compounding Medicines.

A box of small weights and scales, for the weighing of medicines in small portions, as from a grain to two drachms — the weights marked with English characters.

One pair of two-ounce scales; one pair of pound scales, one pound of brass box-weights.

A graduated glass for the measure of fluids, marked with English characters.

One large and one small pestle and mortar.

One marble slab, a foot and a half square, for mixing ointments.

One large and one small ladle.

One large and one small pallet knife—to mix and spread plasters

Articles necessary to be kept for administering and applying Medicines.

Improved Ball Iron.
Drenching horn.
Flannel—for the applying of fomentations and poultices.
Woollen and linen bandages.
Tow, and broad coarse tape.

Instruments.

Stomach-pump,
Elastic tube,
Fleam and blood stick,
Abscess lancet,
Tooth rasp, with a guard,

Seton, and curved needles,
Improved casting hobbles,
Brushes, currycombs, &c., of
course.

CALVING TABLE.

Day Bulled.	Will Calve.	Day Bulled.	Will Calve.	Day Bulled.	Will Calve.	Day Bulled.	Will Calve.
Jan'y 1	Oct'r 8	April 1	Jan'y 6	July 1	April 7	Oct'r 1	July 9
" 7	" 14	" 7	" 12	" 7	" 13	" 7	" 15
" 14	" 21	" 14	" 19	" 14	" 20	" 14	" 22
" 21	" 28	" 21	" 26	" 21	" 28	" 21	" 29
" 28	Nov. 4	" 28	Feb'y 2	" 28	May 4	" 28	Aug. 5
" 21	" 7	" 30	" 4	" 31	" 8	" 31	" 8
Feb'y 1	" 8	May 1	" 5	Aug. 1	" 9	Nov. 1	" 9
" 7	" 14	" 7	" 11	" 7	" 15	" 7	" 15
" 14	" 21	" 14	" 18	" 14	" 22	" 14	" 21
" 21	" 28	" 21	" 25	" 21	" 29	" 21	" 29
" 28	Dec'r 5	" 28	Mar. 4	" 28	June 5	" 28	Sept. 5
Mar. 1	" 6	" 31	" 7	" 31	" 8	" 30	" 7
" 7	" 12	June 1	" 8	Sept. 1	" 9	Dec'r 1	" 8
" 14	" 19	" 7	" 14	" 7	" 15	" 7	" 21
" 21	" 26	" 14	" 21	" 14	" 22	" 14	" 21
" 28	Jan'y 2	" 21	" 28	" 21	" 29	" 21	" 28
" 31	" 5	" 28	April 4	" 28	July 6	" 28	Oct'r 5
		" 30	" 6	" 30	" 8	" 31	" 8

LAMBING TABLE.

When to Ram.	Will Lamb.	When to Ram.	Will Lamb.	When to Ram.	Will Lamb.	When to Ram.	Will Lamb.
Jan'y 1	May 27	April 1	Aug. 26	July 1	Nov. 25	Oct'r 1	Feb. 25
" 14	June 10	" 14	Sept. 8	" 14	Dec'r 9	" 14	Mar. 10
Feb'y 1	" 28	May 1	" 22	Aug. 1	" 26	Nov. 1	" 26
" 14	July 12	" 14	Oct'r 8	" 14	Jan'y 8	" 14	April 9
Mar. 1	" 26	June 1	" 25	Sept. 1	" 26	Dec'r 1	" 25
" 14	Aug. 8	" 14	Nov. 8	" 14	Feb'y 9	" 14	May 9

TO THE PUBLISHERS.

SHOULD you have anywhere a spare corner, please enter a protest in my name, against the cruel practice recommended, of *firing* for the *lampas;* which takes its name from the brutal custom among old farriers, but now abandoned in England, of *burning* the swell-ing down with a red-hot lamp-iron. In most cases, it will soon subside of itself, especially if a few mashes be given, aided by a gentle alterative. If need be, a few moderate cuts may be made across the bars with a pen-knife.

Founder may be cured, and the traveller pursue his journey the next day, by giving a *table-spoonful of alum !* This I got from Dr. P. Thornton, of Montpelier, Rappahannoc county, Virginia, as founded on his own observation in several cases.

. S. S

TROTTING.

~~~~~~~~~~~~

This is a gait held in high estimation in the northern parts of the United States, and in Canada; especially when a horse can go his mile within three minutes. Then, as he falls by seconds, his value rises by guineas. In the south, gentlemen don't "cotton" to such action; though a passion for this sort of equestrian display is travelling towards the land of the magnolia grandiflora, with some other changes less compatible with their ancient high-born chivalry.

On the good old track at Charleston, among gentlemen who have never let the old Huguenot fires go down, you rarely see a *snafflebridle*, or what is called a "*goer !*" They have an eye and a heart for a good horse; but choose to retain the power of throwing him on his haunches when occasion may demand it.

It is, we believe, a rule on all courses in the United States, that the jockey's weight, in a trotting race, whether in harness or saddle, must be not under 145 pounds.

In harness, simply signifies a sulky, as light as the owner may choose. They generally weigh from 75 to 125 lbs. The weight of a trotting wagon is from 125 to 200 lbs. Hiram Woodruff's weight was about 160 lbs.

An interesting investigation is now going on in England to ascertain whether Tom Thumb, the celebrated American trotter, ever performed 20 miles within the hour. Large bets are pending on the result. If he has ever accomplished such a feat, it has not been, within our knowledge, officially recorded. Many of the parties betting on Tom Thumb having performed the above feat, failing to procure satisfactory proof thereof, have paid their bets.

Fanny Jenks trotted 101 miles in harness, over the Bull's Head course, Albany, in 9 hours, 42 minutes, 57 seconds, on the 5th of May, 1845.

Fanny Murray trotted 100 miles, in harness, in 9 hours, 41 minutes, 26 seconds, on the 15th of May, 1846, over the Bull's Head course, Albany.

(52)

BEST TROTTING TIME, AT MILE HEATS.

Name.	Saddle or Harness.	Time.	Course.	Date.
Aggy Down.....	saddle	2 27, 2 29½, 2 30, 2 30, 2 31,	Beacon Course, N. J.	Sep. 25, 1845
Beppo	"	2 32½, 2 31½, 2 33, 2 38,	Beacon Course, N. J.	June 26, 1843
Confidence......	harness	2 35, 2 37, 2 36 ...	Beacon Course, N. J.	June — 1841
Dutchman	"	2 35, 2 32, 2 35 ...	Beacon Course, N. J.	July — 1839
Dutchman	saddle	2 36, 2 35, 2 33, 2 33, 2 40,	Trenton, N. J.	Sep. — 1836
Edwin Forrest ..	"	2 31½, 2 33	Centreville, L. I. ...	May — 1834
Lady Suffolk....	"	2 28½, 2 28, 2 28, 2 29, 2 32,	Beacon Course, N. J.	July 4, 1843
Lady Suffolk....	"	2 26½, 2 27, 2 27 ..	Beacon Course, N. J	July 12, 1843
Norman Leslie..	"	2 38, 2 36½, 2 38, 2 39, 2 38,	Trenton, N. J.	June — 1836

TWO MILE HEATS.

Americus	harness	5 13, 5 11	Union Course, L. I.	Oct. 8, 1846
Americus........	"	5 17½, 5 17, 5 22 .	Hunting Park, Pa. .	Oct. 17, 1846
Black Maria	saddle	5 19½, 5 12½	Cambridge Park ...	June 18, 1845
Confidence......	harness	5 16½, 5 16½, 5 16, 5 18, 5 25,	Centreville, L. I. ...	May — 1841
D. D. Tompkins.	saddle	5 16½, 5 11	Centreville, L. I. ...	Oct. — 1837
Dutchman	"	5 16, 5 09	Beacon Course, N. J.	April — 1839
Dutchman	harness	5 11, 5 16	Beacon Course, N. J.	Oct. — 1839
Edwin Forrest ..	saddle	5 05, 5 06	Hunting Park, Pa. .	May — 1840
Edwin Forrest ..	harness	5 17, 5 13, 5 17 ..	Hunting Park, Pa. .	Oct. — 1840
Hector	"	5 24, 5 19, 5 17½ .	Hunting Park, Pa. .	June 2, 1846
James K. Polk ..	"	5 16, 5 16½	Union Course, L. I.	Nov. 18, 1846
Lady Suffolk....	saddle	4 59, 5 03½	Centreville, L. I. ...	Sep. — 1840
Lady Suffolk....	harness	5 10, 5 15	Centreville, L. I. ..	May — 1842
Lady Suffolk....	"	5 17, 5 19, 5 13 ..	Beacon Course, N. J.	May 21, 1844
Ripton	"	5 10½, 5 12½	Beacon Course, N. J.	May — 1842
Ripton	"	5 07, 5 15	Hunting Park, Pa. .	May — 1842
Ripton	"	5 07, 5 15, 5 17 ..	Hunting Park, Pa. .	May — 1842

THREE MILE HEATS.

Columbus	saddle	7 58, 8 07	Hunting Park, Pa. .	June — 1834
Dutchman	"	7 32½	Beacon Course, N. J.	Aug — 1839
Dutchman	harness	7 41	Beacon Course, N. J.	July — 1839
Dutchman	saddle	7 51½, 7 50, 8 02, 8 24½,	Beacon Course, N. J.	Oct. — 183
Dutchman	"	7 51, 7 51	Hunting Park, Pa. .	May — 1840
Lady Suffolk....	"	7 40½, 7 56	Hunting Park, Pa. .	May — 1841
Ripton	harness	8 00, 7 56½	Beacon Course, N. J.	Aug — 1842

FOUR MILE HEATS.

Dutchman	saddle	11 19, 10 51	Centreville, L. I. ..	May — 1836
Lady Suffolk....	"	11 15, 11 59	Centreville, L. I.	June — 1840
Lady Suffolk...	"	11 22, 11 34	Cambridge Park ...	Nov. — 1839
Sir Peter	harness	11 23, 11 27	Hunting Park, Pa. .	Oct. — 1829
Ellen Thompson,	saddle	11 55, 11 33	Beacon Course, N. J.	May — 1842

RACING.

BEST TIME ON RECORD AT MILE HEATS.

Name.	Time.	Course.	Date.
Aduella	1 48, 1 50, 1 49	New Orleans, La.	Dec. 25, 1842
Aduella	1 50, 1 47, 1 52¾	New Orleans, La.	Mar. 19, 1843
Bendigo	1 50, 1 48, 1 49	Lexington, Ky.	Sep. 24, 1840
Bendigo	1 48, 1 50, 1 48, 1 49	New Orleans, La.	Mar. 21, 1841
Beta	1 45, 1 45, 1 57, 2 01	Nashville, Tenn.	May 22, 1841
Big Alick	1 57, 1 47½, 1 50, 1 51	Louisville, Ky.	June 4, 1842
Capt. McHeath	1 49, 1 48, 1 50	Columbus, Ga.	May 4, 1839
Cassandra	1 48, 1 49½	Washington, D. C.	June 1, 1841
Colt by Leviathan, D. F. Kenner's,	1 48, 1 47¾, 1 50	New Orleans, La.	April 1, 1846
Creath	1 48, 1 48, 1 46	New Orleans, La.	Mar. 27, 1842
Croton	1 51, 1 54½, 1 49	New Orleans, La.	Dec. 20, 1846
Croton	1 49	New Orleans, La.	Mar. 15, 1846
Dan. McIntyre.	1 50, 1 48, 1 51	Georgetown, Ky.	Apr. 28, 1842
Fred. Kaye	1 50, 1 48½, 1 50, 1 53½, 1 52½,	Louisville, Ky.	Oct. 9, 1846
Fred. Kaye	1 52, 1 47½, 1 52, 1 48, 1 57½, 1 56½,	New Orleans, La.	Dec. 6, 1846
Gildersleeve	1 51, 1 49, 1 53, 1 56	Versailles, Ky.	Sep. 18, 1846
Harden'd Sinner	1 50, 1 48, 1 49½	Jackson, Miss.	Feb. 17, 1844
Houri, (*Imp.*)	1 47, 1 53	New Orleans, La.	Mar. 18, 1840
Jane Adams	1 47½, 1 52	New Orleans, La.	Oct. 29, 1845
Jim Bell	1 51, 1 46	Lexington, Ky.	May 21, 1841
John Hampden	1 48, 1 49, 1 53	Orange C. H., Va.	Sep. 18, 1839
Kitty Harris	1 48, 1 51, 2 02	Baltimore, Md.	May 17, 1842
Leda	1 48, 1 48	Lexington, Ky.	May 19, 1841
Little Trick	1 48	E. Feliciana, La.	Apr. 24, 1844
Lucy c. (Buford's)	1 49, 1 48, 1 51	Bardstown, Ky.	Oct. 12, 1839
Mary Brennan	1 48, 1 49	Cincinnati, Ohio	Oct. 19, 1839
Minstrel	1 48, 1 48	Louisville, Ky.	June 4, 1839
Miss Foote	1 47, 1 49, 1 48, 1 50, 1 50	New Orleans, La.	Dec. 12, 1841
Music	1 50, 1 48	New Orleans, La.	Mar. 17, 1842
Music	1 48½, 1 46½, 1 48	New Orleans, La.	Dec. 29, 1844
Nathan Rice	1 45, 1 52	Louisville, Ky.	Oct. 7, 1844
Prospect	1 50, 1 48, 1 53	Trenton, N. J.	May 25, 1841
Sailor Boy	1 51, 1 49, 1 48	Cynthiana, Ky.	Oct. 25, 1839
Serenade	1 48, 1 55, 2 00	Cynthiana, Ky.	Oct. 25, 1839
St. Pierre	1 47, 1 56, 1 55	Orange C. H., Va.	Sep. 15, 1841
Sunbeam	1 47, 1 48, 1 46½, 1 47, 1 47	New Orleans, La.	Mar. 24, 1844
Susan Hill	1 55, 1 50, 1 48	Havana	Apr, 30, 1843
The Duke	1 48, 1 55½, 1 53½	Trenton, N. J.	May 31, 1839
Uncas	1 45½, 1 48, 1 47½	E. Feliciana, La.	Apr. 27, 1844
Victor	1 50, 1 55, 1 48	Kanawha, Va.	June 7, 1839

BEST TIME ON RECORD AT TWO MILE HEATS.

Name.	Time.	Course.	Date.
Alarick	3 54, 3 39	Lexington, Ky.	Sep. 26, 1845
Ann Hayes	3 43½, 3 42½	New Orleans, La.	Nov. 21, 1844
Ann Stuart	3 50, 3 44, 3 45	Memphis, Tenn.	Nov. 14, 1843
Arraline	3 44½, 3 49, 3 49, 3 50	Louisville, Ky.	June 7, 1843
Attakapas	3 46, 3 52	Columbus, Ga.	May 2, 1839
Balie Peyton	3 54, 3 45	Broad Rock, Va.	Apr. 26, 1839
Bee's-Wing	3 44, 3 47	New Orleans, La.	Mar. 26, 1839
Betsey Archy, filly,	3 53, 3 44	Washington, D. C.	May 31, 1841
Black-Nose	3 49½, 3 45,	Georgetown, Ky.	Sep. 18, 1841
Brown Kitty	3 49½, 3 44, 3 45	New Orleans, La.	Dec. 1, 1846
Buck-Eye	3 56, 3 40, 3 47	New Orleans, La.	Mar. 18, 1841
Butterfly filly	3 48½, 3 50, 3 40	Lexington, Ky.	Sep. 27, 1845

Continued on page 55

BEST TIME ON RECORD AT TWO MILE HEATS.

Name.	Time.	Course.	Date.
Churchill	3 49, 3 46, 3 47	Lexington, Ky.	Sep. 20, 1843
Consol Junior	3 46, 3 53, 3 47	Louisville, Ky.	June 8, 1843
Creath	3 41, 3 41	New Orleans, La.	Mar. 24, 1842
Creath	3 46, 3 42	Louisville, Ky.	June 1, 1842
Creath	3 40, 3 45	Havana	Apr. 26. 1843
Croton	3 50, 3 44½, 3 50	New Orleans, La.	Apr. 13, 1845
Croton	3 44½, 3 43½, 3 43½	New Orleans, La.	Apr. 2, 1846
Croton	3 47½, 3 46	New Orleans, La.	Apr. 16, 1846
Cub	3 45½, 3 44	Louisville, Ky.	Oct. 1, 1839
Earl of Margrave	3 46, 3 40½	New Orleans, La.	Mar. 23, 1842
Gazan	3 45, 3 45	Lexington, Ky.	May 23, 1840
George W. Kendall	3 50, 3 47, 3 45, 4 07	New Orleans, La.	Dec. 10, 1841
Governor Butler	3 57, 3 46	Camden, S. C.	Nov. 21, 1840
Grey Medoc	3 46, 3 49, 3 55	New Orleans, La.	Dec. 27, 1840
Grey Medoc	3 45, 3 55	New Orleans, La.	Mar. 18, 1839
Hero	3 45, 3 55	Pineville, S. C.	Feb. — 1843
Hornblower	3 46, 3 51	Union Course, L. I.	May 8, 1839
La Bacchante	3 41, 4 03	New Orleans, La.	Apr. 3, 1845
Laneville	3 50, 3 45, 3 51½	Fairfield, Va.	May 29, 1841
Maid of Northampt'n	3 45	Washington, D. C.	Oct. 2, 1845
Midas	3 46, 3 46	Baltimore, Md.	May 9, 1844
Miss Clash	3 46, 3 43	Louisville, Ky.	June 15, 1844
Motto	3 48½, 3 43	Lexington, Ky.	Sep. 26, 1844
Motto	3 46, 3 48½	Bardstown, Ky.	Sep. — 1842
Music	3 49, 3 46, 3 51	New Orleans, La.	Dec. 22, 1842
Music	3 51, 3 46, 3 55	New Orleans, La.	Jan. 4, 1844
Music	3 49, 3 45	New Orleans, La.	Dec. 25, 1844
Nancy Clark	3 46, 3 46	Augusta, Ga.	Dec. 9, 1840
Nanny Rogers	3 48, 3 46, 3 51	Lexington, Ky.	May 22, 1846
Ol' Bee	3 49, 3 46	Richmond, Va.	Apr. 16, 1845
Passenger, (Imp.)	4 10, 3 53, 3 44	Trenton, N. J.	Oct. 25, 1839
Purity	3 50, 3 44	Jackson, Miss.	Jan. 28, 1846
Ralph	3 51, 3 45	Louisville, Ky.	June 5, 1839
Richard of York	3 49, 3 46	New Orleans, La.	Mar. 13, 1839
Richard of York	3 46, 3 44	New Orleans, La.	Mar. 24, 1839
Robert Bruce	3 43, 3 43, 3 47	Cincinnati, Ohio.	Oct. 14, 1839
Rocker	3 48, 3 46	Trenton, N. J	May 20, 1839
Ruffin	3 49, 3 45½, 4 42½	Lexington, Ky.	Sep. 21, 1843
Ruffin	3 49, 3 46	Natchez, Miss.	Nov. 19, 1845
Sally Shannon	3 50, 3 43	Frankfort, Ky.	Sep. 7, 1842
Sally Ward	3 50, 3 41½	New Orleans, La.	Dec. 3, 1846
Sarah Bladen	3 46	New Orleans, La.	Mar. 17, 1842
Sarah Washington	3 45	Orange C. H. Va.	Sep. 16, 1841
Senator	3 46, 3 46	Baltimore, Md.	May 8, 1844
Snag	3 48, 3 43	Terre Haute, Ind.	Sep. — 1844
Sorrow, (Imp.)	3 55, 3 43	Springfield, Ill	Apr. 24, 1839
Stanley Eclipse	3 44, 3 45½	Trenton, N. J.	Oct. 30, 1845
Sthreshley	3 43, 3 45	New Orleans, La.	Mar. 19, 1840
Susan Hill	3 45, 3 51	Havana	Apr. 28, 1843
Taglioni	3 49, 3 46	Pineville, S. C.	Jan. 30, 1844
Tarantula	3 48, 3 46	Nashville, Tenn.	Oct. 4, 1844
The Colonel	3 45, 3 50	Baltimore, Md.	May 4, 1844
Treasurer	3 47, 3 45½	Trenton, N. J.	May 28, 1840
Trenton	3 46, 3 45	Union Course, L. I.	Oct. 8, 1839
Vertner	3 46, 3 48	Lexington, Ky.	Sep. 21, 1839
Viola	3 47, 3 45	E. Feliciana, La.	Apr. 25, 1844
Warfield's Too Soon colt,	3 49, 3 45	Lexington Ky.	May 23, 1846
Wellington	3 56, 3 52, 3 43, 3 50	Camden, N. J.	Oct. 26, 1841
West Florida	3 51½, 3 46, 3 53	Georgetown, D. C.	Apr. 12, 1839
Will Go	3 46	Broad Rock, Va.	Oct. 2, 1839
Wilton Brown	3 52, 3 45	Alexandria, D. C.	June 5, 1842
Young Whig	3 53, 3 44	Oakley, Miss.	Dec. 7, 1844

BEST TIME ON RECORD AT THREE MILE HEATS.

Name.	Time.	Course.	Date.
Ailsey Scroggins....	5 57, 5 46, 5 54½	Bardstown, Ky.	Sep. — 1842
Andrewetta	5 48, 5 42½	Trenton, N. J.	May 29, 1340
Argentile...........	5 42, 5 51	Louisville, Ky.	June 6, 1844
Astor	5 45, 5 44	Washington, D. C. .	June 3, 1841
Black-Nose	5 48, 5 46	Frankfort, Ky.....	Sep. 24, 1840
Black-Nose	5 45, 5 46	Lexington, Ky.....	Sep. 28, 1840
Blue Dick	5 44, 5 38½	Trenton, N. J.	May 19, 1842
Blue Dick	5 42, 5 39½	Alexandria, D. C..	June 3, 1842
Blue Dick	5 50, 5 46.........	Baltimore, Md.	May 6, 1844
Bob Letcher	5 52, 5 46, 6 12, 5 51 ...	Lexington, Ky.	May 26, 1843
Boston	5 46	Broad Rock, Va. ...	Apr. 27, 1839
Clarion...........	5 45½, 5 57	Union Course, L. I.	Oct. 9, 1839
Creath	5 57, 5 43	Louisville, Ky. ...	Oct. 15, 1841
Creath	5 45, 5 44½	New Orleans, La. ..	Mar. 28, 1843
Creath	5 44, 5 53	Havana	Apr. 29, 1843
Eliza Calvert	6 00½, 5 59, 5 46	Camden, N. J.	Oct. 29, 1841
Fashion	5 43	Baltimore, Md. ...	Oct. 16, 1846
George Martin.....	5 40, 5 46	New Orleans, La. ..	Mar. 25, 1842
George Martin.....	5 45½, 5 49, 5 52	New Orleans, La. ..	Mar. 17, 1843
Glorvina	5 45, 5 51	Natchez, Miss.....	Apr. 25, 1839
Hard Cider	5 41, 6 14, 5 55, 5 50	Fredericksburg, Va.	May 28, 1840
Isola	6 04½, 5 45, 6 02½, 6 44 ...	Lexington, Ky.....	Sep. 19, 1843
James F. Robinson .	5 46, 5 55	Lexington, Ky.....	May 20, 1841
Jeannetton	5 45, 5 38½	New Orleans, La...	Dec. 27, 1844
Joe Chalmers	5 48, 5 45	Memphis, Tenn. ...	Nov. 15, 1843
Kate Aubray	5 40, 5 41	New Orleans, La. ..	Dec. 23, 1842
Liz Hewitt	5 44½	Peoria, Ill.........	Oct. 28, 1843
Louisa Jordan.....	5 39, 5 40	New Orleans, La...	Dec. 4, 1846
Maria	5 57, 5 44	E. Feliciana, La....	Apr. 26, 1844
Mariner...........	5 46, 5 56	Camden, N. J......	May 21, 1841
Master Henry	5 47½, 5 40, 5 56, 6 01 ...	Baltimore, Md.	May 16, 1839
Midas	5 45, 5 58	Washington, D. C. .	May 16, 1844
Miss Foote	5 59, 5 46..........	Mobile, Ala.	Mar. 10, 1842
Polly Green	5 46, 5 48	Columbus, Ga.	May 2, 1839
Queen Mary	5 37, 5 40, 5 40	Cincinnati, Ohio...	Oct. 17, 1839
Red Bill	5 40, 5 48, 5 49	Lexington, Ky.	May 22, 1840
Register..........	5 45, 5 49	Baltimore, Md.	Oct. 20, 1842
Ripple............	5 51, 5 47, 5 44, 5 52 ...	Louisville, Ky. ...	Oct. 8, 1840
Rover	5 47, 5 48, 5 46, 5 52 ...	Lexington, Ky. ...	Sep. 24, 1844
Ruffin............	5 40½, 5 36	New Orleans, La..	Mar. 22, 1844
Sally Shannon	5 41½, 5 50, 5 57, 6 01 ...	Lexington, Ky.	Sep. 21, 1842
Santa Anna	5 43½, 5 48	Pineville, S. C.....	Feb. 8, 1843
Sarah Washington.	5 51, 5 45	Broad Rock, Va. ...	Apr. 21, 1842
Sarah Washington..	5 40, 5 45	Baltimore, Md.	May 19, 1842
Tazewell	5 46	Rome, Ga.........	Sep. 16, 1840
Ten Broeck........	6 01, 5 41, 5 49	Louisville, Ky. ...	June 2, 1842
The Colonel	5 42, 5 54, 5 56	Camden, N. J.	Nov. 27, 1845
Treasurer	5 42	Union Course, L. I.	June 5, 1840
Treasurer	5 55½, 5 46..........	Union Course, L. I.	Oct. 5, 1842
Wilton Brown.....	5 45, 6 05	Alexandria, D. C..	June 4, 1842

BEST TIME ON RECORD AT FOUR MILE HEATS.

Name.	Time.	Course.	Date.
Andrewetta	7 46	Raleigh, N. C.	Nov. 7, 1839
Ann Hayes	7 36½, 7 42	New Orleans, La.	Mar. 23, 1844
Bandit	8 02, 7 44	Baltimore, Md.	May 15, 1840
Boston	8 13, 7 46, 7 58½	Union Course, L. I.	May 13, 1842
Eutaw	8 01, 7 43	Washington, D. C.	May 6, 1842
Eclipse	7 37½, 7 49, 8 24	Union Course, L. I.	May 27, 1843
Fashion	7 42, 7 48	Camden, N. J.	Oct. 28, 1841
Fashion	7 32½, 7 45	Union Course, L. I.	May 10, 1842
Fashion	7 38, 7 52½	Camden, N. J.	Oct. 29, 1842
Fashion	7 36, 7 49	Trenton, N. J.	Nov. 4, 1842
Fashion	7 35½	Baltimore, Md.	Oct. 20, 1843
Fashion	7 43½	Union Course, L. I.	Oct. 23, 1845
Fashion	7 36, 7 51	Baltimore, Md.	May 14, 1846
George Martin	7 33, 7 43	New Orleans, La.	Mar. 20, 1843
Greyhead	7 45½, 7 50	Lexington, Ky.	Sep. 23, 1843
Grey Medoc	7 35, 8 19, 7 42, 8 17	New Orleans, La.	Mar. 20 1841
Iago	7 45, 7 58	St. Louis, Mo.	June 24, 1844
Jerry Lancaster	7 43, 7 40	New Orleans, La.	Apr. 5, 1845
Jerry Lancaster	7 38, 8 14	New Orleans, La.	Apr. 12, 1845
Jerry Lancaster	7 55, 7 45	St. Louis, Mo.	Oct. 21, 1846
Jerry Lancaster	7 51, 7 43, 8 08	New Orleans, La.	Dec. 5, 1846
Jim Bell	7 37, 7 40	New Orleans, La.	Mar. 19, 1842
Miss Foote	8 02, 7 35	New Orleans, La.	Mar. 26, 1842
Miss Foote	7 42, 7 40	Lexington, Ky.	Sep. 25, 1842
Miss Foote	7 36½, 7 39, 7 51½	New Orleans, La.	Dec. 24, 1842
Omega	7 57, 7 45	Augusta, Ga.	Dec. 11, 1840
Peytona	7 45, 7 48	New Orleans, La.	Jan. 5, 1844
Peytona	7 39½, 7 45½	Union Course, L. I.	May 13, 1845
Reel	7 40, 7 43	New Orleans, La.	Dec. 11, 1841
Reel	7 43½, 7 41	New Orleans, La.	Mar. 18, 1843
Rover	7 39, 7 39½, 7 51, 8 29	New Orleans, La.	Dec. 28, 1841
Sarah Bladen	7 45, 7 40	New Orleans, La.	Mar. 17, 1841
Vashti	7 53, 7 46, 8 19	Baltimore, Md.	May 15, 1839

THE ST. LEGER.

THE Doncaster St. Leger (in England prono'.nce Sellenger,) is the most important stake in Great Britain, amou ing to from eighteen to twenty-four thousand dollars, and is run for, annually, by three year old colts and fillies: the former carry '19 pounds, the latter 114.

With these tables in view, a comparison of the speed of English and American horses can easily be made, having due regard to weight, age, and the distance run. The St. Leger is a race of one straight heat, and the horse has only to do his ' est for that single run.

J. S. S.

The following table will show the read the distance per second averaged by horses running at any dista ce:

Time of running one mile.	Distance per second Yds.	Ft.	In.
1 40	17	1	$9\frac{3}{5}$
1 41	17	1	$3\frac{1}{5}$
1 42	17	0	$9\frac{1}{8}$
1 43	17	0	$3\frac{1}{7}$
1 44	16	2	$9\frac{1}{4}$
1 45	16	2	$3\frac{3}{7}$
1 46	16	1	$9\frac{3}{4}$
1 47	16	1	$4\frac{1}{7}$
1 48	16	0	$10\frac{2}{3}$
1 49	16	0	$5\frac{2}{7}$
1 50	16	0	0
1 51	15	2	$6\frac{9}{11}$
1 52	15	2	$1\frac{2}{3}$
1 53	15	1	$8\frac{2}{13}$
1 54	15	1	$3\frac{5}{8}$
1 55	15	0	11
1 56	15	0	$6\frac{4}{15}$
1 57	15	0	$1\frac{21}{30}$
1 58	14	2	$8\frac{56}{59}$
1 59	14	2	$4\frac{52}{119}$
2 00	14	2	0

AVERAGE SPEED FOR THE DONCASTER ST. LEGER.

Distance 1 mile 6 furlongs 132 yards.

Year.	Name of Horse.	Time. M. s.	Yds. in a minute.
1818	Reveller	3 15	988
1846	Sir Tatton Sykes	3 16	983
1838	Don John	3 17	978
1819	Antonio	3 18	973
1842	Blue Bonnet	3 19	968
1835	Queen of Trumps	3 20	963
1836	Elis	3 20	963
1840	Launcelot	3 20	963
1843	Nutwith	3 20	963
1847	Van Tromp	3 20	963
1834	Touchstone	3 22	954
1841	Satirist	3 22	954
1837	Mango	3 23	949
1844	Faugh-a-ballagh	3 23	949
1823	Barefoot	3 23¼	948
1825	Memnon	3 23½	947
1827	Matilda	3 24	945
1826	Tarrare	3 25	940
1839	Charles XII.	3 25	940
1845	The Baron	3 25	940
1820	St. Patrick	3 26	935
1822	Theodore	3 26	935
1824	Jerry	3 29	922
1810	Octavian	3 30	918
1812	Otterington	3 31	913
1833	Rockingham	3 38	884
	Mean speed	3 24	945

PEDIGREES

OF

WINNING HORSES,

SINCE 1839.

Being an Appendix to Mason's Farrier

~~~~~~~~~~~~~~~~~~~~~~~~~~~~

A.

AARON, b. h. by Tennessee Citizen, dam by Timoleon.

ABBEVILLE, b. h. by Nullifier, dam by Gallatin.

ABNER HUNTER, b. h. by Medoc, dam by Blackburn's Whip.

ACALIA, b. m. by Luckless.

ACHILLES, gr. h. by Boxer.

ADELA, b. m. by The Colonel, dam [*Imp.*] Variella by Blacklock.

ADELAIDE, b. m. by [*Imp.*] Leviathan, dam by Napoleon.

ADELIA, b. m. by Mons. Tonson, dam by Sir Archy.

ADRIAN, ch. h. by [*Imp.*] Luzborough, dam Phenomena, by Sir Archy.

ADUELLA, ch. m. by [*Imp.*] Glencoe, dam Giantess by [*Imp.*] Leviathan.

ÆSOP, ch. h. by [*Imp.*] Priam, dam Trumpetta by Mons. Tonson.

ÆTNA, b. m. by Volcano, dam Rebecca by Palafox.

AHIRA, b. h. by Medoc, dam by Tiger.

AILSEY SCROGGINS, ch. m. by Giles Scroggins, dam by Pirate.

AJARRAH HARRISON, ch. m. by Eclipse, dam by Gallatin.

AJAX, gr. h. by [*Imp.*] Leviathan, dam by Pacolet.

A. J. LAWSON, b. h. by [*Imp.*] Hedgford, dam Kitty Fisher by Gallatin.

ALAMODE, ch. h. by [*Imp.*] Margrave, dam by Timoleon.

ALARIC, b. h. by Mirabeau, dam by [*Imp.*] Tranby.

ALATOONA, b. m. by Argyle, dam Viola by Gallatin.

ALBION, [*Imp.*] bl. h. by Cain or Actæon, dam by Comus or Blacklock.

ALBORAC, b. h. by Telegraph, dam by Monday.

ALDERMAN, ch. g. by [*Imp.*] Langford, dam by Sir Charles.

ALLEGRA, b. m. by Stockholder, dam by Pacolet.

ALLEN BROWN, ch. h. by Stockholder, dam by [*Imp*] Eagle.

ALEXANDER CAMPBELL, b. h. by Collier, dam by Kosciusko.

ALEXANDER CHURCHILL, b. h. by [*Imp.*] Zinganee, dam by Bertrand.

ALICE, b. m. by Conqueror, dam by Wild Medley.

———— b. m. by [*Imp.*] Sarpedon, dam Rowena by Sumpter.

ALICE ANN, gr. m. by Director, dam by Gallatin.

ALMIRA, gr. m. by Eclipse dam by Stockholder.

ALTORF, b. h. by [*Imp.*] Fylde, dam by Virginian.

ALWILDA, gr. m. by Monmouth Eclipse, dam by John Richards.

AMBASSADOR, ch. h. by Plenipotentiary, dam [*Imp.*] Jenny Mills by Whisker.

AMELIA, br. m. by Bluster, dam by Messenger.

AMERICA, b. m. by Stockholder, dam by Democrat.

———— b. m. by [*Imp.*] Trustee, dam Di Vernon by Florizel.

AMERICAN CITIZEN, b. h. by Marion, dam by Harwood.

———— EAGLE, gr. h. by Grey Eagle, dam by Waxy.

———— STAR, ch. h. by Cramp, dam by Pulaski.

AMY THE ORPHAN, ch. m. by [*Imp.*] Nonplus, dam by Comet.

ANDREWANNA, b. m. by Andrew, dam by Gallatin.

ANDREWETTA, gr. m. by Andrew, dam by Oscar.

ANDREW HAMET, b. h. by Sidi Hamet, dam by Trumpator.

ANN BARROW, b. m. by Cock of the Rock, dam by Virginian.

ANN BELL, ch. m. by Frank, dam Jonquil by Little John.

ANN BLAKE, b. m. by Lance, dam by Blackburn's Whip.

ANN CALENDAR, ch. m. by Eclipse, dam Grand Duchess by [*Imp.*] Gracchus.

ANN GILLESPIE, br. m. by McCarty's Henry Clay, dam Susan by Sir William.

ANN HARROD, ch. m. by Hickory John, dam by King William.

ANN HAYES, b. m. by [*Imp.*] Leviathan, dam by Pacific.

ANN INNIS, ch. m. by Eclipse, dam (the dam of Mary Morris) by Sumpter.

ANN KING, b. m. by [*Imp.*] Sorrow, dam Lady of the Lake by Henry Tonson.

ANN STEVENS, ch. m. by [*Imp.*] Trustee, dam (an imported mare) by Muley.

ANN STEWART, ch. m. by Eclipse, dam Kitty Hunter by Paragon

ANNE ROYALE, br. m. by Stockholder, dam Alice Lee by Sir Henry Tonson.

ANTOINETTE, ch. m. by [*Imp.*] Leviathan, dam Multiflora by Director.

ANTIPATOR, ch. h. by Tychicus, dam Club Foot by Napoleon.

ANVIL, b. h. by [*Imp.*] Contract, dam by Eclipse.

ARAB, b. h. by Arab, dam by Sir Archy.

ARABELLA, b. m. by Collier, dam by Gallatin.

ARABIAN MARK, b. h. by [*Imp.*] Fylde, dam by Sir Charles.

ARGENTILE, b. m. by Bertrand, dam Allegrante by [*Imp.*] Truffle

ARGYLE, br. h. by Mons. Tonson, dam Thistle by Ogle's Oscar.

ARILLA, gr. m. by O'Kelly, dam by Medley.

ARKALUKA, ch. h. by [*Imp.*] Leviathan, dam Sally McGehee

AROOSTOOK, b. h. by Wheeling Rodolph, dam by Moses.

ARRALINE, ch. m. by [*Imp.*] Leviathan, dam by Stockholder.

ARRAH NEAL, ch. m. by [*Imp.*] Leviathan, dam Martha Washington by Sir Charles.

ARSENIC, ch m. by [*Imp.*] Leviathan, dam Mary Farmer by Conqueror.

ASHLAND, ch. h. by Medoc, dam Lady Jackson by Sumpter.

ASTOR, b. h. by Ivanhoe, dam Tripit by Mars.

ATTAKAPAS, ch. h. by [*Imp.*] Luzborough, dam by Arab.

ATTILA LECOMTE, b. m. by [*Imp.*] Glencoe, dam Extant by [*Imp.*] Leviathan.

ATLANTIC, b. m. by Blood and Turf, dam Old Fly.

AUNT PONTYPOOL, ch. m. by Bertrand Junior, dam Gold Finder by Virginius.

AUSTER, br. h. by Westwind, dam by [*Imp.*] Leviathan.

AUTHENTIC, ch. h. by [*Imp.*] Leviathan, dam Timoura by Timoleon.

B.

BALD HORNET, ch. g. by Bald Hornet, dam by Bertrand.

BALIE PEYTON, b. h. by Andrew, dam (Master Henry's dam) by Eclipse.

BALTIMORE, b. h. by [*Imp.*] Luzborough, dam by Gohanna.

BAND BOX, gr. m. by O'Kelly, dam Lucy Brooks by Bertrand.

BANDIT, b. h. by [*Imp.*] Luzborough, dam by Virginian.

BANJO BILL, b. h. by [*Imp.*] Sarpedon, dam by Darnaby's Diomed.

BAND OF MUSIC, ch. m. by O'Kelly, dam by Oscar.

BARBARA ALLEN, ch. m. by Collier, dam Lady Jackson by Sumpter.

BASSINGER, bl. h. by [*Imp.*] Fylde, dam by Randolph's Roanoke.

BAYWOOD, b. h. by Editor, dam by Pacolet.

BEACON LIGHT, ch. m. by [*Imp.*] Glencoe, dam Giantess by [*Imp.*] Leviathan.

BEATRICE OF FERRARA, m. by Stockholder, dam by Duroc.

LEAU-CATCHER, ch. m. by Leopold, dam Cranberry.

BEE'S-WING, ch. m. by [*Imp.*] Leviathan, dam Black Sophia by Topgallant.

BELFIELD, b. h. by [*Imp.*] Priam, dam [*Imp.*] Bustle by Whalebone.

BELLISSIMA, b. m. by [*Imp.*] Belshazzar, dam Wingfoot by Rattler.

BELLE OF WINCHESTER, ch. m. by Stockholder, dam by Sir Archy.

————————————, ch. m. by [*Imp.*] Shakspeare, dam Cado by Sir Archy.

BELLE TAYLOR, b. m. by Medoc, dam by Sumpter.

BEN BARKLEY, b. h. by Push Pin, dam Miss Wakefield by Sir Hal.

BEN BUSTER, b. h. by Cherokee, dam by Whip.

BEN FRANKLIN, ch. h. by Flagellator, dam Medora by Eclipse.

———————————— ch. h. by Woodpecker, dam by Franklin Beauty

BEN FRANKLIN, ch. h. by [*Imp*] Leviathan, dam by Stockholder
BENDIGO gr. h. by Timoleon, dam by Sir Charles.
——————— b. h. by Medoc, dam by Sir Archy.
BENGAL, ch. h. by Gohanna, dam Sportsmistress (or Gulnare) by
    Duroc.
BERENICE, ch. m. by Skylark, dam Kathleen by [*Imp.*] Leviathan.
BETA, ch. m. by [*Imp.*] Leviathan, dam by Kosciusko.
BETHESDA, b. m. by Pacific, dam by Sir Henry Tonson.
BETHUNE, br. h. by Sidi Hamet, dam Susette by Aratus.
BETSEY COLEMAN, ch. m. by Goliah, dam Melinda.
BETSEY COODEY, ch. m. by [*Imp.*] Leviathan, dam by Sir Charles.
BETSEY HUNTER, ch. m. by Sir Clinton, · am by Hamiltonian.
BETSEY LAUDERDALE, ch. m. by [*Imp.*] Leviathan, dam by Sir
    Richard.
BETSEY MILLER, gr. m. by [*Imp.*] Leviathan, dam Jane Shore
    by Oscar.
BETSEY RED, ch. m. by Red Rover, dam Betsey West by [*Imp.*]
    Buzzard.
BETSEY SHELTON, b. m. by Jackson, dam Harriet Haxall by Sir
    Hal.
BETSEY WATSON, br. m. by Jefferson, dam by Sir Henry Tonson.
BETSEY WHITE, ch. m. by Goliah, dam by Sir Charles.
BIG ALECK, ch. h. by Medoc, dam by Tiger.
BIG ELLEN, b. m. by Medoc, dam by Old Whip.
BIG JOHN, ch. h. by Bertrand, dam by Hamiltonian.
BIG NANCY, ch. m. by Jackson, dam by Gallatin.
BILLY AYNESWORTH, ch. h. by Traveller, dam Helen by Timo-
    leon.
BILLY BLACK, b. h. by Volcano.
BILLY BOWIE, b. h. by Drone, dam Agility by Sir James.
BILLY GAY, b. h. by [*Imp.*] Hedgford, dam Mary Francis by Di
    rector.
BILLY TONSON, gr. h. by Mons. Tonson, dam by Cherokee.
BILLY TOWNES, b. h. by [*Imp.*] Fylde, dam by Virginian.
BILLY WALKER, ch. h. by [*Imp.*] Valparaiso, dam by Sir Richard.
BILLET, ch. h. by Mingo, dam by Mambrino.
BILOXE, ch. h. by Dick Chinn, dam Extio by [*Imp.*] Leviathan.
BLACK BOY, bl. h. by [*Imp.*] Chateau Margaux, dam by [*Imp.*]
    Chance.
——————— bl. h. by [*Imp.*] Chateau Margaux, dam Lady Mayo
    by Van Tromp.
BLACK DICK, bl. h. by [*Imp.*] Margrave, dam by Pamunky.
BLACK FOOT, ch. h. by Medoc, dam by Blackburn's Whip.
BLACK HAWK, bl. h. by Industry.
——————— bl. h. by Mucklejohn.
BLACK JACK, bl. h. by Tom Fletcher, dam by Baronet.
BLACK LOCUST, bl. h. by [*Imp.*] Luzborough, dam by Sir Archy.
BLACK NOSE, ch. h. by Medoc, dam Lucy by Orphan.
BLACK PRINCE, bl. h. by [*Imp.*] Fylde, dam Fantail by Sir Archy

BLACK RABBIT, bl. h. by [Imp.] Nonplus, dam (Fair Ellen's dam) by Virginius.
BLACK ROSE, bl. m. by [Imp.] Leviathan, dam by Arab.
BLAZING STAR, b. h. by Henry, dam by Eclipse.
BLOODY NATHAN, gr. h. by [Imp.] Leviathan, dam by Pacolet.
BLOOMFIELD RIDLEY, b. h. by Bell-Air, dam Cedar Snags.
BLOOMSBURY, ch. m. by [Imp.] Fylde, dam by Giles Scroggins.
BLUE BONNET, gr. m. by [Imp.] Hedgford, dam Grey Fanny by Bertrand.
BLUE DICK, gr. h. by [Imp.] Margrave, dam by Lance.
BLUE JIM, ch. h. by Mucklejohn.
BLUE SKIN, h. by Marmion, dam by Tecumseh.
BOB BUSH, ch. h. by Medoc, dam by Bertrand.
BOB LETCHER, b. h. by Medoc, dam by Rattler.
BOB LOGIC, br. h. by [Imp.] Langford, dam by Mambrino.
BOB RUCKER, ch. h. by Eclipse, dam by Sir Charles.
BOIS D'ARC, ch. h. by Eclipse, dam Hortensia by Contention.
BONNY BLACK, bl. m. by [Imp.] Valentine, dam Helen Mar by Rattler.
BORAC, ch. h. by Pacific, dam by Bagdad.
BOSTON, ch. h. by Timoleon, dam (Robin Brown's dam) by Ball's Florizel.
BOSTON FILLY, m. by Boston, dam by [Imp.] Priam.
BOWDARK, b. h. by Anvil, dam by Bagdad.
BOXER, b. h. by Mingo, dam by Eclipse.
BOYD M'NAIRY, ch. h. by [Imp.] Leviathan, dam Morgiana by Pacolet.
BRACELET, ch. m. by Eclipse, dam [Imp.] Trinket.
BREAN, ch. h. by Goliah.
BRILLIANT, b. h. by Sidi Hamet, dam Miss Lancess by Lance.
BRITANNIA, [Imp.] b. m. by Actæon, dam by Scandal.
BROCKLESBY, ch. h. by [Imp.] Luzborough, dam by Roanoke.
BROKER, b. h. by [Imp.] Rowton, dam Jane Bertrand by Bertrand.
BROTHER TO HORNBLOWER, b. h. by Monmouth Eclipse, dam Music by John Richards.
BROTHER TO PEYTONA, ch. h. by [Imp.] Glencoe, dam Giantess by [Imp.] Leviathan.
BROTHER TO VICTOR, b. h. by [Imp.] Cetus, dam [Imp.] My Lady by Comus.
BROWN ELK, b. h. by Buck Elk, dam by Whip.
BROWN GAL, br. m. by [Imp.] Leviathan, dam by Virginian.
BROWN KITTY, br. m. by Birmingham, dam by Tiger.
BROWN LOCK, br. h. by Pacific, dam by Sir Hal.
BROWN STOUT, br ... by [Imp.] Sarpedon, dam Feathers by Mons. Tonson.
BROWNLOW, br. h. by [Imp.] Merman, dam (Glenare's dam) by [Imp.] Leviathan.
BRUCE, ch. h. by [Imp.] Nonplus, dam La..nbal'e by Kosciuskc
BUBB, b m. by Bertrand, dam by Whig.

BUCK-EYE, b. h. by Critic, dam Ann Page by Ogle's Oscar.

———————— b. h. by Lafayette Stockholder, dam Old Squaw by Indian.

———————— BELLE, ch. m. by Medoc, dam by Sumpter.

———————— LAD, ch. h. by Bertrand, dam by a Son of Spread Eagle.

BUCK RABBIT, b. h. by [*Imp.*] Nonplus, dam (Fair Ellen's dam) by Virginius.

BULGER BROWN, b. h. by Lance, dam by Jenkins' Sir William.

BUNKUM, ch. g. by Hyazim, dam by Gallatin.

BURLEIGH, b. h. by Sir Archie Montorio, dam Mary Lee by Contention.

BUSTAMENTE, ch. h. by Whalebone, dam Sarah Dancy by Timoleon.

BUZ FUZ, gr. h. by Medley, dam by [*Imp.*] Luzborough.

C.

CADMUS, b. h. by Cadmus.

CALANTHE, b. m. by [*Imp.*] Leviathan, dam by Jackson.

CALANTHE, ch. m. by Medoc, dam by Sumpter.

CAMANCHE, ch. h. by Grey Eagle, dam by Rattler.

CAMDEN, br. h. by Shark, dam [*Imp.*] Invalid by Whisker.

CAMEO, b. m. by [*Imp.*] Tranby, dam by Buzzard.

CAMEL, ch. h. by Birmingham, dam by Whip or Sumpter.

CAMILLA, br. m. by [*Imp.*] Hedgford, dam (Picayune's dam) by Sir William of Transport.

CAPTAIN BURTON, br. h. by Cherokee, dam by Green Oak.

CAPTAIN M'HEATH, ch. h. by [*Imp.*] Leviathan, dam Miss Bailey by [*Imp.*] Boaster.

CAPTAIN THOMAS HOSKINS, b. h. by [*Imp.*] Autocrat, dam by Tom Tough.

CAPTAIN (The) b. h. by Sir Archy Montorio, dam Ophelia by Wild Medley.

CAPTAIN WHITE-EYE, bl. h. by Chifney, dam by Sumpter.

CAROLINE MALONE, (Col. Thomas Watson's), ch. m. by [*Imp.*] Leviathan, dam Proserpine by Oscar.

———————————— (Col. J. C. Guild's), b. m. by [*Imp.*] Leviathan, dam by Sir Richard.

———————————— (Col. Thomas Watson's), b. m. by [*Imp.*] Leviathan, dam by Jerry.

CASHIER, ch. h. by Goliah, dam by Sir Charles.

CASKET, b. m. by [*Imp.*] Priam, dam by Constitution.

CASETTA CHIEF, ch. h. by Andrew, dam by Wildair.

CASSANDRA, b. m. by [*Imp.*] Priam, dam Flirtilla Jr. by Sir Archy.

CASTIANIRA, ch. m. by [*Imp.*] Leviathan, dam by Stockholder.

CATALPA, b. m. by Frank, dam by John Richards.

CATARACT, b. h. by Monmouth Eclipse, dam by John Richards.

CATHERINE, b. m. by Bertrand, dam Black-eyed Susan by Tiger.

CATHERINE FENWICK, gr. m. by Mucklejohn, dam by Saxe Weimar.

CATHERINE RECTOR, ch. m. by Pacific, dam Mary Tonson.
CAVALIER SERVANTE, gr. h. by Bertrand, dam by Andrew.
CEDRIC, b. h. by [Imp.] Priam, dam Countess Plater by Virginian.
CELERITY, ch. m. by [Imp.] Leviathan, dam Patty Puff by Pacolet.
CHAMPAGNE, b. h. by Eclipse, dam by Sir Archy.
CHARLES, b. h. by [Imp.] Rowton, dam Leocadia.
CHARLES ARCHY, ch. h. by Sir Charles, dam by Eclipse.
CHARLES MALCOLM, ch. h. by Malcolm, dam by Albert Gallatin.
CHARLEY ANDERSON, ch. h. by Medoc, dam by Mercury.
CHARLEY FOX, b. h. by Waxy, dam by Buckner's Leviathan.
CHARLEY NAYLOR, b. h. by Medoc, dam by Tiger.
CHARLOTTE BARNES, b. m. by Bertrand, dam by Sir Archy.
CHARLOTTE CLAIBORNE, b. m. by Havoc, dam by Conqueror
CHARLOTTE HILL, b. m. by Hephestion, dam by Cook's Whip.
CHARITY GIBSON, ch. m. by [Imp.] Leviathan, dam by Sir Charles.
CHATEAU, [Imp.] b. m. by Chateau Margaux. dam Cuirass by
Oiseau.
CHEMISETTE, b. m. by [Imp.] Glencoe, dam by Arab.
CHEROKEE MAID, gr. m. by Marmion, dam by Tecumseh.
CHESAPEAKE, b. or br. h. by [Imp.] Leviathan, dam by Thaddeus.
CHICOMAH, ch. m. by [Imp.] Leviathan, dam White Feather by
Conqueror.
CHICOPA, ch. m. by Tuscahoma, dam Fortuna by Pacolet.
CHIEFTAIN, b. h. by Godolphin, dam Young Lottery by Sir Archy.
CHOTAUK, br. h. by Pamunky, dam by Arab.
CHURCHILL, b. h. by [Imp.] Zinganee, dam by Buzzard.
CINDERELLA, b. m. by Pacific, dam Mary Vaughan by Pacolet.
CLARA BOARDMAN, b. m. by [Imp.] Consol, dam Sally Bell by
Sir Archy.
CLARION, ch. h. by Monmouth Eclipse, dam by Ogle's Oscar.
CLARISSA, ch. m. by Monmouth Eclipse, dam (Clarion's dam) by
Ogle's Oscar.
CLEAR THE TRACK, ch. h. by [Imp.] Luzborough, dam by Stock-
holder.
CLEOPATRA, b. m. by [Imp.] Leviathan, dam by Pacolet.
CLEVELAND, gr. h. by [Imp.] Emancipation, dam by [Imp.] Levia-
than.
COAL BLACK ROSE, bl. m. by [Imp.] Leviathan, dam by Arab.
COLUMBUS, Junior, b. h. by Columbus, dam by Bertrand.
COMPROMISE, b. m. by Nullifier, dam by Anti-Tariff.
CONCHITA, ch. m. by [Imp.] Leviathan, dam Miss Bailey by [Imp.
Boaster.
CONSOL, Junior, br. h. by [Imp.] Consol, dam [Imp.] The Nun's
Daughter by Filho da Puta.
CORA, [Imp.] ch. m. by Muley Moloch, dam by Champion.
CORA MUNRO, ch. m. by Hugh L. White, dam by Crusher.
CORDELIA, ch. m. by [Imp.] Leviathan, dam by Sir Archy.
CORK, b. h. by [Imp.] Leviathan, dam Caledonia by Jerry.
CORNELIA, b. m. by Skylark, dam by Arab.
CORONATION, ch. h. by Laplander, dam by Oscar.

COTTON PLANT, gr. m. by Bertrand, dam by Pacolet.
COWBOY, ch. h. by Medoc, dam by Virginian.
CRACKAWAY, ch. h. by Marmaduke.
CRACOVIENNE, gr. m. by [Imp.] Glencoe, dam [Imp.] Gallopade
by Catton.
CREATH, b. h. by [Imp.] Tranby, dam by Sir Archy Montorio.
CRICHTON, ch. h. by Bertrand, dam by Phenomenon.
CRIPPLE, gr. m. by [Imp.] Philip, dam (Gamma's dam) by Sir Ri-
chard.
CROCKETT, b. h. by Crockett, dam by Sir Archy.
CROTON, gr. h. by Chorister, dam by Mucklejohn.
CRUCIFIX, ch. m. by [Imp.] Leviathan, dam Virginia by Sir Archy.
CUB, ch. m. by Medoc, dam by Sumpter.
CUMBERLAND, b. h. by [Imp.] Leviathan, dam by Sir William.
CURCULIA, ch. m. by Medoc, dam by Sumpter.
CZARINA, gr. m. by [Imp.] Autocrat, dam Aurora by Arab.

D.

DANDRIDGE, b. h. by Garrison's Zinganee, dam by Walnut or La-
fayette.
DAN MARBLE, ch. h. by Woodpecker, dam (a sister to West Flo-
rida's dam) by Potomac.
DAN M'INTYRE, ch. h. by Medoc, dam by Sumpter.
DAN TUCKER, ch. h. by [Imp.] Belshazzar, dam by Pulaski.
DANIEL BUCK, ch. h. by Collier, dam by Pacolet.
DARIUS, b. h. by Orphan Boy, dam by Cumberland.
DARKNESS, bl. m. by Wagner, dam Sally Shannon's dam) by Sir
Richard.
DARNLEY, ch. h. by John Richards, dam Lady Gray by Sir Richard.
DART, b. h. by [Imp.] Doncaster, dam Jane Gray by Orphan Boy.
DAVE PATTON, ch. h. by Sumpter, dam by Hamiltonian.
DAVID FYLDE, b. h. by [Imp.] Fylde, dam by Clay's Sir William.
DAY DREAM, br. m. by [Imp] Luzborough, dam by Sir Archy.
DAYTON, ch. h. by Tormentor, dam by Tuckahoe.
DECATUR, ch. h. by Henry, dam Ostrich by Eclipse.
DECEPTION, b. h. by Stockholder, dam by [Imp.] Leviathan.
DE LATTRE, br. h. by [Imp.] Consol, dam [Imp.] Design by Tramp.
DELAWARE, b. h. by Mingo, dam by John Richards.
DELPHINE, ch. m. by Sumpter.
DEMOCRAT, ch. h. by [Imp.] Luzborough, dam by Eagle.
DENMARK, br. h. by [Imp.] Hedgford, dam Betsey Harrison by
Aratus.
DENIZEN, [Imp.] b. h. by Actæon, dam Design by Tramp.
DEVIL JACK, ch. h. by [Imp.] Leviathan, dam Lady Burton by
Timoleon.
DIANA CROW, bl. m. by Mark Antony, dam by Botts' Lafayette.
DIANA SYNTAX, br. m. by Doctor Syntax, dam [Imp.] Diana by
Catton.
DICK COLLIER, ch. h. by Collier, dam by Whip.
DICK MENIFEE, br. h. by Lance, dam by Sir William of Transport

DOCTOR DUDLEY, b. h. by Bertrand, dam by Robin Gray.
DOCTOR DUNCAN, ch. h. by Cadmus, dam by Old Court.
DOCTOR FRANKLIN, ch. h. by Frank, dam Althea by Big Archy.
DOCTOR WILSON, ch. h. by John Bascombe, dam Bolivia by Bolivar.
DOLLY DIXON, b. m. by [Imp.] Tranby, dam Sally House by Virginian.
DOLLY MILAM, b. m. by [Imp.] Sarpedon, dam by Eclipse.
DONCASTER, [Imp.] bl. h. by Longwaist, dam by Muley.
DONNA VIOLA, b. m. by [Imp.] Luzborough, dam (Jack Downing's dam) by Mons. Tonson.
DUANNA, gr. m. by [Imp.] Sarpedon, dam Goodloe Washington by Washington.
DUBLIN, gr. h. by [Imp.] Leviathan, dam by Jerry.
DUCKIE, b. m. by [Imp.] Sarpedon, dam Mary Jones by Kosciusko.
DUKE SUMNER, gr. h. by Pacific, dam by Grey Archy.
DUNGANNON, b. h. by Mingo, dam by John Stanley.
DUNVEGAN, b. h. by [Imp.] Trustee, dam Jemima by Rattler.

E.

EARL OF MARGRAVE, b. h. by [Imp.] Sarpedon, dam Duchess of Marlborough by Sir Archy.
ECLIPTIC, ch. h. by Eclipse, dam (Rodolph's dam) by Moses.
EDISTA, b. h. by [Imp.] Rowton, dam Empress.
EDWARD EAGLE, ch. h. by Grey Eagle, dam by Director.
EFFIE, b. m. by [Imp.] Leviathan, dam by Stockholder.
EL BOLERO, br. h. by Stockholder, dam by [Imp.] Leviathan.
EL FURIOSO, b. h. by [Imp.] Hedgford, dam Rattlesnake by Bertrand.
ELIAS RECTOR, b. h. by [Imp.] Luzborough, dam Kate Blair.
ELI ODOM, br. h. by [Imp.] Leviathan, dam Chuckfahila by Bertrand.
ELIZA CULVERT (or Calvert), ch. m. by Cymon, dam Lady Sumner by Shawnee.
ELIZA HUGHES, b. m. by Marmion, dam by Whip.
ELIZA JANE, b. m. by [Imp.] Monarch, dam Big Jinny by Rattler.
ELIZA ROSS, b. m. by Marmion, dam by Tiger or Whip (or Tiger Whip).
ELIZABETH GREATHOUSE, b. m. by Masaniello, dam by Waxy.
ELIZABETH JONES, m. by Pacific, dam by Mons. Tonson.
ELLA, ch. m. by Young Virginian, dam by Harwood.
ELLEN HUTCHINSON, ch. m. by [Imp.] Leviathan, dam by Bertrand.
ELLEN CARNELL, ch. m. by [Imp.] Belshazzar, dam by [Imp.] Leviathan.
ELLEN JORDAN, b. m. by (Imp.) Jordan, dam Ellen Tree by Henry.
ELLEN PERCY, ch. m. by Godolphin, dam by (Imp.) Bedford.
——————————— ch. m. by Godolphin, dam by Financier.
ELLEN WALKER, b. m. by (Imp.) Consol, dam (Imp.) Plenty by Emilius.
ELLISIF, b. m. by Platoff, dam by Mucklejohn.

ELLIPTIC, ch. h. by Monmouth Eclipse, dam Amanda by Revenge

ELOISE, ch. m. by (*Imp.*) Luzborough, dam Mary Wasp by Don Quixotte.

ELVIRA, ch. m. by Red Gauntlet, dam by Rob Roy.

EMERALD, b. m. by (*Imp.*) Leviathan, dam (*Imp.*) Eliza by Rubens.

EMIGRANT, gr. h. by Cadet, dam by (*Imp.*) Contract.

EMILY, ch. m. by Medoc, dam Spider by Almanzar.

———— br. m. by (*Imp.*) Priam, dam by Tom Tough.

———— (*Imp.*) b. m. by Emilius, dam Elizabeth by Rainbow.

EMILY SPEED, ch. m. by (*Imp.*) Leviathan, dam by Pacolet.

EMMET, b. h. by Bertrand, dam by Gallatin.

ESMERALDA, b. m. by Pressure, dam by Murat.

ESPER SYKES, (*Imp.*) br. h. by Belshazzar, dam Capsicum by Emilius.

ESTA, gr. m. by Bolivar, dam by (*Imp.*) Barefoot.

ESTHER WAKE, gr. m. by (*Imp.*) Luzborough, dam by Stockholder.

ETHIOPIA, bl. m. by Dashall, dam by (*Imp.*) Expedition.

EUDORA, br. m. by Jefferson, dam by Oscar.

EUCLID, br. h. by (*Imp.*) Luzborough, dam by Sir Archy.

EUTAW, b. h. by (*Imp.*) Chateau Margaux, dam by Sir Charles.

EVERGREEN, ch. m. by Wild Bill, dam by Sir Charles.

EXTIO, b. m. by (*Imp.*) Leviathan, dam (*Imp.*) Refugee by Wanderer.

F.

FANCY, br. m. by (*Imp.*) Fylde, dam by Sir Archy.

FANDANGO, gr. m. by (*Imp*) Leviathan, dam (*Imp.*) Gallopade by Catton.

FANNY, ch. m. by Eclipse, dam Maria West by Marion.

———— (J. Guildersleeve's), bl. m. by Sidi Hamet, dam by Sumpter.

———— (Joseph Alston's), b. m. by Woodpecker, dam Fan by Trumpator.

FANNY BAILEY, ch. m. by Andrew, dam by Bertrand.

FANNY FORESTER, b. m. by (*Imp.*) Emancipation, dam by Industry.

FANNY GREEN, b. m. by (*Imp.*) Trustee, dam Betsey Archy by Sir Archy.

FANNY KING, b. m. by (*Imp.*) Glencoe, dam Mary Smith by Sir Richard.

FANNY LIGHTFOOT, b. m. by Stockholder, dam by Sumpter.

FANNY ROBERTSON, b. m. by (*Imp.*) Priam, dam Arietta by Virginian.

FANNY STRONG, ch. m. by (*Imp.*) Leviathan, dam Sally Bell by Sir Archy.

FANNY WYATT, ch. m. by Sir Charles, dam by Sir Hal.

FANTAIL, ch. m. by Waxy, dam by Sumpter.

FAIRLY FAIR, ch. m. by (*Imp.*) Luzborough, dam by Peter Teazle.

FAITH, b. m. by (*Imp.*) Tranby, dam Lady Painter by Lance

FASHION, ch. m. by (*Imp.*) Trustee, dam Bonnets O' Blue by Sir Charles.

FEATHERS, ch. m. by (*Imp.*) Leviathan, dam (George Kendall's dam) by Stockholder.

FESTIVITY, b. h. by (*Imp.*) Leviathan, dam Magnolia by Mons. Tonson.

FIAT, b. m. by (*Imp.*) Hedgford, dam Lady Tompkins by Eclipse.

FIFER, b. h. by Monmouth Eclipse, dam Music by John Richards.

FILE-LEADER, ch. h. by (*Imp.*) Barefoot, dam Saluda by Timoleon.

FINANCE, b. m. by Davy Crocket, dam by Sir Henry Tonson.

FLASH, b. m. by (*Imp.*) Leviathan, dam by Conqueror.

FLAXINELLA, gr. m. by (*Imp.*) Leviathan, dam by Virginian.

FLEETFOOT, gr. m. by (*Imp.*) Barefoot, dam Dove by Duroc.

FLETA (James L. French's) br. m. by (*Imp.*) Sarpedon, dam by Rasselas.

——— (G. B. Williams's), ch. h. by (*Imp.*) Leviathan, dam by Clay's Sir William.

FLIGHT, ch. m. by (*Imp.*) Leviathan, dam by Sir Charles.

FLORA HUNTER, gr. m. by Sir Charles, dam by Duroc.

FORDHAM, ch. h. by Eclipse, dam Janette by Sir Archy.

FORTUNATUS, ch. h. by Carolinian, dam by Sir Charles.

FORTUNE, b. m. by (*Imp.*) Tranby, dam by Maryland Eclipse.

FRANCES AMANDA, ch. m. by Pennoyer, dam Sally McGrath.

FRANCES TYRREL, b. m. by Bertrand, dam by Rockingham.

FRED KAYE, b. h. by Grey Eagle, dam by Moses.

FRESHET, ch. m. by Tom Fletcher, dam Caroline (or Catherine) by Pacific.

FREE JACK, br. h. by (*Imp.*) Luzborough, dam (*Imp.*) Tinsel by Napoleon.

FROSTY, ch. h. by Eclipse, dam Martha Holloway by Rattler.

FURY, bl. m. by Terror, dam by Smith's Bedford.

——— (Col. Wade Hampton's), ch. m. by (*Imp.*) Priam, dam (*Imp.*) sister to Ainderby by Velocipede.

G.

GABRIEL, ch. h. by Napoleon, dam Harpalyce by Collier.

GALANTHA, b. m. by (*Imp.*) Leviathan, dam by Jackson.

GAMMA, gr. m. by Pacific, dam (Melzare's dam) by Sir Richard.

GANO, b. h. by Eclipse, dam Betsey Richards by Sir Archy.

GARRICK, gr. h. by (*Imp.*) Shakspeare, dam by Eaton's Columbus

GARTER, b. m. by (*Imp.*) Glencoe, dam by Trumpator.

GAS-LIGHT, br. h. by (*Imp.*) Merman, dam by Mercury.

GAZAN, b. h. by Sir Leslie, dam Directress by Director.

GENERAL DEBUYS, ch. h. by (*Imp.*) Leviathan, dam (*Imp.*) Nanny Killham by Voltaire.

GENERAL RESULT, b. h. by (*Imp.*) Consol, dam by Timoleon.

GENEVA, ch. m. by Medoc, dam by Arab.

GEORGE BURBRIDGE, b. h. by (*Imp.*) Chateau Margaux, dam by Mons. Tonson.

GEORGE ELLIOTT, br. h. by (*Imp.*) Leviathan, dam by Lawrence

GEORGE LIGHTFOOT, b. h. by Eclipse Lightfoot, dam Mary Logan by Arab.

GEORGE MARTIN, b. h. by Garrison's Zinganee, dam **Gabriella** by Sir Archy.

GEORGE W. KENDALL, ch. h. by Medoc, dam Jenny Devers **by** Stockholder.

GEROW, ch. h. by Henry, dam Vixen by Eclipse.

GERTRUDE, b. m. by (*Imp.*) Leviathan, dam Parasol by Napoleon

GIFT, ch. m. by Dick Chinn, dam Milch Cow.

GIPSEY, b. m. by Nullifier, dam by Anti-Tariff.

GLENARA, b. h. by (*Imp.*) Rowton, dam Nell Gwynne by Tramp.

·———— (Davis & Ragland's,) ch. m. by (*Imp.*) Glencoe, dam Kitty Clover by Sir Charles.

———— (Dr. Thos. Payne's,) b. m. by (*Imp.*) Leviathan, dam Jane Shore by Sir Archy.

GLIDER, ch. h. by (*Imp.*) Valparaiso, dam by Clifton.

GLIMPSE, b. h. by Medoc, dam by Tiger.

GLORVINA, ch. m. by Industry, dam by Bay Richmond.

GLOVER ANN, gr. m. by (*Imp.*) Autocrat, dam by Bolivar.

GOLD EAGLE, ch. h. by Grey Eagle, dam Eliza **Jenkins** by Sir William.

GOLD FRINGE, ch. h. by (*Imp.*) Glencoe, dam (*Imp.*) Gold **Wire.**

GONE AWAY, b. h. by (*Imp.*) Leviathan, dam by Virginian.

GOSPORT, br. h. by (*Imp.*) Margrave, dam Miss Valentine by (*Imp.*) Valentine.

GOVERNOR BARBOUR, b. h. by (*Imp.*) Truffle, dam by Holmes' Vampire.

GOVERNOR BUTLER, ch. h. by Argyle, dam Mary Frances by Director.

GOVERNOR CLARK, ch. h. by Medoc, dam by Old Court.

GOVERNOR POINDEXTER, ch. h. by (*Imp.*) Leviathan, dam Eliza Clay (the dam of Giantess,) by Mons. Tonson.

GRACE DARLING, ch. m. by (*Imp.*) Trustee, dam Celeste by ' Henry.

GRAMPUS, b. h. by (*Imp.*) Whale, dam by Timoleon.

———— br. h. by Shark, dam by Mons. Tonson.

GRATTAN, b. h. by (*Imp.*) Chateau Margaux, dam Flora by Maryland Eclipse.

GREY ELLA, (A. G. Reed,) gr. m. by Big Archy, dam by Bertrand.

———— (A. G. Reed,) gr. m. by Collier, dam by Gallatin

GREY FRANK, gr. h. by Frank, dam by Buzzard.

GREY-HEAD, (J. L. Bradley's,) b. h. by Chorister, dam by Sumpter

———— (J L. Bradley's,) b. h. by Chorister, dam by Muckle john.

GREY MARY, gr. m. by Ben. Sutton, dam by Hamiltonian.

GREY MEDOC, gr. h. by Medoc, dam Grey Fanny by Bertrand.

GREY MOMUS, gr. h. by Hard Luck, dam by Mons. Tonson.

GUINEA-COCK, br. h. by Merlin, dam by Grey-tail Florizel.

GULNARE, b. m. by (*Imp.*) Sarpedon, dam by Sir William of Transport.

GUSTAVUS, b. h. by Sussex, dam by Thornton's Rattler

GUY OF WARWICK, ch. h. by Frank, dam by Hamiltonian.

19 *

H.

HANNAH HARRIS, b. m. by Bertrand, dam Grey Goose oy Pacolet
HANNIBAL, b. h. by O'Kelly, dam Roxana by Sir Charles.
HA'-PENNY, b. m. by Birmingham, dam Picayune by Medoc.
HARDENED SINNER, b. h. by (*Imp.*) Philip, dam by (*Imp.*)
  Bluster.
HARD CIDER, b. h. by (*Imp.*) Tranby, dam by Sir Charles.
HARK-AWAY, ch. h. by Emilius, dam (*Imp.*) Trapes.
HARPALYCE, ch. m. by Collier, dam by Sea-Serpent.
HARRIET, ch. m. by Eclipse, dam by Shylock.
HARRY BLUFF, bl. h. by (*Imp.*) Autocrat, dam by Pakenham.
HARRY CARGILL, ch. h. by (*Imp.*) Leviathan, dam (*Imp.*) Flo-
  rentine by Whisker.
HARRY HILL, b. h. by (*Imp.*) Chateau Margaux, dam (*Imp.*) Anna
  Maria by Truffle.
HARRY WHITEMAN, ch. h. by Orphan Boy, dam by Sir Archy.
HAWK-EYE, ch. h. by Sir Lovell, dam Eliza Jenkins by Sir Wil-
  liam.
HEAD 'EM, b. h. by (*Imp.*) Trustee, dam Itasca by Eclipse.
HEALER, ch. m. by Monmouth Eclipse, dam by Sir Archy of
  Transport.
HEBE, ch. m. by Collier, dam by Bertrand.
HECTOR BELL, gr. h. by Drone, dam Mary Randolph by Gohanna.
HEIRESS, (THE) ch. m. by (*Imp.*) Trustee, dam by Henry.
HELEN, (*Imp.*) b. m. by (*Imp.*) Priam, dam Malibran by Rubens.
HENRY A. WISE, br. h. by Dashall, dam by Hickory.
HENRY CLAY, br. h. by Cock of the Rock, dam by Virginian.
HENRY CROWELL, b. h. by Bertrand Junior, dam sister to Muckle-
  john Junior.
HERALD, ch. h. by Plenipotentiary, dam (*Imp.*) Delphine by
  Whisker.
HERMIONE, ch. m. by (*Imp.*) Non Plus, dam Leocadia by Virginian.
HERO, ch. h. by Bertrand Junior, dam (*Imp.*) Mania by Figaro.
HIT-OR-MISS, b. m. by (*Imp.*) Somonocodrom, dam (*Imp.*) Baya-
  dere. [These horses are owned in Canada.]
HOOSIER-GIRL, ch. m. by (*Imp.*) Langford.
HOPE, ch. h. by the Ace of Diamonds, dam (The Captain's dam,)
  by Oscar.
HORNBLOWER, br. h. by Monmouth Eclipse, dam Music by John
  Richards.
HOURI, (*Imp.*) ch. m. by Langar, dam Annot Lyle by Ashton.
HUGUENOT, ch. h. by Convention, dam (*Imp.*) Marigold.
HUMMING-BIRD, br. m. by Industry, dam Virginia by Thornton's
  Rattle .
HUNTSMAN, gr. h. by (*Imp.*) Leviathan, dam by Pacolet.
HYDE PARK, ch. h. by (*Imp.*) Barefoot, dam Saluda by Timoleon

I.

IAGO, bl. h. by Othello, dam (Sartin's dam,) by Timoleon.
ICELAND, ch. h. by Medoc, dam Lady Jackson by Sumpter
ILLINOIS, b. h. by Medoc, dam by Bertrand.

IOWA, ch. h. by (*Imp.*) Barefoot, dam (*Imp.*) Woodbine.
IRENE, ro. m. by Printer, dam McKinney's Roan.
ISEE TURNER, ch. m. by (*Imp.*) Leviathan, dam by Stockholder
ISIDORA, b. m. by (*Imp.*) Blacklock.
ISOLA, ch. m. by Bertrand, dam Susette.

J.

JACK DOWNING, b. h. by Pacific, dam by Mons. Tonson.
JACK PENDLETON, ch. h. by Goliah, dam (Philip's dam,) by Tra falgar.
JACK WALKER, ch. h. by Cymon, dam by (*Imp.*) Luzborough.
JAMES ALLEN, ch. h. by (*Imp.*) Leviathan, dam Donna Maria by Sir Hal.
JAMES CROWELL, br. h. by Bertrand, dam by Sir Charles.
JAMES JACKSON, ch. h. by (*Imp.*) Leviathan, dam Parasol by Tiger.
JAMES F. ROBINSON, ch. h. by Medoc, dam by Potomac.
JAMES K. POLK, b. h. by (*Imp.*) Luzborough, dam Oleana by Telegraph.
——————————— b. h. by Telegraph, dam by Buzzard.
——————————— ch. h. by Buck-eye, dam by Medoc.
JANE ADAMS, b. m. by (*Imp.*) Tranby.
JANE FRANCIS, b. m. by Granby, dam by Tecumseh.
JANE MITCHELL, ch. m. by (*Imp.*) Leviathan, dam by Conqueror.
JANE ROGERS, ch. m. by (*Imp.*) Leviathan, dam by Sir Charles.
JANE SMITH, b. m. by John Dawson, dam by Pacolet.
JANE SPLANE, gr. m. by (*Imp.*) Autocrat, dam Helen McGregor by Mercury.
JEANETTE BERKELEY, ch. m. by Bertrand jr., dam Carolina by Young Buzzard.
JEANNETTON, ch. m. by (*Imp.*) Leviathan, dam by Stockholder.
JENNY-ARE-YOU-THERE, ro. m. by Sir Archy Montorio, dam by Potomac.
JENNY RICHMOND, ch. m. by Medoc, dam by Hamiltonian.
JENNY ROBERTSON, b. m. by (*Imp.*) Luzborough, dam by Marcus.
JEROME, b. h. by (*Imp.*) Luzborough, dam by Sir Charles.
JERRY, gr. h. by Jerry, dam by Blackburn's Sir William.
JERRY LANCASTER, ch. g. by Mark Moore, dam Maid of Warsaw by Gohanna.
JIM BELL, b. h. by Frank, dam Jonquil by Little John.
JIM ROCK, ch. h. by Young Eclipse, dam by Potomac.
JOB, b. h. by Eclipse, dam Jemima by Rattler.
JOE, ch. h. by Medoc, dam by Sir Archy Montorio.
JOE ALLEN, ch. h. by Goliah, dam by Sir Charles.
JOE CHALMERS, ch. h. by (*Imp.*) Consol, dam (*Imp.*) Rachel by Partisan (or Whalebone).
JOE DAVIS, b. h. by Eclipse, dam Virginia Washington by Saxe Weimar.
JOE GATES, ch. h. by Marlborough, dam by Eclipse.
JOE MURRAY, br. h. by Waxy, dam by Hamiltonian.
JOE STURGES, ch. h. by John Bascombe, dam by Thomas's Sir Andrew.
JOE WINFIELD, b. h. by John Dawson, dam Sally Dillard.

JOHN ANDERSON, b. h. by (*Imp.*) Luzborough, dam by Bagdad.
———————————— ch. h. by Cadmus, dam (Kate Anderson's dam,)
    by (*Imp.*) Eagle.
JOHN ARCHY, ch. h. by John Richards, dam by Old Whip.
JOHN BELL, b. h. by Shark, dam Kate Kearney.
JOHN BENTON, gr. h. by (*Imp.*) Leviathan.
JOHN BLEVINS, ch. h. by The Colonel, dam (*Imp.*) Trinket.
JOHN B. JONES, b. h. by Bertrand, dam by Director.
JOHN BLUNT, b. h. by Marion, dam (Mary Blunt's dam,) by Alfred.
JOHN CAUSIN, b. h. by (*Imp.*) Zinganee, dam Attaway by Sir
    James.
JOHN C. STEVENS, ch. h. by Medoc, dam by Sumpter.
JOHN DUNKIN, b. h. by Mucklejohn, dam Coquette.
JOHN FRANCIS, ch. h. by Francis Marion, dam Mary Doubleday by
    Sir Henry.
JOHN HAMPDEN, ch. h. by Goliah, dam by Director.
JOHN HUNTER, b. h. by Shark, dam Coquette by Sir Archy.
JOHN KIRKMAN, ch. h. by Birmingham, dam by Sir Henry Tonson.
JOHN LEMON, ch. h. by Uncas, dam by Oscar.
JOHN MALONE, ch. h. by (*Imp.*) Leviathan, dam Proserpine by
    Tennessee Oscar.
JOHN MARSHALL, b. h. by (*Imp.*) Luzborough, dam Lady Bass
    by Conqueror.
JOHN R. GRYMES, gr. h. by (*Imp.*) Leviathan, dam Alice Grey by
    Pacolet.
——————————— (Col. A. L. Bingaman's,) gr. h. by (*Imp.*) Levia-
    than, dam Fanny Jarman by Mercury.
JOHN ROSS, bl. h. by Waxy, dam by Topgallant.
——————————— ch. h. by (*Imp.*) Leviathan, dam by Oscar.
JOHN VALIANT, bl. h. by Valiant, dam by King's Archer.
JOHN YOUNG, b. h. by John Richards, dam by Trumpator.
JOHNSON, br. h. by Star, dam Vanity by Grigsby's Potomac.
JOSHUA BELL, ch. h. by Frank, dam Jonquil by Little John.
JOYCE ALLEN, b. m. by (*Imp.*) Emancipation, dam Leannah by
    Seagull.
JULIA, b. m. by (*Imp.*) Rowton, dam by Roscius.
JULIA BURTON, ch. m. by Gohanna, dam by Tom Tough.
JULIA DAVIE, ch. m. by (*Imp.*) Rowton, dam by Kosciusko.
JULIA FISHER, ro. m. by (*Imp.*) Luzborough, dam Polly Bellew by
    Timoleon.
JULIUS, ch. h. by (*Imp.*) Luzborough, dam by Jackson.
JUMPER, ch. h. by Timoleon, dam Diana Vernon by Herod.

K.

KANAWA, ch. h. by Medoc, dam by Rattler.
KATE, b. f. by Monmouth Eclipse, dam Shepherdess by Apollo.
KATE ANDERSON, b. m. by Columbus, dam Eaglet by (*Imp.*)
    Eagle.
KATE AUBREY, gr. m. by Eclipse, dam Grey Fanny by Bertrand.
KATE CONVERSE, b. m. by (*Imp.*) Non Plus, dam Daisy by
    Kosciusko.
KATE COY, b. m. by Critic, dam Nancy Bone by Sussex.
KATE HAUN, br. m. by Stockholder, dam by Timoleon

KATE LUCKETT, b. m. by Monmouth Eclipse, dam Shepherdess by Apollo.
KATE NICKLEBY, br. m. by (*Imp.*) Trustee, dam by Teniers.
———————————— b. m. by (*Imp.*) Glencoe, dam by (*Imp.*) Levia than.
KATE SEYTON, br. m. by Argyle, dam Pocahontas by Sir Archy.
KATE SHELBY, ch. m. by (*Imp.*) Leviathan, dam Maria Shelby by Stockholder.
KAVANAGH, b. or ch. h. by Bertrand, dam by Director.
KEWANNA, b. m. by (*Imp.*) Cetus, dam (*Imp.*) My Lady by Comus
KITTY HARRIS, gr. m. by (*Imp.*) Priam, dam Ninon de l'Enclos by Rattler.
KITTY THOMPSON, gr. m. by (*Imp.*) Margrave, dam Ninon de l'Enclos by Rattler.

L.

LA BACCHANTE, ch. m. by (*Imp.*) Glencoe, dam by Bertrand.
LA BELLA COMBS, ch. m. by Andrew, dam by Director.
LADY CANTON, gr. m. by (*Imp.*) Tranby, dam Mary Randolph b. Gohanna.
LADY CAVA, ch. m. by Bertrand, dam Betsey Echols by Archy Montorio.
LADY FRANCIS, b. m. by Trumpator, dam (Pressure's grandam.)
LADY FRANKLIN, b. m. by (*Imp.*) Luzborough, dam Sting by Con queror.
LADY HARRISON, b. m. by Sir Henry, dam by Mucklejohn.
LADY JACKSON, -. m. by Sumpter.
LADY JANE, gr. m. by (*Imp.*) Leviathan, dam Lady Grey by Orphan Boy.
LADY PLAQUEMINE, ch. m. by Little Red, dam by (*Imp.*) Eagle.
LADY PLYMOUTH, b. m. by Flagellator, dam Black Sophia by Eclipse.
LADY SKIPETH, m. by (*Imp.*) Leviathan, dam by Truxton.
LADY SLIPPER, ch. m. by (*Imp.*) Leviathan.
LADY STOCK, ch. m. by Stockholder, dam by Potomac.
LADY SUSAN, b. m. by Cramp, dam by Pantaloon.
LAFITTE, gr. h. by O'Kelly, dam Caroline Wilson by Timoleon
LANDSCAPE, b. h. by (*Imp.*) Margrave, dam by Sir Archy.
LANGHAM, ch. h. by Medoc, dam by Cumberland.
LANEVILLE, ch. h. by Eclipse, dam by Arab.
LASSO, b. m. by Mucklejohn, dam by Gallatin.
LAURA, b. m. by Medoc, dam by Moses.
LAURA LECOMTE, b. m. by Tarquin, dam Sarah by (*Imp.*) Sar. pedon.
LAURETTE, ch. m. by Jerseyman, dam Maria Harrison.
LAVINIA PIPER, ch. m. by (*Imp.*) Leviathan, dam by Murphy s Pacolet.
LAVOLTA, b. m. by Medoc, dam by Blackburne's Buzzard.
LAWYER McCAMPBELL, b. h. by Lord Byron, dam Warping Bars by Rattle the Cash.
LEDA, ch. m. by Tiger, dam by Sumpter.
LEESBURG, ch. h. by Red Rover, dam by Tuckahoe.
LEG-BAIL, ch. h. by Jackson, dam by Marshal Ney.
LEG-TREASURER, ch. h. by Medoc, dam by Cumberland.

LEHIGH, ch. h. by (*Imp.*) Skylark, dam Nelly Webb by Industry.
LENNOX, b. h. by (*Imp.*) Trustee, dam (*Imp.*) Rosalind by Pawlowitz
LESLIE, ch. h. by (*Imp.*) Leviathan, dam by Stockholder.
LETTY FLOYD, ch. m. by (*Imp.*) Rowton, dam Palmetto by Rob Roy
LEVI, -. h. by Star, dam by Walnut.
LEVITHA, ch. m. by (*Imp.*) Leviathan.
LEXPIHILI, ch. m. by Hugh L. White, dam by Pacolet.
LIATUNAH, ch. m. by (*Imp.*) Ainderby, dam (*Imp.*) Jenny Mills by Whisker.
LIBERALITY, ch. h. by Maryland Eclipse, dam by Sir Alfred.
LIBERTAS, ch. h. by Eclipse, dam by Director.
LIEUTENANT BASSINGER, br. h. by (*Imp.*) Fylde, dam by Roanoke.
LIKENESS, (*Imp.*) ch. m. by Sir Peter Lely, dam Worthless by Walnut.
LILY, gr. m. by Tychicus, dam Laura by Rob Roy.
LIMBER JOHN, ch. h. by Kosciusko, dam by Moses.
LINWOOD, ch. h. by Wild Bill, dam by Pacolet.
LITTLE BARTON, b. h. by Bertrand, dam by Hamiltonian.
LITTLE BLUE, gr. h. by Marmion, dam by Tecumseh.
LITTLE MISERY, b. m. by Anvil, dam (*Imp.*) Anna Maria by Truffle.
LITTLE PRINCE, gr. h. by John Bascombe, dam Bolivia by Bolivar.
LITTLE RED, ch. h. by Medoc, dam by Sumpter.
LITTLE TRICK, b. h. by (*Imp.*) Tranby, dam (Occident's dam,) by Florizel.
LIVE OAK, b. h. by (*Imp.*) Luzborough, dam by Pacific.
LIVINGSTON, gr. h. by Medley, dam by Van Tromp.
——————— b. h. by (*Imp.*) Trustee, dam by Henry.
LIZ LONG, br. m. by (*Imp.*) Merman, dam by Alpheus.
LIZ TILLETT, ch. m. by Frank, dam by Medoc.
LIZZY HEWITT, b. m. by Ivanhoe, dam Princess Ann by Mons Tonson.
LOG-CABIN, ch. h. by Frank, dam by Hamiltonian.
LONG TOM, ch. h. by Pacific, dam by Jerry.
LORD OF LORN, br. h. by Argyle, dam Maria by Virginian.
——————— br. h. by Argyle, dam Duck Filly by Virginius.
LORD OF THE ISLES, gr. h. by Pacific, dam by Jerry.
LORENZO, b. h. by Bertrand, dam by Whip.
LORINDA, ch. m. by Havoc, dam by Conqueror.
LOUISA JORDAN, ch. m. by (*Imp.*) Jordan, dam Betsey Marshal by John Richards.
LOUISA WINSTON, b. m. by Waxy.
LUCRETIA NOLAND, br. m. by (*Imp.*) Hedgford, dam Frances Ann by Frank.
LUCY A. MEYER, b. m. by Pacific, dam by Sir Richard.
LUCY BENTON, br. m. by Hugh L. White, dam by Moloch.
LUCY DASHWOOD, gr. m. by (*Imp.*) Leviathan, dam Miss Bailey by (*Imp.*) Boaster.
LUCY FULLER, ch. m. by Eclipse, dam by Pakenham.
LUCY LONG, m. by John Richards, dam by Diomed.
——————— b. m. by Latitude, dam by Whip.
LUCY WEBB, ch. m. by Medoc, dam by Sumpter.

LUDA, b. m. by Medoc, dam Duchess of Marlborough by Sir Archy
LUNA DOE, ch. m. by (*Imp.*) Leviathan, dam Telic Doe by Pacific
LYNDHURST, ch. h. by (*Imp.*) Leviathan, dam by Wonder.
LYNEDOCH, ch. h. by (*Imp.*) Leviathan, dam by Wonder.

M.

MABEL WYNNE, b. m. by (*Imp.*) Rowton, dam by Sir Archy.
MADAME ARRALINE, ch. m. by Medoc, dam by Cadmus.
MAFFIT, b. h. by Frank, dam by Aratus.
MAGNATE, ch. h. by Eclipse, dam Cherry Elliott by Sumpter
MAID OF ATHENS, b. m. by (*Imp.*) Priam, dam by Arab.
MAID OF NORTHAMPTON, gr. m. by (*Imp.*) Autocrat, dam by
    Rattler.
MAJOR BOOTS, br. h. by (*Imp.*) Merlin, dam by Alborak.
MANALOPAN, gr. h. by Medley, dam by John Richards.
MANGO, (*Imp.*) ch. m. by Taurus, dam Pickle by Emilius.
MARCHIONESS, ch. m. by (*Imp.*) Rowton, dam (Fancy's dam) by
    Sir Archy.
MARCO, b. h. by Sir Leslie, dam by Lance.
MARGARET CARTER, b. m. by Medoc, dam Lady Whip by Sir
    Archy.
MARGARET BLUNT, b. m. by Eclipse, dam by Contention.
MARGARET WOOD, b. m. by (*Imp.*) Priam, dam Maria West by
    Marion.
MARIA, ch. m. by (*Imp.*) Jordan, dam Polly Powell by Virginian.
MARIA BLACK, (*Imp.*) br. m. by Filho da Puta, dam by Smolensko
MARIA BROWN, br. m. by (*Imp.*) Luzborough, dam Brunette by Sir
    Hal.
MARIA COLLIER, br. m. by Collier, dam by Gallatin.
MARIA MILLER, br. m. by Stockholder, dam by Madison.
MARIA PEYTON, ch. m. by Balie Peyton, dam by Tariff.
MARIA SHELTON, ch. m. by Andrew, dam (Ajarrah Harrison's
    dam) by Gallatin.
MARIA SPEED, ch. m. by (*Imp.*) Leviathan, dam by Pacific.
MARIA WILLIAMS, ch. m. by (*Imp.*) Leviathan, dam by Napoleon
MARINER, bl. h. by Shark, dam Bonnet's o' Blue by Sir Charles.
MARION, b. m. by (*Imp.*) Autocrat, dam by Rob Roy.
MARTHA BICKERTON, b. m. by Pamunky, dam by Tariff.
MARTHA CARTER, ch. m. by Bertrand, dam Sally Naylor by Gal-
    latin.
———————————— ch. m. by Bertrand, dam by Oscar.
MARTHA CALVIN, b. m. by Agrippa, dam by Walnut.
MARTHA MALONE, b. m. by (*Imp.*) Leviathan, dam Tatcheeana
    by Bertrand.
MARTHA RANEY, b. m. by (*Imp.*) Luzborough, dam by Sumpter.
MARTHA ROWTON, ch. m. by (*Imp.*) Rowton, dam Martha Griffin
    by Phenomenon.
MARTHAVILLE, b. m. by Dick Singleton, dam Black-Eyed Susan
MATCHEM, ch. h. by (*Imp.*) Luzborough, dam by P'nd Jackson.
MARTIN'S JUDY, br. m. by Young's Mercury, dam by Eclipse.
MARTIN VAN BUREN, b. h. by Lafayette Stockholder, dam by In-
    diar
MARY, gr. m. by Old Saul, dam by Free Mulatto.

MARY, gr. m. by (*Imp.*) Consol, dam Sally Bell by Sir Archy.
MARY ANN FURMAN, br. m. by (*Imp.*) Sarpedon, dam by Bertrand.
MARY BEECHLAND, b. m. by Sir Leslie, dam by Potomac.
MARY BELL, b. m. by Seagull, dam (Vidocq's dam) by Stockholder
MARY BRENNAN, b. m. by Richard Singleton, dam by Hamiltonian.
MARY BURNHAM, b. m. by Archy Montorio, dam by Stockholder.
MARY CHASE, b. m. by (*Imp.*) Felt, dam by Sir Archy.
MARY CHURCHILL, b. m. by (*Imp.*) Barefoot.
MARY DOUGLAS, gr. m. by Jerry, dam by Stockholder.
MARY ELIZABETH, ch. m. by Andrew, dam by Gallatin.
MARY ELLEN, b. m. by Woodpecker, dam by Sumpter.
MARY HEDGFORD, br. m. by (*Imp.*) Hedgford, dam Mary Francis by Director.
MARY JONES, ch. m. by (*Imp.*) Barefoot, dam by Eclipse.
MARY LEWIS, ch. m. by (*Imp.*) Leviathan, dam Proserpine by Oscar.
MARY LONG, b. m. by (*Imp.*) Tranby, dam Lady Pest by Carolinian.
MARY LUCKETT, ch. m. by Marion, dam (Charles Archy's dam) by Eclipse.
MARY MASON, br. m. by Pirate, dam by (*Imp.*) Consol.
MARY MEADOWS, ch. m. by Stockholder, dam by Timoleon.
MARY MILLER, ch. m. by Arab, dam by Peacemaker.
MARY MORRIS, b. m. by Medoc, dam Miss Obstinate by Sumpter.
MARY OUSLEY, br. m. by King's Bertrand, dam by Pacolet.
MARY PORTER, ch. m. by Mucklejohn, dam by Printer.
MARY REED, br. m. by Industry, dam by Rattler.
MARY RODGERS, b. m. by (*Imp.*) Hibiscus, dam Ten Broeck's dam.
MARY SCOTT, b. m. by Bertrand, dam by Blackburn's Whip.
MARY SHERWOOD, b. m. by Stockholder, dam by (*Imp.*) Leviathan.
MARY STEWART, b. m. by (*Imp.*) Valentine, dam by Henry.
MARY THOMAS, b. m. by (*Imp.*) Consol, dam Parrot by Roanoke.
MARY TRIFLE, ch. m. by Medoc, dam by Hamiltonian.
MARY VAUGHAN, b. m. by Waxy, dam by (*Imp.*) Bluster.
MARY WALTON, ch. m. by (*Imp.*) Leviathan, dam Miss Bailey by (*Imp.*) Boaster.
MARY WATSON, gr. m. by Robin Hood, dam Bolivia by Bolivar.
MARY WELLER, ch. m. by Sterling, dam Discord by (*Imp.*) Luzborough.
MARY WICKLIFFE, b. m. by Medoc.
MARY WYNNE, b. m. by Eclipse, dam Flirtilla Jr. by Sir Archy.
MASTER HENRY, b. h. by Henry, dam (Balic Peyton's dam) by Eclipse.
MAT. MURPHY, ch. h. by Pete Whetstone, dam by Rattler.
MEDINA, b. m. by (*Imp.*) Barefoot, dam by Director.
MEDOCA, ch. m. by Medoc, dam by Doublehead.
MEDORA WINSTON, b. m. by Telegraph, dam by Pacolet.
MELISSE BYRON, b. m. by Cherokee, dam by Barnett's Diomed.
MELODY, ch. m. by Medoc, dam (Randolph's dam) by Haxall's Moses
MERCER, cn. h. by Woodpecker, dam by Hamiltonian

MERIDIAN, ch. h. by (*Imp.*) Barefoot, dam by Eclipse.
METARIE, ch. m. by Frank, dam (Musedora's dam) by Kosciusko.
METEOR, ch. h. by (*Imp.*) Priam, dam (Baltimore's dam) by Gohanna.
McINTYRE, ch. h. by Medoc, dam by Sumpter.
MIDNIGHT, bl. m. by Shark, dam Meg Dods, by Sir Archy.
MIDAS, b. h. by (*Imp.*) Rowton, dam by Roanoke.
MILTON HARRISON, b. h. by Orange Boy, dam by Quicksilver.
MINERVA ANDERSON, ch. m. by (*Imp.*) Luzborough, dam by Sir Charles.
MINERVA PROFFIT, ch. m. by (*Imp.*) Luzborough, dam Sophia Bess.
MINISTER, b. h. by Medoc, dam by Alexander.
MINSTREL, b. m. by Medoc, dam by Bedford's Alexander.
MINT JULEP, br. h. by Godolphin, dam Isora by Dockon.
MIRABEAU, b. h. by Medoc, dam Ann Merry by Sumpter.
MIRIAM, b. m. by (*Imp.*) Autocrat, dam Laura by Rob Roy.
MIRTH, b. m. by Medoc, dam (Minstrel's dam) by Bedford's Alexander.
MISKWA, ch. m. by Dick Chinn, dam Linnet by (*Imp.*) Leviathan.
MISSISSIPPI, b. h. by John Dawson, dam by Partnership.
MISSOURI, ch. m. by Eclipse, dam by Director.
MISTAKE, b. m. by Eclipse, dam by Timoleon.
MISS ACCIDENT, (*Imp.*) b. m. by Tramp, dam Florestine by Whisker.
MISS ANDREW, ch. m. by Andrew, dam by Gallatin.
MISS BELL, b. m. by (*Imp.*) Consol, dam (*Imp.*) Amanda by Morisco.
MISS CHESTER, b. m. by (*Imp.*) Sarpedon, dam Delilah by Tiger.
MISS CLARK, ch. m. by Birmingham, dam by Cumberland.
MISS CLASH, ch. m. by Birmingham, dam by Stockholder.
MISS CLINKER, (*Imp.*) b. m. by Humphrey Clinker, dam Mania by Maniac.
MISS FOOTE, b. m. by (*Imp.*) Consol, dam (*Imp.*) Gabriella by Oscar (or Oiseau).
MISS JACKSON, ch. m. by Oakland, dam by Diomed.
MISS LETTY, b. m. by (*Imp.*) Priam, dam Patty Burton by Marion.
MISS MACARTY, b. m. by Waxy.
MISS RIDDLE, ch. m. by (*Imp.*) Riddlesworth, dam Lady Jackson by Sumpter.
MISS WILLS, gr. m. by (*Imp.*) Zinganee, dam Sorrow by Rob Roy.
MOBILE, b. h. by (*Imp.*) Consol, dam (*Imp.*) Sessions by Whalebone.
MOLLY LONG, ch. m. by Tom Fletcher, dam by (*Imp.*) Janus.
MOLLY WARD, b. m. by (*Imp.*) Hedgford, dam by Bertrand.
MOLOCH, (*Imp.*) b. h. by Muley Moloch, dam Sister to Puss by Teniers.
MONARCH, (*Imp.*) b. h. by Priam, dam Delphine by Whisker.
MONGRELIA, ch. m. by Medoc, dam Brownlock by Tiger.
MONKEY DICK, b. h. by Dick Singleton, dam by Sumpter.
MORDAC, ch. h. by Eclipse, dam by Whip.
MORGAN, ch. h. by John Bascombe, dam Amy Hamilton.
MORGIANA, ch. m. by Red Gauntlet, dam by Joe Kent.
MORTIMER, ch. h. by Monmouth Eclipse, dam by Ogle's Oscar

MOSELLE (Colonel Gavan's), b. m. by Telegraph, dam (*Imp.*) Jane Shore.

MOSELLE (E. P. Dave's), b. m. by (*Imp.*) Luzborough, dam (*Imp.*) Jane Shore.

MOTH, ch. m. by (*Imp.*) Glencoe, dam (*Imp.*) Jessica by Velocipede.

MOTTO, ch. m. by (*Imp.*) Barefoot, dam Lady Tompkins by Eclipse.

MOUNTAINEER, ch. h. by Yorkshire, dam by Rattler.

MOUNTJOY, b. h. by (*Imp.*) Tranby, dam by Sir Charles.

MUD, gr. h. by (*Imp.*) Leviathan, dam by Pacolet.

MUSEDORA, ch. m. by Medoc, dam by Kosciusko.

MUSE SANDFORD, b. h. by Hickory, dam by (*Imp.*) Contract.

MUSIC, gr. h. by (*Imp.*) Philip, dam Piano by Bertrand.

N.

NANCY BUFORD, ch. m. by Medoc, dam by Thornton's Rattler.

NANCY CLARK, b. m. by Bertrand, dam Morocco Slipper by Timoleon.

NANCY DAWSON, ch. m. by Frank, dam by Voltaire.

NANCY O., ch. m. by Flagg, dam Milly Tonson by Mons. Tonson.

NANCY ROWLAND, b. m. by (*Imp.*) Rowton, dam by Rob Roy.

NANNY, b. m. by (*Imp.*) Trustee, dam Miss Mattie by Sir Archy.

NARCISSA PARISH, ch. m. by Stockholder, dam by (*Imp.*) Eagle.

NARINE, ch. m. by (*Imp.*) Jordan, dam Louisianaise.

NAT BRADFORD, gr. h. by Bertrand, dam Morocco Slipper by Timoleon.

NATHAN RICE, br. h. by Birmingham, dam by Whipster.

NED WELLS, b. h. by O'Connell, dam by Stockholder.

NEPTUNE, ch. m. by (*Imp.*) Jordan, dam Louisianaise.

NIAGARA, ch. h. by (*Imp.*) Trustee, dam Gipsey by Eclipse.

NICK BIDDLE, b. h. by Score Double, dam Highland Mary.

NICK DAVIS, ch. h. by (*Imp.*) Glencoe.

NICON, ch. h. by Pacific, dam by Jackson.

NOBLEMAN, ch. h. by (*Imp.*) Cetus, dam (*Imp.*) My Lady by Comus.

NORFOLK, br. h. by (*Imp.*) Fylde, dam Polly Peachem by John Richards.

NORMA, ch. m. by Longwaist, dam (*Imp.*) Novelty by Blacklock.

NORTH STAR, ch. h. by Emilius, dam Polly Hopkins by Virginian.

O.

OCTAVE, b. m. by (*Imp.*) Emancipation, dam Polly Kennedy.

OGLENAH, ch. h. by Medoc, dam Maria by Hamiltonian.

OH SEE, ch. h. by (*Imp.*) Foreigner, dam by Mons. Tonson.

OLD DOMINION, ch. h. by Eclipse, dam Isabella by Sir Archy.

OLD MISTRESS, ch. m. by Count Badger, dam Timoura by Timoleon.

OLEAN, ch. m. by (*Imp.*) Leviathan, dam by Truxton.

OLEANDER, ch. m. by (*Imp.*) Glencoe, dam Aranetta by Bertrand.

OLIVIA WAKEFIELD, gr. m. by Patrick Henry.

OLYMPUS, ch. h. by Eclipse, dam Flirtilla Junior, by Sir Archy.

OMEGA, gr. m. by Timoleon, dam Daisy Cropper by Ogle's Oscar.

OMOHONDRO, ch. h. by Robin Brown, dam by Mason's Rattler.

OREGON b. h. by (*Imp.*) Philip, dam by (*Imp.*) Luzborough.

ORIANA, br. m. by (*Imp.*) Longwaist, dam (*Imp.*) Orleana by Buzzard.

ORIFLAMME, ch. h. by Mons. Tonson, dam by Sir Hal.
ORIOLE, b. m. by (*Imp*) Leviathan, dam Object by Marshal Ney.
ORLEANS, ch. h. by Cock of the Rock, dam by Timoleon.
ORSON, ch. h. by (*Imp.*) Valentine, dam Ethelinda by Marshal Bertrand.
OSTRICH, ch. h. by Collier, dam by Shakspeare.
OSCAR, (Josiah Chambers's), ch. h. by Ulysses, dam by Bertrand.
OSCEOLA, b. h. by Pacific, dam by Oliver II. Perry.
————— ch. h. by Wild Bill, dam by Timoleon.
————— ch. h. by Collier, dam by Sumpter.
OTHELLO, ch. h. by Waxy, dam by Hickory.
OUR MARY, br. m. by (*Imp.*) Langford, dam Ostrich by Eclipse.

P.

PAIXHAN, b. h. by (*Imp.*) Felt, dam Mary Hutton.
PALMERSTON, b. h. by (*Imp.*) Merman, dam (*Imp.*) by Cadmus.
PANIC, ch. h. by Eclipse, dam Aggy-up by Timoleon.
PARIS, bl. h. by (*Imp.*) Priam, dam Water-Witch.
PARTNER, ch. h. by Medoc, dam by Doublehead.
PASSENGER, (*Imp.*) b. h. by Langar, dam My Lady by Comus.
————— b. h. by Balie Peyton, dam by Pamunky.
PASSAIC, (*Imp.*) ch. h. by Reveller, dam Rachel by Moses.
PATRICK H. GALWEY, ch. h. by (*Imp.*) Jordan, dam Duchess of Ashland by Shakspeare.
PATSEY ANTHONY, b. m. by (*Imp.*) Priam, dam (Josephus's dam) by Virginian.
PATSEY BUFORD, b. m. by Mazeppa, dam by Rattler.
PATSEY CROWDER, gr. m. by Patrick Henry, dam Hillon by Antelope.
PATSEY DAVIS, ch. m. by Count Badger, dam Timoura by Timoleon.
PATSEY STUART, b. m. by Bertrand, dam by Redgauntlet.
PEDLAR, ch. h. by (*Imp.*) Leviathan, dam by Pizarro.
PEGGY HALE, ch. m. by (*Imp.*) Skylark, dam by Sir Charles.
PENELOPE, (*Imp.*) ch. m. by Plenipo, dam Brazil by Ivanhoe.
PENSEE, gr. m. by Lauderdale, dam by Lightning.
PEORIA, ch. m. by Medoc, dam by Whip.
PETER PINDAR, ch. h. by (*Imp.*) Daghee, dam by (*Imp.*) Barefoot.
PETER SPYKE, ch. h. by Eclipse, dam by (*Imp.*) Jack Andrews.
PETWAY, b. h. by (*Imp.*) Glencoe, dam Kitty Clover by Sir Charles.
————— br. h. by (*Imp.*) Glencoe, dam by Sir Archy.
PETWORTH, b. h. by (*Imp.*) Philip, dam (Kinlock's dam) by Shawnee.
PEYTONA, ch. m. by (*Imp.*) Glencoe, dam Giantess by (*Imp.*) Leviathan.
PHANTOM, b. h. by (*Imp.*) Contract, dam by Potomac.
PHIL. BROWN, (*Imp.*) ch. h. by Glaucus, dam Bustle by Whalebone.
PICKWICK, b. h. by Pacific, dam by Pacolet.
PICOLO, br. h. by Lord Byron, dam Highland Mary (Nick Biddle's dam).
PILOT, b. h. by Wild Bill, dam by Oscar.
PLENIPO, (*Imp.*) b. h. by Plenipo, dam Polly Hopkins by Virginian
POKEROOT, gr. h. by William Tell, dam by Citizen

POLLARD BROWN, b. h. by Wild Bill, dam Hippy by Pacolet.
POLLY ELLIS, m. by (*Imp.*) Trustee, dam Rosalind by Ogle's Oscar
POLLY GREEN, br. m. by Sir Charles, dam Polly Peachem by John
    Richards.
POLLY HUNTER, ch. m. by Andrew, dam by Crusader.
POLLY MILAM, b. m. by (*Imp.*) Sarpedon, dam by Escape.
POLLY PILLOW, b. m. by (*Imp.*) Leviathan, dam by Sir Archy.
POLLY PIPER, ch. m. by Count Piper, dam by Consul or Sumpter
PONEY, ch. h. by (*Imp.*) Leviathan, dam by Stockholder.
PONOLA, ch. h. by Hannibal, dam by Sir Archy.
PORTSMOUTH, br. h. by (*Imp.*) Luzborough, dam Polly Peachem
    by John Richards.
POSTMASTER, (The) b. h. by (*Imp.*) Consol, dam Country Maid by
    Pacific.
POWELL, ch. h. by Medoc, dam by Alexander or Virginian.
PRENTISS, (S. S.) b. h. by (*Imp.*) Fylde, dam by Washington.
PRESTON, br. h. by Telegraph, dam (Olivia's dam).
———— b. h. by (*Imp.*) Leviathan, dam Parrot by Roanoke.
PRIMA DONNA, b. m. by (*Imp.*) Priam, dam Lady Rowland by Te
    riff.
PRINCE ALBERT, ch. h. by (*Imp.*) Margrave, dam (Eutaw's dam,
    by Sir Charles.
PRINCESS, ch. m. by (*Imp.*) Priam, dam Sally Hope by Sir Archy.
PRINCESS ANN, b. m. by (*Imp.*) Leviathan, dam by Stockholder.
PRISCILLA MARTIN, ch. m. by (*Imp.*) Leviathan, dam by Arab.
PROMISE, ch. m. by Wagner, dam by Lance.
PROSPECT, ch. h. by Monmouth Eclipse, dam by (*Imp*) Expedition.
———— ch. m. by (*Imp.*) Luzborough, dam Anvilina Smith by
    Stockholder.
PRYOR, b. h. by (*Imp.*) Priam, dam Queen of Clubs by Virginian.
PURITY, b. m. by (*Imp.*) Ainderby, dam Betty Martin by Giles
    Scroggins.
PUSS, b. m. by (*Imp.*) Priam, dam by Virginian.

Q.

QUEEN ANNE, (*Imp.*) bl. m. by Camel, dam by Langar.
QUEEN ELIZABETH, br. m. by (*Imp.*) Leviathan, dam by Sir **Archy**
QUEEN MARY, ch. m. by Bertrand, dam by Brimmer.
QUININE, ch. m. by Red Tom, dam by Bertrand.

R.

RAGLAND, ch. h. by (*Imp*) Leviathan, dam by Stockholder.
RALPH, b. h. by Woodpecker, dam Brown Mary by Sumpter.
RANCOPUS, ch. m. by Flagellator, dam Molly Longlegs.
RAN PEYTON, ch. h. by (*Imp.*) Leviathan, dam by Stockholder.
RAPIDES, ch. h. by (*Imp.*) Skylark, dam Margaret May by Pacific.
RASP, gr. h. by (*Imp.*) Fylde, dam by Director.
REBECCA KENNER, b. m. by (*Imp.*) Skylark, dam Lady Halston
    by Bertrand.
REBEL, ch. h. by Gohanna, dam (Ohio's dam).
RED BILL, ch. h. by Medoc, dam Brown Mary by Sumpter.
RED BREAST, ch. h. by (*Imp.*) Priam, dam Fanny Wyatt by Sir
    Charles.

RED BUCK, ch. h. by (*Imp.*) Rowton, dam Lady Deerpond.
RED EAGLE, br. h. by Grey Eagle, dam by Moses.
RED FOX, ch. h. by (*Imp.*) Luzborough.
RED GAUNTLET, ch. h. by (*Imp.*) Trustee, dam (*Imp.*) Vaga.
RED HAWK, ch. h. by Medoc, dam by Sumpter.
RED HEAD, b. h. by Woodpecker, dam by Whipster.
RED MOROCCO, ch. m. by Medoc, dam Brownlock by Tiger.
RED ROSE, br. m. by (*Imp.*) Leviathan, dam by (*Imp.*) Bagdad.
RED TOM, ch. h. by Bertrand, dam Duchess of Marlborough by Sir
    Archy.
REEL, gr. m. by (*Imp.*) Glencoe, dam (*Imp.*) Gallopade by Catton.
REGENT, b. h. by (*Imp.*) Priam, dam Fantail by Sir Archy.
REGISTER, gr. h. by (*Imp.*) Priam, dam Maria Louisa by Mons.
    Tonson.
RELIANCE, b. h. by (*Imp*) Autocrat, dam Lady Culpeper by Caroli-
    nian.
RESCUE, br. h. by (*Imp.*) Emancipation, dam Louisa Lee by Medley.
REVEILLE, b. m. by Bertrand, dam Sally Melville by Virginian.
———————— b. or br. h. by Young Virginian, dam by Harwood.
REVERIE, b. or br. m. by (*Imp.*) Ainderby, dam by Giles Scroggins.
RHYNODINO, gr. h. by Pacific, dam by Hamiltonian.
RICHARD OF YORK, b. h. by Star, dam by Shylock.
RICHARD ROWTON, b. h. by (*Imp.*) Rowton, dam by Falstaff.
RIENZI, b. h. by (*Imp.*) Autocrat, dam by Sir Charles.
——————— b. h. by (*Imp.*) Autocrat, dam Peggy White by (*Imp.*) Sy
    phax (or Diomed).
RINGDOVE, b. m. by (*Imp.*) Merman.
RIPPLE, b. m. by Medoc, dam Belle Anderson by Sir William.
ROANNA, ro. m. by Archy Montorio, dam by Potomac.
ROBERT BRUCE, b. h. by Clinton, dam by Sir Archy.
ROBIN COBB, ch. h. by (*Imp.*) Felt, dam Polly Cobb.
ROCKER, b. h. by Eclipse, dam by Virginian.
ROCKETT, b. h. by Sir Leslie, dam Miss Lancess by Lance.
RODERICK DHU, gr. h. by Merlin, dam by (*Imp.*) Bagdad.
RODNEY, br. h. by (*Imp.*) Priam, dam Medora.
ROSABELLA, b. m. by (*Imp.*) Shakspeare, dam by Timoleon.
ROSA VERTNER, b. m. by Sir Leslie, dam Directress by Director.
ROSCOE, b. h. by Pacific, dam by Grey Archy.
ROTHSCHILD, b. h. by (*Imp.*) Zinganee, dam by Tiger.
ROVER, b. h. by Woodpecker, dam Sally Miller by Cherokee.
ROWTONELLA, ch. m. by (*Imp.*) Rowton, dam Sally Hopkins by
    Kosciusko.
RUBY, b. m. by (*Imp.*) Rowton, dam Bay Maria by Eclipse.
——— ch. m. by Duke of Wellington, dam Lively by Eclipse
RUFFIN, b. h. by (*Imp.*) Hedgford, dam Duchess of Marlborough by
    Sir Archy.

S.

SAILOR BOY, b. h. by Jim Cropper, dam by Marshal.
SALADIN, b. h. by John Richards, dam by Henry.
SALKAHATCHIE, b. m. by Vertumnus, dam Sally Richardson by
    Kosciusko.
SAL STRICKLAND, ch. m. by (*Imp.*) Leviathan, dam by Pacolet

SALLY BARTON, ch. m. by Jackson, dam by Gallatin.
SALLY BROWN, b. m. by Jackson, dam by Gallatin.
SALLY CARR, b. m. by Stockholder, dam by ————
SALLY CRESSOP, ch. m. by Eclipse, dam by Arab.
SALLY DILLIARD (or HILLIARD), gr. m. by O'Kelly, dam by Shawnee.
SALLY HARDIN, b. m. by Bertrand, dam Peggy Stewart by Whip
SALLY HART, m. by (*Imp.*) Luzborough, dam Clear-the-Kitchen by Shakspeare.
SALLY McGHEE, ch. m. by Gascoigne, dam Thisbe.
SALLY MORGAN, b. m. by (*Imp.*) Emancipation, dam Lady Morgan by John Richards.
SALLY SHANNON, b. m. by Woodpecker, dam (Darnley's dam,) by Sir Richard.
SALLY WARD, m. by John R. Grymes, dam by ————
SAMBO, ch. h. by Equinox, dam by Aratus.
SAM HOUSTON, b. h. by (*Imp.*) Autocrat, dam by (*Imp.*) Major.
SANDY YOUNG, b. h. by Medoc, dam Natchez Bell by Seagull.
SANTA ANNA, ch. h. by Bertrand Junior, dam Daisy by Kosciusko.
SANTEE, ch. h. by Wild Bill, dam Sally McGhee by Timoleon.
SARAH BLADEN, ch. m. by (*Imp.*) Leviathan, dam Morgiana by Pacolet.
SARAH BURTON, m. by Pacific, dam by Timoleon.
SARAH CHANCE, ch. m. by Lafayette, dam by Sir Archy.
SARAH JACKSON, JUNIOR, b. m. by Piamingo, dam by Arab.
SARAH MORTON, b. m. by Sidi Hamet, dam Rowena by Sumpter.
SARAH WASHINGTON, b. m. by Garrison's Zingance, dam by Contention.
SARTIN, br. h. by (*Imp.*) Luzborough, dam Julia Fisher by Timoleon.
SCARLET, ch. h. by Uncas, dam by Pacolet.
SENATOR, ch. h. by (*Imp.*) Priam, dam Ariadne by Gohanna.
SERENADE, b. h. by Woodpecker, dam by Cook's Whip.
SEVEN-UP, b. m. by (*Imp.*) Chateau Margaux, dam by Arab.
SHAMROCK, (*Imp.*) ch. h. by St. Patrick, dam Delight by Reveller.
SHARATOCK, ch. h. by Medoc, dam by Trumpator.
SHEPHERDESS, ch. m. by Lance, dam Amanda by Revenge.
SIGNAL, bl. h. by (*Imp.*) Margrave, dam by Mons. Tonson.
SIMON BENTON, ch. h. by Medoc, dam by Rattler.
SIMON GURTY, ch. h. by Mark Moore, dam by Tiger.
SIMON KENTON, ch. h. by Eclipse, dam by Rattler.
SIR ARISS, gr. h. by Trumpator, dam Ophelia by Wild Medley.
SIR ELLIOTT, b. h. by (*Imp.*) Leviathan, dam Lady Frolic by Sir Charles.
SIR JOSEPH BANKS, b. h. by (*Imp.*) Luzborough, dam by Sir Archy.
SIR WILLIAM, b. h. by Sir William, dam by Rattler.
SISSY, b. m. by (*Imp.*) Leviathan, dam (*Imp.*) Gutty by Whalebone.
SISTER TO THORNHILL, ch. m. by (*Imp.*) Glencoe, dam (*Imp.*) Pickle by Emilius.
SLEEPER, gr. h. by (*Imp.*) Sarpedon, dam Flora by Grand Seignor.

SLEEPER (THE), gr. h. by (*Imp.*) Sarpedon, dam by C.　　　's Mes.
senger.

SLEEPY JOHN, b. h. by John Dawson, dam Sally Dilliard by Virginian, (or Phenomena).

SMOKE, ch. h. by (*Imp.*) Trustee, dam Bianca by Medley.

SNAG, ch. h. by Medoc, dam by Rattler.

SNOWBIRD, gr. h. by (*Imp.*) Chateau Margaux, dam Forsaken Filly
by Jerry.

SOPHIA LOVELL, b. m. by Sir Lovell, dam Eliza Jenkins by Sir
William.

SORROW, (*Imp.*) ch. h. by Defence, dam Tears by Woful.

SPLINT, ch. m. by Hualpa, dam by Phenomenon.

STACKPOLE, ch. h. by (*Imp.*) Leviathan, dam by Stockholder.

STAGE-DRIVER, b. h. by Lance, dam by Bertrand.

STANHOPE, ch. h. by Eclipse, dam Helen Mar by Rattler.

STANLEY, ch. h. by (*Imp.*) Leviathan, dam Aronetta by Bertrand.

———— ch. h. by Eclipse, dam by Busiris.

———— ECLIPSE, ch. h. by Busiris, dam by John Stanley.

STAR, b. h. by (*Imp.*) Skylark, dam Betsey Epps by Timoleon.

STAR OF THE WEST, b. m. by Bertrand, dam by Whip.

——————————— ch. m. by (*Imp.*) Luzburough, dam by Ber
trand.

STEEL, b. h. by (*Imp.*) Fylde, dam Dimont by Constitution.

STHRESHLEY, ch. h. by Medoc, dam by Paragon.

STRANGER, b. h. by Lance, dam by Whip.

STOCKBOROUGH, ch. h. by (*Imp.*) Luzborough, dam by Stockholder.

ST. CHARLES, ch. h by (*Imp.*) Jordan, dam by Mercury.

ST. CLOUD, ch. h. by (*Imp.*) Belshazzar, dam by Old Partner.

ST. LOUIS, gr. h. by Altorf, dam Fleta by Jackson's (or Johnson's)
Medley.

ST. PIERRE, bl. h. by Pamunky, dam by Lafayette.

SUFFERER, b. h. by Eclipse, dam Meg Dods by Sir Archy.

SUFFOLK, b. h. by Andrew, dam Ostrich by Eclipse.

SUNBEAM, ch. m. by (*Imp.*) Leviathan, dam Alice Grey by Mercury.

———— ch. h. by (*Imp.*) Langford, dam Gipsey, (sister to Medoc).

SUSAN HILL, ch. m. by (*Imp.*) Glencoe, dam Susan Hill by Timoleon.

SUSAN TYLER, b. m. by (*Imp.*) Sarpedon.

SUSAN VANCE, ch. m. by Saladin, dam by Sir William.

SWALLOW, b. m. by (*Imp.*) Leviathan, dam Object by Marshal Ney

SWEET HOME, ch. m. by Medoc, dam by Hamiltonian.

SWISS BOY, br. h. by (*Imp.*) Swiss, dam by Stockholder.

SYLPHIDE, (*Imp.*) b. m. by Emilius, dam Polly Hopkins by Virgi
nian.

SYMMETRY, b. m. by (*Imp.*) Priam, dam Phenomena by Sir Archy

T.

TABITHA, ch. m. by Hualpa, dam by Phenomenon.

TAGLIONI, ch. m. by (*Imp.*) Priam, dam by Sir Charles.

TALLEY, ch. h. by Talleyrand, dam by Bertrand.

TALLULAH, ch. m. by Hyazim, dam by Gallatin.

TAMERLANE, ch. h. by Cowper, dam by Director.

TAMMANY, b. h. by (*Imp.*) Trustee, dam Camilla by Henry.
TARANTULA, ch. m. by (*Imp.*) Belshazzar, dam Mary Jane Davis by Stockholder.
TARLTON, b. h. by Woodpecker, dam by Robin Grey.
TARQUIN, b. h. by (*Imp.*) Consol, dam Jeannie Deans by Powhattan.
TATTERSALL, ch. h. by (*Imp.*) Emancipation, dam (Volney's dam,) by Sir Archy.
TAYLOE, b. h. by (*Imp.*) Autocrat, dam Peggy White.
TAZEWELL, b. h. by (*Imp.*) Fylde, dam by Gallatin.
TEARAWAY, b. h. by (*Imp.*) Trustee, dam Jemima by Thornton's Rattler.
TELAMON, ch. h. by Medoc, dam Cherry Elliott by Sumpter.
TELIE DOE, b. m. by Pacific, dam Matilda by Greytail.
TELLULA, ch. m. by Eclipse, dam by Whip.
TEMPEST, ch. h. by (*Imp.*) Trustee, dam Jeanette by Sir Archy.
TEMPLAR, b. h. by (*Imp.*) Sarpedon, dam by Timoleon.
TEN BROECK, ch. h. by Eclipse, dam by Bertrand.
TENNESSEE, b. m. by (*Imp.*) Felt, dam Berenice by Archy Junior.
TEXANA, b. m. by (*Imp.*) Hedgford, dam Goodlee Washington by Washington.
TEXAS, b. h. by (*Imp.*) Fylde, dam by Potomac.
THE COLONEL, ch. h. by (*Imp.*) Priam, dam (*Imp.*) My Lady by Comus.
THE COLONEL'S DAUGHTER, b. m. by The Colonel, dam (*Imp.*) Variella by Blacklock.
THE DUKE, ch. h. by Monmouth Eclipse, dam by (*Imp.*) Expedition.
THE MAJOR, b. h. by Othello, dam by Citizen.
THE MERCER COLT, br. h. by (*Imp.*) Mercer, dam Miss Mattie by Sir Archy.
THE PONEY, ch. h. by (*Imp.*) Leviathan, dam by Stockholder.
THE POSTMASTER, b. h. by (*Imp.*) Consol, dam Country Maid by Pacific.
THE QUEEN, (*Imp.*) ch. m. by Priam, dam Delphine by Whisker.
THOMAS HOSKINS, b. h. by (*Imp.*) Autocrat, dam Minerva by Tom Tough.
THOMAS R. ROOTS, b. h. by (*Imp.*) Tranby, dam Eliza Jenkins by Sir William of Transport.
THORNHILL, ch. h. by (*Imp.*) Glencoe, dam (*Imp.*) Pickle by Emilius.
TIBERIAS, b. h. by (*Imp.*) Priam, dam Fanny Wright by Silverheels.
TIPPECANOE, ch. h. by Eclipse, dam by Rattler.
TISHANNA, b. m. by Benbow, dam Fidget by Eclipse.
TISHIMINGO, b. h. by (*Imp.*) Leviathan, dam Maria Shepherd by Sir Archy.
TOBY, b. h. by Bertrand, dam by Eagle.
TOM AND JERRY, ch. h. by Heart of Oak, dam by Lafayette.
TOM BENTON, b. h. by Wild Bill, dam by Pacolet.
TOM BUCK, ro. h. by (*Imp.*) Glencoe, dam Lady Sykes by Timoleon.
TOM CHILTON, ch. h. by (*Imp.*) Leviathan, dam by Childers.
TOM CORWIN, b. h. by (*Imp.*) Emancipation, dam by Lottery.
TOM CRINGLE, ch. h. by Carolinian.
TOM DAY, b. h. by Bertrand, dam Sally Melville by Virginian.
TOM MARSHAL, (Col. Bingaman's.) gr. h. by (*Imp.*) Leviathan, dam Fanny Jarman by Mercury.

TOM MARSHAL, (Col. Buford's,) b. h. by Medoc, dam by Sumpter.
TOM PAINE, bl. h. by (*Imp.*) Margrave, dam (Emily Thomas's dam) by Tom Tough.
TOM THURMAN, b. h. by (*Imp.*) Fylde, dam by Citizen.
TOM WALKER, ch. h. by Marylander, dam by Rattler.
TOMMY WAKEFIELD, ch. h. by Drone, dam by Eclipse.
TORCH-LIGHT, ch. m. by (*Imp.*) Glencoe, dam Wax-light by (*Imp.* Leviathan.
TORNADO, ch. h. by Eclipse, dam Polly Hopkins by Virginian.
TRANBYANNA, m. by (*Imp.*) Tranby dam Lady Tompkins by Eclipse.
TRANSIT, b. h. by (*Imp.*) Hedgford, dam (Molly Ward's dam) by Bertrand.
TREASURER, b. h. by (*Imp.*) Roman, dam Dove by Duroc.
TRENTON, o. h. by Eclipse Lightfoot, dam by Tuckahoe.
TROUBADOUR, bl. h. by (*Imp.*) Luzborough, dam by Stockholder.
TRUXTON, b. h. by (*Imp.*) Barefoot, dam Princess by Defiance.
TUSKENA, b. h. by Mons. Tonson, dam Creeping Kate.
TYLER, b. h. by (*Imp.*) Trustee, dam Kate Kearney by Sir Archy.

U.

UNCAS, ch. h. by Diomed.
———- b. h. by (*Imp.*) Jordan, dam by Pacific.
UNITY, ch. f. by Genito, dam Lady Pest by Carolinian.

V.

VAGABOND, ch. h. by (*Imp.*) Ainderby, dam (*Imp.*) Vaga.
VAGRANT, ch. h. by (*Imp.*) Trustee, dam (*Imp.*) Vaga.
VANITY, b. m. by Traveller.
VAN TROMP, h. by Van Tromp, dam by Mucklejohn.
VASHTI, b. m. by (*Imp.*) Leviathan—Slazy by Bullock's Mucklejohn.
VELASCO, b. h. by Shark, dam by Virginian.
VELOCITY, ch. m. by (*Imp.*) Leviathan, dam Patty Puff by Pacolet.
VERTNER, ch. h. by Medoc, dam Lady Adams by Whipster.
VETO, ch. h. by Eclipse, dam by Diomed.
——— h. by (*Imp.*) Luzborough, dam Lady Washington by Washington.
VICTOR, br. h. by (*Imp.*) Cetus, dam (*Imp.*) My Lady by Comus.
VICTORIA, gr. m. by Sir Kirkland, dam by Tippoo Saib.
——— b. m. by (*Imp.*) Luzborough, dam by Timoleon.
——— ROWTON, ch. m. by (*Imp.*) Rowton, dam by Phenomenon.
VICTRESS, b. m. by Grey Eagle, dam by Royal Charley.
VIDOCQ, br. h. by Medoc, dam by Stockholder.
VIOLA, ch. m. by (*Imp.*) Leviathan, dam Mary Longfit by Pacific.
VIRGINIA, ch. m. by (*Imp.*) Leviathan, dam by Sir Rich'd Tonson.
VIRGINIA ROBINSON, b. m. by (*Imp.*) Luzborough, dam Becky by Marquis (or Marcus).
VOLTAIRE, ch. h. by (*Imp.*) Leviathan, dam by Bertrand.

W.

WACOUSTA, ch. h. by Jerseyman, dam Lady Vixen.
WADDY THOMPSON, ch. h. by (*Imp.*) Emancipation, dam by Trafalgar.

WAGNER, ch. h. by Sir Charles, dam Maria West by Marion.
WALK IN-THE-WATER, b. h. by Collier, dam by Bertrand.
WALTER L. b. h. by (*Imp.*) Fylde, dam by Sir Charles.
WANTON WILL, b. h. by Brunswick, dam by Prince Edward.
WARSAW, ch. h. by Eclipse, dam by Arab.
WARWICK, ch. h. by Stockholder, dam by (*Imp.*) Leviathan.
WASHENANGO, ch. h. by (*Imp.*) Sorrow, dam by (*Imp.*) Leviathan
WATKINS, ro. h. by John Richards, dam by Whip.
WAXETTA, br. m. by Waxy, dam by Kennedy's Diomed.
WEBSTER, b. h. by (*Imp.*) Priam, dam Fairy.
WELLINGTON, b. h. by (*Imp.*) Sarpedon, dam (Volney's dam) by
    Sir Archy.
WESLEY MALONE, b. h. by (*Imp.*) Leviathan, dam by Sir Richard.
WEST FLORIDA, b. m. by Bertram, dam by Potomac.
WEST-WIND, br. h. by (*Imp.*) Chateau Margaux, dam Mambrina
    by Bertrand.
WHALEBONE, b. h. by (*Imp.*) Cetus, dam by Gohanna.
WHISKER, b. h. by (*Imp.*) Emancipation, dam by Walnut.
WILL-GO, b. or br. h. by (*Imp.*) Luzborough, dam by Eclipse.
WILLIAM R., b. h. by Goliah, dam by Sir Alfred.
WILD BURK, ch. h. by Medoc, dam by (*Imp.*) Bluster.
WILLIS, ch. h. by Sir Charles, dam by (*Imp.*) Merryfield.
WILLIS P. MANGUM, b. h. by Shark, dam Aggy Down.
WILTON BROWN, gr. h. by (*Imp.*) Priam, dam Ninon de l'Enclos
    by Rattler.
WINCHESTER, ch. h. by Clifton, dam by Contention.
WINFIELD (or WINFIELD SCOTT), ch. h. by Andrew, dam by
    Eclipse.
WONDER, b. h. by Tychicus, dam Nancy Marlborough by Rob Roy.
WOODCOCK, b. h. by (*Imp.*) Emancipation, dam by Shylock.
WORKMAN, ch. h. by (*Imp.*) Luzborough, dam by Timoleon.

Y.

YAZOO TRAPPER, ch. h. by Sir William.
YELLOW ROSE, ch. m. by Andrew, dam Tuberose by Arab.
YORKSHIRE, b. h. by St. Nicholas, dam Moss Rose by Tramp.
YOUNG DOVE, gr. m. by (*Imp.*) Trustee, dam Dove by Duroc.
YOUNG FRAXINELLA, gr. m. by (*Imp.*) Autocrat, dam by Virgi-
    nian.
YOUNG MEDOC, ch. h. by Medoc.

Z.

ZAMPA, ch. h. by (*Imp.*) Priam, dam Celeste by Henry.
ZEBA, ro. m. by Eclipse, dam Miss Walton by Mendoza.
ZEMMA (or ZAMOUR), ch. h. by Ulysses, dam by Stockholder
ZENITH, b. h. by Eclipse, dam Belle Anderson by Sir William of
    Transport.
ZENOBIA, c.. m. by (*Imp.*) Roman, dam Dove by Duroc.
ZOE, ch. m. by (*Imp.*) Kowton, dam (Little Venus's dam,) by Sir
    William.
ZORAIDA, b m. by Virginius, dam by Comet.

CELEBRATED STALLIONS AND BROOD MARES.

~~ ~~~~~~~ ~~~~~~ ~~~~~ ~~~

A.

ABDALLAH, b. h. by Mambrino, dam Amazonia.

ABJER, [*Imp.*] got by Old Truffle, dam Briseis by Beningbrough, gr. dam Lady Jane by Sir Peter Teazle — Paulina by Florizel, &c.—foaled 1817, died 1828.—Alabama.   James Jackson.

ADMIRAL, [*Imp.*] b. h. got by Florizel, dam the Spectator mare, (who was also the dam of Old imp. Diomed)—foaled 1779.— New York.   J. Delancy.

AFRICAN, bl. h. by [*Imp.*] Valentine, dam by Marshal Bertrand.

ALLEN BROWN, ch. h. by Stockholder, dam by [*Imp.*] Eagle.

ALL FOURS, [*Imp.*] got by All Fours, son of Regulus—Blank— Bolton Starling—Miss Meynell by Partner—Greyhound—Curwin's Bay Barb, &c. imp. into Massachusetts or Connecticut.

ALONZO, ch. h. by Eclipse, dam by Sir Archy.

AINDERBY, [*Imp.*] ch. h. by Velocipede, dam Kate by Catton.

ALTORF, b. h. by [*Imp.*] Fylde, dam Countess Plater by Virginian.

AMBASSADOR, [*Imp.*] b. h. by Emilius, dam [*Imp.*] Trapes by Tramp.

AMERICUS, [*Imp.*] b. h. got by Babraham—Creeping Molly by Second—General Evans' Arabian Cartouch—foaled 1775.
                                         William Macklin.

ANDREW, ch. h. by Sir Charles, dam by Herod.

ANDREW JACKSON, b. h. by Timoleon, dam by [*Imp.*] Whip.

ANN PAGE, m. by Maryland Eclipse, dam by Tuckahoe.

ARAMINTA, b. m. by May-Day, dam Tripit by Mars.

ARGYLE, br. h. by Mons. Tonson, dam Thistle by Ogle's Oscar.

AUTOCRAT, [*Imp.*] gr. c. got by Grand Duke, dam Olivetta by Sir Oliver—Scotina by Delphi—Scota by Eclipse—foaled 1822.— New York.   William Jackson.

B.

BABRAHAM, [*Imp.*] b. h. got by Old Fearnought (son of Godolphin Ar.) — Silver — imported into Virginia by William Evans of Surrey county, and got by the Belsize Arabian in England, and foaled 1759.—Va. 1765.   William and George Evans.

——————— [*Imp.*] b. h. got by Wildair—Babraham—Sloe—Bartlett's Childers — Counsellor — Snake, &c. — foaled 1775. —Va 1783.   Augustine Willis

BALIE PEYTON, b. h. by Andrew, dam Pocahontas by Eclipse
BAY MIDDLETON, b. h. by [Imp.] Fylde, dam by Potomac.
BELLE ANDERSON, m. by William of Transport, dam Butterfly
BELSHAZZAR, [Imp.] ch. by Blacklock, dam Manuella by Dick
    Andrews.
BERNER'S COMUS, [Imp.] b. h. by Comus, dam Rotterdam by
    Juniper.
BERTRAND Junior, ch. h. by Bertrand, dam Transport by Virgi-
    nius.
BETSEY MALONE, m. by Stockholder, dam by Potomac.
BIANCA, m. by Medley, dam Powancey by Sir Alfred.
BIG JOHN, ch. h. by Bertrand, dam by Hamiltonian.
BILL AUSTIN, b. h. by Bertrand, dam by Timoleon.
BIRMINGHAM, br. h. by Stockholder, dam Black Sophia by Top
    gallant.
BLACK ARABIAN, [Imp.]—Presented by the Emperor of Morocco
    to the United States' Government.
BLACK PRINCE, b. h. by [Imp.] Fylde, dam Fantail by Sir Archy.
————————————— [Imp.] bl. h. got by Babraham — Riot by Regulus
    —Blaze—Fox, &c.—foaled 1760.—New York.   A. Ramsay.
BLOODY NATHAN, ch. h. by [Imp.] Valentine, dam Daphne by
    Duroc.
BOHEMOTH, Junior, b. h. by Old Bohemoth.
BONNYFACE, [Imp.] (also called Master Stephen) dk. b. h. got by
    a son of Regulus out of the Fen mare, got by Hutton's Royal
    colt—Blunderbuss, &c.—foaled 1768.—Va.   French.
BOSTON, ch. h. by Timoleon, dam (Robin Brown's dam) by Ball's
    Florizel.
BRITANNIA, [Imp.] m. by Muley, dam Nancy by Dick Andrews.
BUFF COAT, [Imp.] dun h. got by Godolphin Arabian — Silver
    Locks by the Bald Galloway—Ancaster Turk—Leeds Arabian,
    &c.—foaled 1742.—Va. 1761.   Joseph Wells.
BULLE ROCK, [Imp.] got by the Darley Arabian — Byerly Turk,
    out of a natural Arabian mare, &c.—foaled 1718.—Virginia,
    1735–6.   Samuel Patton.
BUSIRIS, ch. h. by Eclipse, dam Grand Duchess by [Imp.] Grac-
    chus.
BUSSORAH ARABIAN — Imported by Abraham Ogden, Esq., of
    New York.
BUTTERFLY, m. by Sumpter, dam by [Imp.] Buzzard.

C.

CADMUS, ch. h. by Eclipse, dam Di Vernon by Ball's Florizel
CAMDEN, b. h. by [Imp.] Sarpedon, dam by Old Cherokee.
CAMILLA, m. by [Imp.] Philip, dam Roxana by Timoleon.
CANNON, [Imp.] br. h. got by Dungannon—Miss Spindleshanks by
    Omar—Staring, &c.—foaled 1789.—Boston.   Gen. Lyman.
CAROLET, ch. m. by [Imp.] Leviathan, dam Peg Caruthers by
    Arab.
CAROLINE, m. by Eclipse, dam Miss Mattie.

CARVER, [Imp.] b. h. got by Young Snap — Blank — Babraham — Ancaster Starling — Grasshopper, &c.— foaled 1770.— Norfolk county, Va. Dr. Charles Mayle.

CETA, m. by [Imp.] Cetus, dam Harriet Heth by Mons. Tonson.

CETUS, [Imp.] b. h. by Whalebone, dam Lamea by Gohanna.

CHARLEY NAILOR, b. h. by Medoc, dam by Tiger.

CHARLOTTE PAGE, m. by Sir Archy, dam by [Imp.] Restless.

CHATEAU MARGAUX, [Imp.] dk. br. h. got by Whalebone, (best son of Waxy,) dam Wasp by Gohanna — Highflyer — Eclipse, &c.—foaled 1822.—Va. 1835. J. J. Avery & Co.

CHEROKEE, h. by Sir Archy, dam Roxana by Hephestion.

CHESTERFIELD, b. h. by Pacific, dam by Wilkes' Madison.

CHIFNEY, ch. h. by Sir Charles, dam by Sir Archy.

CHILTON, b. h. by Seagull, dam by Hazard.

CINDERELLA, b. m. by Saladin, dam by Aratus.

CIPPUS, bl. h. by Industry, dam by Randolph's Mark Antony.

CIVIL JOHN, gr. h. by Tariff, dam by Pakenham.

CLARET, [Imp.] got by Chateau Margaux, dam by Partisan—Silver Tail by Gohanna—Orville, &c.—foaled 1830.—N. Carolina. Wyatt Cardwell.

CLARINET, ch. m. by Kentucky Sir Charles, dam Mary Grindle by Eclipse.

CLARION, ch. h. by Monmouth Eclipse, dam by Ogle's Oscar.

COCK OF THE ROCK, b. h. by Duroc, dam by Romp.

COLORADO, h. by Eclipse, dam by Sir Archy.

COMMENCEMENT, m. by Arab, dam by Francisco.

COMMODORE, b. h. by Mambrino, dam by True American.

CORONET, [Imp.] b. h. by Catton, dam by Paynator.

CORTES, b. by Old Rattler, dam by Jack Andrews.

COUNT BADGER, ch. h. by Eclipse, dam Arabella by Hickory.

COUNT ZALDIVAR, ch. h. by Andrew, dam by Timoleon.

COUNTESS BERTRAND, m. by Bertrand, dam Nancy Dawson by Platt's Alexander.

CRIPPLE, b. h. by Medoc, dam Grecian Princess by Whip.

CRITIC, ch. h. by Eclipse, dam by Eclipse Herod.

CUSSETA CHIEF, ch. h. by Andrew, dam Virago by Wildair or Wonder.

CYMON, ch. h. by Marion, dam Fair Forester by [Imp.] Chance.

D.

DAGHEE, [Imp.] b. h. by Muley, dam by Arabian Sheik.

DAMASCUS, h. by [Imp.] Zilcadi, dam Dido by [Imp.] Expedition.

DANCING MASTER, [Imp.] b. h. got by Woodpecker—Madcap by Snap—Miss Meredith by Cade, &c.—foaled 1788.—S. Carolina.

DANIEL O'CONNELL, gr. h. by Sir Henry Tonson, dam by [Imp.] Sir Harry.

DAVY CROCKETT, h. by Constitution, dam by Sutton's Whip.

DEBASH, [Imp.] b. h. got by King Fergus—Highflyer—Madcap by Snap—Miss Meredith by Cade, &c.—foaled 1772. Imported into Massachusetts. Jones.

DECATUR, ch h. by Henry, dam Ostrich by Eclipse.

DERBY, [*Imp.*] dr. b. h. got by Peter Lely out of Urganda. formerly
Lady Eleanor, she by Milo, dam by Sorcerer out of Twins, &c
—foaled 1831.   R. D. Shepherd.

DIANA, m. by Mons. Tonson, dam by Conqueror.

DIANA, [*Imp.*] m. by Catton, dam Trulla by Sorcerer.

DIANA, m. by Mercury, dam Rarity.

DONCASTER, [*Imp.*] b. h. by Longwaist, dam by Muley, grandam
Lady Ern by Stamford.

DON QUIXOTE, [*Imp.*] ch. h. by O'Kelly's Eclipse—Grecian Princess by Forester — Coalition colt — Bustard, &c.— foaled 1784.
Imported into Va.

DORMOUSE, [*Imp.*] dk. b. h. got by Old Dormouse, dam by White-
foot — Silverlocks by Bald Galloway, &c. — foaled 1753.—Va.
1759.

DOSORIS, ch. h. by Henry, dam (Goliah's dam) by Mendoza.

DRONE, [*Imp.*] b. h. got by King Herod—Lily by Blank—Peggy by
Cade—Croft's Partner—Bloody Buttocks, &c.—foaled 1777.—
Duchess county, New York.

———— ch. h. by Mons. Tonson, dam Isabella by Sir Archy.

DUANE, br. h. by [*Imp.*] Hedgford, dam Goodloe Washington by
Washington.

DUCHESS, b. m. by [*Imp.*] Coronet, dam by Tariff.

DUCHESS OF YORK, [*Imp.*] ch. m. got by Catton, dam by Sancho
—Coriander—Highflyer, &c.—foaled 1821.—Va.
                                                      R. D. Shepherd.

DUKE SUMNER, gr. h. by Pacific, dam by Grey Archy.

DUNGANNON, ch. h. by Sumpter, dam by Duke of Bedford.

E.

ECLIPSE, (American,) ch. h. by Duroc, dam Miller's Damsel by
[*Imp.*] Messenger.

ECLIPSE THE SECOND, b. h. by Eclipse, dam Lady Nimble by
Sir William.

ELIZA ARMSTRONG, m. by Flying Childers, dam Gipsey by Florizel.

ELIZA MILLER, m. by Miller's Bertrand, dam Lucy Forester by
Marshal Ney.

ELLEN GRANVILLE, b. m. by [*Imp.*] Tranby, dam by Contention.

EMANCIPATION, [*Imp.*] br. h. by Whisker, dam by Ardrossan.

ENGLISHMAN, (*Imp.* by Mr. Walter Bell of Va., in his dam,) by
Eagle (also imported)—Pot8os—Pegasus—Small Bones by Justice, &c.—foaled 1812.

ENTERPRISE, h. by John Richards, dam by Don Quixote.

EUGENIUS, [*Imp.*] ch. h. by Chrysolite, dam Mixbury by Regulus
—Little Bowes by a brother to Mixbury—Hutton's Barb, &c.—
foaled 1770.

EXILE, h. by [*Imp.*] Leviathan, dam [*Imp.*] Refugee by Wanderer

F.

FAIRFAX ROANE, [*Imp.*] (alias Strawberry Roan) ro. h. got by Adolphus, dam by Smith's Tartar (a son of Croft's Partner) g. dam by Midge (son of Snake) — Hip, &c. — foaled 1764—Va. —Fairfax.

FANNY WRIGHT, m. by Silverheels, dam Aurora by Governor Wright's Vingtun.

FELT, [*Imp.*] b. h. by Langar, dam Steam by Waxy Pope.

FESTIVAL, ch. h. by Eclipse, dam by Timoleon.

FIFER, b. h. by Monmouth Eclipse, dam Music by John Richards.

FLATTERER, [*Imp.*] b. h. by Muley, dam Clari by Marmion.

FLORANTHE, m. by John Richards, dam Fanny Wright.

FOP, [*Imp.*] gr. h. by Stumps, dam by Fitz James.

FRANCIS MARION, ch. h. by Marion, dam Malvina by Sir Archy

FRANK, ch. h. by Sir Charles, dam Betsey Archy by Sir Archy.

G.

GANDER, gr. h. by Wild Bill, dam Grey Goose by Pacolet.

GANO, b. h. by Eclipse, dam Betsey Richards by Sir Archy.

GENERAL MABRY, h. by [*Imp.*] Leviathan, dam Galen by Pacific.

GEROW, ch. h. by Henry, dam Vixen by Eclipse.

GIFT, [*Imp.*] b. h. got by Cadormus, dam by Old Crab — Second Starling, &c.—foaled 1768.—New Kent county, Va.

Colonel Dangerfield.

GILES SCROGGINS, b. h. by Sir Archy, dam Lady Bedford by [*Imp.*] Bedford.

GLENCOE, [*Imp.*] ch. h. by Sultan, dam Trampoline by Tramp.

GLOSTER, b. h. by Sir Charles, dam by Alfred.

GOHANNA, h. by Sir Archy, dam Merino Ewe by [*Imp.*]·Bedford.

GOLD BOY, b. h. by Industry, dam (Buck Eye's dam) by Medoc.

GOLDWIRE, [*Imp.*] br. m. by Whalebone, dam Young Amazon by Gohanna.

GOVERNOR HAMILTON, gr. h. by Sir Andrew, dam by Bonaparte.

GRANBY, [*Imp.*] b. h. got by Blank—Old Crab—Cyprus Ar.—Commoner—Makeless — Brimmer, &c.—foaled 1759. — Powhatan county, Va.   Samuel Watkins.

GRECIAN PRINCESS, m. by Virginian, dam Calypso by Bell-Air —Dare Devil—Old Wilda.r — Piccadilla by Fearnought — Godolphin—Hob or Nob, &c.

GREY EAGLE, gr. h. by Woodpecker, dam Ophelia by Wild Medley.

GREY MEDOC, gr. h. by Medoc, dam Grey Fanny by Bertrand.

GROUSE, br. h. by Eclipse, dam by Erie.

GUM ELASTIC, b. h. by Waxy, dam by Read's Spread Eagle.

H.

HALO, h. by Sir Archy Montorio, dam Semiramis.

HARD LUCK, gr. h. by Randolph's Roanoke, dam Lady Washington.

HAYWOOD, h. by [*Imp.*] Leviathan, dam Black Sophia by Topgallant.

HECTOR, [*Imp.*] bl. h. got by Lath — Childers—Basto — Curwin's Bay Barb, &c.—foaled 1745.   Colonel Marshall.

HEDGFORD, [*Imp.*] br. h. by Filho da Puta, dam Miss Craigie by Orville.

———————— (Young) h. by [*Imp.*] Hedgford, dam by [*Imp.*] Eagle.

HERCULES—a grey draft horse, imported into Louisville, Ky.

HERO, [*Imp.*] b. h. got by Blank—Godolphin Ar. &c.—foaled 1747. —Va.   John S. Wilson.

HIAZIM, ch. h. by Sir Archy, dam Janey by [*Imp.*] Archduke.

HIBISCUS, [*Imp.*] b. h. by Sultan, dam Duchess of York by Waxy.

HICKORY JOHN, ch. h. by John Richards, dam Kitty Hickory by Hickory.

HIGHLAND HENRY, ch. h. by Henry, dam Highland Mary by Eclipse.

HORNBLOWER, br. h. by Monmouth Eclipse, dam Music by John Richards.

HUGH LUPUS, [*Imp.*] b. h. by [*Imp.*] Priam, dam Her Highness by Moses.

I.

IBARRA, b. h. by [*Imp.*] Hedgford, dam by Virginian.

IBRAHIM PACHA, [*Imp.*]—a pure Bedouin Arabian—imported by Captain James Riley.

J.

JACK OF DIAMONDS, [*Imp.*] dk. b. h. by Cullen's Arabian—Darley Ar. — Byerly Turk, &c. — Va. 1763. Imported by Colonel Spottswood.   Solomon Dunn.

JACK PENDLETON, ch. h. by Goliah, dam by Trafalgar.

JANE GRAY, m. by Orphan Boy, dam Rosalind by Ogle's Oscar.

JEROME, br. h. by [*Imp.*] Luzborough, dam by Sir Charles.

JESSICA, [*Imp.*] ch. m. by Velocipede, dam by Sancho.

JIM JACKSON, ch. h. by [*Imp.*] Leviathan, dam by Conqueror.

JOB, b. h. by Eclipse, dam Jemima by Thornton's Rattler.

JOHN BASCOMBE, ch. h. by Bertrand, dam Grey Goose by Pacolet

JOHN BULL, [*Imp.*] b. h. by Chateau Margaux, dam by Woful.

JOHN DAWSON, b. h. by Pacific, dam by Grey Archy.

JOHN GASCOIGNE, h. by Randolph's Gascoigne, dam by Virginian.

JOHN RICHARDS, b. h. by Sir Archy, dam by Rattler, gr. dam by [*Imp.*] Medley.

JORDAN, [*Imp.*] ch. h. by Langar, dam Matilda by Comus.

JUNIUS, [*Imp.*] bl. h. got by Old Starling — Old Crab — Monkey— Curwin's Bay Barb—Spot, &c.—foaled 1754.—Va. 1759.

JUSTICE, [*Imp.*] b. h. got by Blank, dam Aura by Stamford Turk, gr. dam by a brother to Conqueror—Childers, &c.—Va. 1780.   George Gould.

JUSTICE, [*Imp.*] got by Old Justice (son of King Herod) — Old
Squirt mare — Mogul — Camilla by Bay Belton, &c. — foaled
1782.—S. Carolina.   Major Butler.

K

KANGAROO, ch. h. by Uncas, dam by Orphan.

KATE NICKLEBY, m. by [*Imp.*] Trustee, dam Lady Mostyn by
Teniers.

KING WILLIAM, [*Imp.*] red sor. h. got by Florizel, dam Milliner
by Matchem —Cassandria by Blank, &c.—foaled 1781.—Chester county, Pa.   Dr. Norriss.

—————————— [*Imp.*] b. h. by King Herod, dam Madcap by
Snap — Miss Meredith by Old Cade, &c.—foaled 1777. — Connecticut.   Skinner.

KITTY BRIM, b. m. by Old Conqueror, dam by Gallatin ; gr. dam
by Highflyer.

L.

LADY CLIFDEN, m. by Sussex, dam Betsey Wilson.

LADY CULPEPER, m. by Carolinian, dam Flora by Ball's Florizel.

LADY MORGAN, m. by John Richards, dam Matchless by [*Imp.*]
Expedition.

LADY MOSTYN, [*Imp.*] m. by Teniers, dam Invalid by Whisker.

LADY NIMBLE, m. by Eclipse, dam Transport by Kosciusko.

LADY SCOTT, [*Imp.*] br. m. got by Ardrosson, dam Dido by Vis-
count—Brilliant by Whiskey, &c.   R. D. Shepherd.

LADY WHIP, m. by Whip, dam by Alonzo, gr. dam by [*Imp.*] Buzzard.

LAFAYETTE, b. h. by Conqueror, dam Julia by Sir Arthur.

LANGFORD, [*Imp.*] br. h. got by Starch, out of Peri by Wanderer,
her dam Thalistris by Alexander, out of Rival by Sir Peter—
Horne by Drone—Manilla by Goldfinder—foaled 1833.
                                                      F. P. Corbin.

LAPLANDER, ch. or br. h. by Flagellator, dam Medora.

LEOPARDESS, m. by Medoc, dam by Haxall's Moses.

LEVIATHAN, [*Imp.*] ch. h. by Muley, dam by Windle.

—————————— Junior, ch. h. by [*Imp.*] Leviathan, dam by Young
Diomed.

LILY, m. by Eclipse, dam Garland by Duroc.

LIMBER JOHN, ch. h. by Kosciusko, dam by Moses.

LOFTY, [*Imp.*] b. h. by Godolphin Arabian—Croft's Partner—Bloody
Buttocks—Greyhound, &c.—foaled 1753.—Virginia, Chesterfield
county.   Thomas Goode.

LOUISA, ch. m. by [*Imp.*] Bluster, dam by Hamiltonian.

LURCHER, [*Imp.*] gr. h by Grey Leg, dam Harpalyce by Gohanna

LUZBOROUGH, [*Imp.*] br. h. by Williamson's Ditto, dam by Dick
Andrews.

—————————— Junior, b. h. by [*Imp.*] Luzborough, dam by Sumpter.

LYCURGUS, [*Imp.*] ch. h. by Blank—Snip—Lath, &c.—foaled 176·
—Va. 1776   Geo. H. Harrison
                                                      44 *

LYNEDOCH, ch. h. by [*Imp.*] Leviathan, dam Rosetta by Wilkes Wonder.

M.

MAGNUM BONUM, [*Imp.*] ro. h. by Matchem—Swift—Regulus— Dairy Maid by Bloody Buttocks, &c. — foaled 1774. — Hartford, Conn.   F. Kilborne.

MANALOPAN, gr. h. by Medley, dam by John Richards.

MARIA DAVIESS, ch. m. by Sir Charles, dam Mary Grindle by Eclipse.

MARIA VAUGHAN, m. by Pacific, dam Mary Vaughan by Pacolet.

MARION, b. h. by Sir Archy, dam by [*Imp.*] Citizen.

MARGRAVE, [*Imp.*] ch. h. by Muley, dam by Election.

MARK MOORE, ch. h. by Eclipse, dam Lalla Rookh by Gabriel Oscar.

MARMION, br. h. by [*Imp.*] Merman, dam by Crusader.

MARPLOT, [*Imp.*] by Highflyer—Omar—Godolphin Arabian, &c.

MARTHA BICKERTON, b. m. by Pamunky, dam by Tariff.

MARSHAL NEY, h. by Pacolet, dam Virginia by Dare Devil.

MARY BIDDLE, m. by [*Imp.*] Priam, dam Flora by Mons. Tonson.

MARY VAUGHAN, gr. m. by Old Pacolet, dam by Old Chanticleer.

MASTER ROBERT, [*Imp.*] ch. h. by Star, dam a young Marske mare—foaled 1793.

MASTER SOLOMON, b. h. by Reveller, dam by Lord Berners.

MATCHEM, [*Imp.*] b. h. by Matchem—Lady by Sweepstakes—Patriot—Old Crab, &c.—foaled 1773.—S. Carolina.   Gibbs.

MATCHLESS, [*Imp.*] b. h. by Godolphin Arabian — Soreheel — Makeless, &c.—S. Carolina.

MATILDA, gr. m. by Greytail Florizel, dam by [*Imp.*] Jonah.

MAXIMUS, b. h. by Bertrand, dam Miss Dance by [*Imp.*] Eagle.

MAYZOUBE — a gr. horse imported from Arabia by Captain James Riley.

MELZARE, br. h. by Bertrand, dam by Sir Richard.

MENDOZA, [*Imp.*] b. h. by Javelin — Paymaster — Pamona by King Herod.

MERMAN, [*Imp.*] br. h. by Whalebone, dam by Orville.

MERRY PINTLE, [*Imp.*] gr. h. by Old England, dam by Old Merry Pintle—Skipjack, &c.—foaled 1752.—Va. 1775.   J. Strong.

MERRY TOM, [*Imp.*] b. h. by Regulus — Locust — a son of Flying Childers—Croft's Old Partner, &c.—foaled 1758.

MERCER, [*Imp.*] b. h. by Emilius, dam Young Mouse by Godolphin.

MERWICK BALL, [*Imp.*] ch. h. by Regulus—dam a Traveller mare —Hartley's blind horse—foaled 1762.

MINOR, b. h. by Mons. Tonson, dam by Topgallant.

MISS ANDREWS, [*Imp.*] b. m. by Catton, dam by Dick Andrews.

MISS MATTIE, m. by Sir Archy, dam Black Ghost by Pantaloon.

MISS ROSE, [*Imp.*] b. m. by Tramp, dam by Sancho, gr. dam by Coriander, &c.—foaled 1826.   R. D. Shepherd.

MISS VALENTINE, m. by [*Imp.*] Valentine, dam by John Richards.

MONARCH, [*Imp.*] b. h. by Priam, dam Delphine by Whisker.

MONMOUTH, b. h. by John Richards, dam by Duroc.

MONMOUTH ECLIPSE, ch. h. by Eclipse, dam Honesty by [*Imp.*] Expedition.

MONS. TONSON, gr. h. by Pacolet, dam Madame Tonson by Topgallant.

MORDECAI, [*Imp.*] b. h. by Lottery, dam by Welbeck.

MORVEN, [*Imp.*] ch. h. by Rowton, dam Nanine by Selim.

MOSES MARE (Chas. Buford's) by Haxall's Moses, dam by Cook's or Blackburn's Whip.

N.

NANCY THATCHER, m. by Medoc, dam by Archy of Transport.

NELL GWYNNE, [*Imp.*] m. by Tramp, dam by Beningbrough.

NETTY, [*Imp.*] ch. m. by Velocipede, dam Miss Rose.

NICHOLAS, [*Imp.*] h. by St. Nicholas, dam Miss Rose.

NIMROD, [*Imp.*] b. h. by King Fergus — O'Kelly's Eclipse — Old Marske, &c.—Philadelphia, 1788.

NON PLUS, [*Imp.*] b. h. by Catton, dam Miss Garforth by Walton—Hyacinthus, &c.—foaled 1824.—S. Carolina. R'd. Singleton.

NORTH BRITAIN, [*Imp.*] b. h. by Alcock's Arabian—Northumberland Arabian—Hartley's blind horse.—Philad. 1768. Crow.

NOVELTY, [*Imp.*] m. by Blacklock, dam Washerwoman by Walton.

O.

O'KELLY, ch. h. by Eclipse, dam by Oscar.

OLIVER, h. by May-Day, dam Young Betsey Richards by John Richards.

ONUS, [*Imp.*] br. h. by Camel, dam The Etching by Rubens.

ORLEANA, [*Imp.*] m. by Bustard, dam Laureola by Orville.

OROONOKO, [*Imp.*] bl. h. by Old Crab, dam Miss Slammerkin by Young True Blue—Bloody Shouldered Arabian, &c.—foaled 1745—S. Carolina. J. Mathews.

OTHELLO, br. h. by [*Imp.*] Leviathan, dam by Sir Archy.

P.

PACIFIC, b. h. by Sir Archy, dam Eliza by [*Imp.*] Bedford.

PACOLET, [*Imp.*] h. by Sparke, dam Queen Mab—Hampton Court Childers—Harrison's Arabian, &c.—Va. 1791. Thos. Goode.

PACTOLUS, ch. h. by Pacific, dam Mary Vaughan by Pacolet.

PAMUNKY, b. h. by Eclipse, dam Bellona by Sir Harry.

PAUL CLIFFORD, h. by Eclipse, dam Betsey Richards by John Richards.

PETE WHETSTONE, b. h. by [*Imp.*] Leviathan, dam by Stockholder.

PHARAOH, [*Imp.*] b. h. by Moses, dam by Godolphin Arabian Smockface by Old Snail, &c.—foaled 1753.—S. Carolina.

PHILIP, [*Imp.*] br. h. by Filho da Puta, dam Treasure by Camillus
——— h. by Randolph's Janus, dam (Jack Pendleton's dam) by Trafalgar.

PICTON, br. h. by [*Imp.*] Luzborough, dam Isabella by Sir Archy.

PLATOFF, b. h. by Kosciusko, dam by Hephestion.

PONEY, (The) ch. h. by [*Imp.*] Leviathan, dam by Stockholder.
PORTLAND, [*Imp.*] ch. h. by Recovery, dam by Walton.
PORTSMOUTH, br. h. by [*Imp.*] Luzborough, dam Polly Peachem by John Richards.
POST BOY, ch. h. by Henry, dam Garland by Duroc.
POWHATTAN, b. h. by Arab, dam by Whip.
PRESTO, b. h. by [*Imp.*] Leviathan, dam by Stockholder.
PRIAM, [*Imp.*] b. h. by Emilius, dam Cressida by Whisker.
—— Junior, h. by [*Imp.*] Leviathan, dam by Sir Archy.
PRINCE, [*Imp.*] b. h. by Herod, dam Helen by Blank—Crab, &c.— foaled 1773.—S. Carolina.
PRINCE FERDINAND, [*Imp.*] by Herod, dam by Matchem — gr dam the Squirt mare, &c.
PRUNELLA, [*Imp.*] m. by Comus, dam by Partisan.
PUZZLE, [*Imp.*] b. h. by Reveller, dam by Juniper.

Q.

QUEEN OF THE WEST, br. m. by Shark, out of Lady Mostyn by Teniers, gr. dam Invalid by Whisker.

R.

RATTLER, ch. h. by Sir Archy, dam by [*Imp.*] Robin Red Breast.
RED BILL, b. h. by Medoc, dam Brown Mary by Sumpter.
RED BUCK, b. h. by [*Imp.*] Leviathan, dam Sally Bell by Contention.
RED TOM, ch. h. by Bertrand, dam Duchess of Marlborough by Sir Archy.
REINDEER, ch. h. by Henry, dam Sportsmistress by Hickory.
———— ch. h. by Sussex, dam by Oscar.
REPUBLICAN, [*Imp.*] ch. h. by Wentworth's Ancaster — Old Royal Changeling—Bethel's Arabian, &c.—Va. 1797.

Charles Young.

RICHARD SINGLETON, b. h. by Bertrand, dam Black-Eyed Susan by Tiger.
RIDDLESWORTH, [*Imp.*] ch. h. by Emilius, dam Filagree by Soothsayer.
ROANOKE, b. h. by Sir Archy, dam by Cœur de Lion.
ROBIN BROWN, ch. h. by Mons. Tonson, dam (Boston's dam) by Ball's Florizel.
RODOLPH, b. h. by Archy of Transport, dam by Haxall's Moses.
RODERICK DHU, [*Imp.*] by Sir Peter Teazle, dam by Young Marske —Matchem—Tarquin, &c.   Imported into New York.
ROSALBA, m. by Old Trafalgar, dam Rosalba by Spread Eagle.
ROSIN THE BOW, b. h. by Bertrand, dam Lady Grey by Robin Grey
RUBY, [*Imp.*] b. h. by Emilius, dam Eliza by Rubens.
RUSHLIGHT, ch. m. by Sir Archy, dam Pigeon by Pacolet.

S.

SALLY BARBOUR, m. by [*Imp.*] Truffle, dam by Ball's Florizel.
SALLY HYDE, m. by Sumner's Grey Archy, dam by Medley.

SAM HOUSTON, ch. h. by Barney O'Lynn, dam Judy Bakewell by Eagle.

SANTEE, b. h. by Rob Roy, dam Betty by [*Imp.*] Buzzard.

SARACEN, b. h. by Eclipse, dam Sally Slouch by Virginian.

SARPEDON, [*Imp.*] br. h. by Emilius, dam Icaria by The Flyer— Parma by Dick Andrews, &c.

SCIPIO, b. h. by [*Imp.*] Leviathan, dam Kitty Clover by Sir Charles.

SCOUT, [*Imp.*] br. h. by St. Nicholas, dam by Blacklock.

SEAGULL, b. h. by Sir Archy, dam Nancy Air by [*Imp.*] Bedford.

SHADOW, bl. h. by Eclipse Lightfoot, dam Sally Slouch by Virginian.

SHADOW, [*Imp.*] b. h. got by Babraham—Bolton Starling—Cough ing Polly by Bartlett's Childers, &c.—foaled 1759.—Va. 1771.
T. Burwell.

SHAKSPEARE, [*Imp.*] br. h. by Smolensko, dam Charming Molly by Rubens.

SHARK, bl. h. by Eclipse, dam Lady Lightfoot by Sir Archy.

SHAMROCK, [*Imp.*] ch. h. St. Patrick, dam Delight by Reveller.

SHEPHERDESS, [*Imp.*] b. m. by Young Blacklock, dam Spermaceti by Sligo Waxy.

SHERIFF PACHA, b. h. Nedji bred—imported by Com. Elliott.

SHOCK, [*Imp.*] got by Shock—Partner—Makeless—Brimmer, &c.— Va. Caroline county.   Jno. Baylor.

SIDI HAMET, b. h. by Eclipse, dam Princess by Defiance.

SIDNEY, b. h. by Sir Charles, dam Virginia by Thornton's Rattler.

SIR CHARLES, ch. h. by Saladin, dam by Cultivator

SIR JOSEPH, br. h. by (*Imp.*) Luzborough, dam Sally Maclin by Sir Archy.

SIR LESLIE, b. h. by Sir William, dam by (*Imp.*) Buzzard.

SIR MEDLEY, ch. h. by Medley, dam by Sir Charles.

SIR PETER TEAZLE, (*Imp.*) ro. h. got by Sir Peter Teazle—Mer cury—Cythera by King Herod—Blank, &c.—foaled 1802.—S. Carolina.   Gen. Jno. McPherson.

SIR ROBERT, (*Imp.*) b. h. by Bobadil, dam Fidalma by Waxy Pope.

SIR WILLIAM, h. by Sir William, dam by Tiger.

SKYLARK, (*Imp.*) br. h. by Waxy Pope, dam Skylark by Musician.

SLOUCH, (*Imp.*) ch. h. by Cade, dam the little Hartley mare by Bartlett's Childers—Flying Whig by Woodstock, &c.—foaled 1747.—S. Carolina.

SOURKROUT, (*Imp.*) b. c. by Highflyer, dam Jewel by Squirrel, Sophia by Blank, &c.—foaled 1786.

SOVEREIGN, (*Imp.*) b. h. by Emilius, dam Fleur de Lis by Bourbon.

STARLING, (*Imp.*) by Young Starling—Regulus---Snake, Partner &c.—foaled 1756.—Va. 1762.   Carlisle & Dalton.

STANHOPE, ch. h. by Eclipse, dam Helen Mar by Rattler.

STEEL, b. h. by (*Imp.*) Fylde, dam Diamond by Constitution.

STOCKHOLDER, b. h. by Sir Archy, dam by (*Imp.*) Citizen

ST. LEGER, gr. h. by Eclipse, dam (Ariel's dam,) by Financier.

ST. PAUL, (*Imp.*) sor. h. by Old Saltram, dam Purity by Matchem, Pratt's famous Squirt mare, &c.—foaled 1789.—Va. 1804.

Wm. Lightfoot.

STRAWBERRY ROAN, (see Fairfax Roan).

SWISS, (*Imp.*) b. h. by Whisker, dam by Shuttle.

SYMMETRY, ch. m. by (*Imp.*) Ainderby, dam Ellen Douglass by Bertrand.

T.

TARGET, ch. h. by (*Imp.*) Luzborough, dam Becky by Marquis.

TARLTON, b. h. by Woodpecker, dam by Robin Gray.

TARQUIN, br. h. by Henry, dam Ostrich by Eclipse.

—————— h. by (*Imp.*) Luzborough, dam Hackabout by Timoleon.

—————— (*Imp.*) h. by the Hampton-Court Chesnut Ar. out of Fair Rosamond by Cade—Traveller, &c.—foaled 1720.

TELIE DOE, m. by Pacific, dam Matilda by Grey-tail Florizel.

TENNESSEE CITIZEN, ch. h. by Stockholder, dam Patty Puff by Pacolet.

THOMAS H. BENTON, br. h. by Waxy, dam Virginia by Matapone.

TITRY, (*Imp.*) ch. m. by Langar, dam Zephyrina by Middlethorpe.

TOBACCONIST, b. h. by Gohanna, dam Yankee Maid by Ball's Florizel.

TOM MOORE, h. by Contention, dam Pocahontas by Virginian.

TORNADO, ch. h. by Eclipse, dam Polly Hopkins by Virginian.

TRANBY, (*Imp.*) br. h. by Blacklock, dam by Orville—Miss Grimstone by Weazle—Ancaster, &c.—foaled 1826.—Va. 1835.

J. J. Avery & Co.

TRIPIT, br. m. by Mars, dam by Post Boy.

TRUFFLE, (*Imp.*) b. h. by Truffle, dam Helen by Whiskey.

TRUSTEE, (*Imp.*) ch. h. by Catton, dam Emma by Whisker.

V.

VALPARAISO, (*Imp.*) ch. h. by Velocipede, dam Julianna by Gohanna.

VERTNER, ch. h. by Medoc, dam Lady Adams by Whipster.

VERTUMNUS, b. h. by Eclipse, dam Princess by Defiance.

VICEROY, ch. h. by Eclipse, dam Saluda by Timoleon.

VOLCANO, b. h. by Stockholder, dam Forest Maid by Ratray.

VOLNEY, b. h. by Mons. Tonson, dam by Sir Archy.

—————— (*Imp.*) b. h. by Velocipede, dam (Voltaire's dam,) by Phantom.

W.

WACOUSTA, ch. h. by (*Imp.*) Leviathan, dam Lady Lightfoot by Oscar.

WAGNER, ch. h. by Sir Charles, dam Maria West by Marion.

WASHENANGO, ch. h. by Timoleon, dam Ariadne by (*Imp.*) Citizen.

WHALE, (*Imp.*) by Whalebone, (who was by Waxy,) dam Rectory by Octavius—Catharine by Woodpecker.—N. Carolina.

Edward Townes.

WHALEBONE, b. h. by Sir Archy, dam by Pacolet.

WILD BILL, b. h. by Sir Archy, dam Maria by Gallatin.
WILLIAM H. HARRISON, gr. h. by Trumpator, dam by Double head.
WILLIS, ch. h. by Sir Charles, dam by (*Imp.*) Merryfield.
WANDER, ch. h. by Monmouth Eclipse, dam Powancey by Alfred.
WOODPECKER, b. h. by Bertrand, dam by (*Imp.*) Buzzard.

Y.

ORKSHIRE, (*Imp.*) got by St. Nicholas, dam Miss Rose.

R. D. Shepherd.

YOUNG GOHANNA, h. by Gohanna, dam by Pacolet.
YOUNG MEDLEY, h. by Potomac, dam by Medley.
YOUNG TRAMP, (*Imp.*) h. by Barefoot, dam Isabella by Comus.
YOUNG WONDER, h. by Cock of the Rock, dam Nell Sanders.

Z.

ZINGANEE, (*Imp.*) b. h. by Tramp, dam Folly by Young Drone.
————— (Garrison's,) b. h. by Sir Archy, dam Atalanta by (*Imp.*) Chance.

THE END.